# Optical Properties of 2D Systems with Interacting Electrons

# NATO Science Series

*A Series presenting the results of scientific meetings supported under the NATO Science Programme.*

The Series is published by IOS Press, Amsterdam, and Kluwer Academic Publishers in conjunction with the NATO Scientific Affairs Division

*Sub-Series*

| | |
|---|---|
| I. Life and Behavioural Sciences | IOS Press |
| II. Mathematics, Physics and Chemistry | Kluwer Academic Publishers |
| III. Computer and Systems Science | IOS Press |
| IV. Earth and Environmental Sciences | Kluwer Academic Publishers |
| V. Science and Technology Policy | IOS Press |

The NATO Science Series continues the series of books published formerly as the NATO ASI Series.

The NATO Science Programme offers support for collaboration in civil science between scientists of countries of the Euro-Atlantic Partnership Council. The types of scientific meeting generally supported are "Advanced Study Institutes" and "Advanced Research Workshops", although other types of meeting are supported from time to time. The NATO Science Series collects together the results of these meetings. The meetings are co-organized bij scientists from NATO countries and scientists from NATO's Partner countries – countries of the CIS and Central and Eastern Europe.

**Advanced Study Institutes** are high-level tutorial courses offering in-depth study of latest advances in a field.
**Advanced Research Workshops** are expert meetings aimed at critical assessment of a field, and identification of directions for future action.

As a consequence of the restructuring of the NATO Science Programme in 1999, the NATO Science Series has been re-organised and there are currently Five Sub-series as noted above. Please consult the following web sites for information on previous volumes published in the Series, as well as details of earlier Sub-series.

http://www.nato.int/science
http://www.wkap.nl
http://www.iospress.nl
http://www.wtv-books.de/nato-pco.htm

Series II: Mathematics, Physics and Chemistry – Vol. 119

# Optical Properties of 2D Systems with Interacting Electrons

edited by

## Wolfgang J. Ossau
Physikalisches Institut der Universität Würzburg,
Würzburg, Germany

and

## Robert Suris
Ioffe Physico-Technical Institute, RAS,
St. Petersburg, Russia

**Kluwer Academic Publishers**

Dordrecht / Boston / London

Published in cooperation with NATO Scientific Affairs Division

Proceedings of the NATO Advanced Research Workshop on
Optical Properties of 2D Systems with Interacting Electrons
St. Petersburg, Russia
13–16 June 2002

A C.I.P. Catalogue record for this book is available from the Library of Congress.

ISBN 1-4020-1548-8

Published by Kluwer Academic Publishers,
P.O. Box 17, 3300 AA Dordrecht, The Netherlands.

Sold and distributed in North, Central and South America
by Kluwer Academic Publishers,
101 Philip Drive, Norwell, MA 02061, U.S.A.

In all other countries, sold and distributed
by Kluwer Academic Publishers,
P.O. Box 322, 3300 AH Dordrecht, The Netherlands.

*Printed on acid-free paper*

All Rights Reserved
© 2003 Kluwer Academic Publishers
No part of this work may be reproduced, stored in a retrieval system, or transmitted
in any form or by any means, electronic, mechanical, photocopying, microfilming,
recording or otherwise, without written permission from the Publisher, with the exception
of any material supplied specifically for the purpose of being entered
and executed on a computer system, for exclusive use by the purchaser of the work.

Printed in the Netherlands.

# Contents

Preface 3

I.V. Kukushkin, J.H. Smet, K. von Klitzing, and W. Wegscheider/ Cyclotron resonance of composite fermions 5

A.V. Larionov and V.B. Timofeev/ Collective behavior of interwell excitons in GaAs/AlGaAs double quantum wells 13

B.D. McCombe, A.B. Dzyubenko, H.A. Nickel, T. Yeo, C.J. Meining, T. Sander, A. Petrou, and A.Yu. Sivachenko/ Internal transitions of negatively charged excitons and many-electron effects in GaAs quantum wells 25

W. Ossau, G.V. Astakhov, D.R. Yakovlev, W. Faschinger, V.P. Kochereshko, J. Puls, F. Henneberger, S.A. Crooker, Q. McCulloch, and A. Waag/ Positively and negatively charged trions in ZnSe-based quantum wells 41

G. Aichmayr, M.D. Martin, L. Viña, R. André, V. Ciulin, J.D. Ganiere, and B. Deveaud/ Non-linear effects on the spin dynamics of polaritons in II–VI microcavities 63

M. Sénès, B.L. Liu, X. Marie, T. Amand, and J.M. Gérard/ Spin dynamics of neutral and charged excitons in InAs/GaAs quantum dots 79

M. Combescot, O. Betbeder-Matibet, and C. Tanguy/ Novel many-body procedure for interacting close-to-boson excitons 89

A. Esser and R. Zimmermann/ Density-matrix theory of trionic spectra in semiconductor nanostructures 99

R.A. Suris/ Correlation between trion and hole in Fermi distribution in process of trion photo-excitation in doped QWs 111

V.P. Kochereshko, D.R. Yakovlev, R.A. Suris, G.V. Astakhov, W. Faschinger, W. Ossau, G. Landwehr, T. Wojtowicz, G. Karczewski, J. Kossut/ Combined exciton-electron optical processes in optical spectra of modulation doped QWs 125

D.R. Yakovlev, G.V. Astakhov, W. Ossau, S.A. Crooker and A. Waag/ II–VI quantum wells with high carrier densities and in high magnetic fields 137

P. Hawrylak/ Magneto-optics of inhomogeneous two-dimensional electron gas 151

R.T. Cox, V. Huard, K. Saminadayar, C. Bourgognon, S. Tatarenko/ Filling factor dependence of optical spectra for CdTe quantum wells under magnetic field ... 159

F.X. Bronold/ Cluster meanfield approximation for the optical response of weakly doped semiconductor quantum wells ... 169

M.M. Glazov and E.L. Ivchenko/ D'yakonov–Perel's spin relaxation under electron–electron collisions in QWs ... 181

B.M. Ashkinadze, V. Voznyy, E. Cohen, Arza Ron, L.N. Pfeiffer/ Evolution of the 2DEG-free hole to charged exciton photoluminescence in GaAs/AlGaAs quantum wells ... 193

M.T. Portella-Oberli, V. Ciulin, J.-D. Ganière, B. Deveaud, P. Kossacki, M. Kutrowski, and T. Wojtowicz/ Radiative decay, diffusion, localization and dephasing of trions and excitons in CdTe quantum wells ... 205

D. Keller, G.V. Astakhov, D.R. Yakovlev, L. Hansen, W. Ossau, and S.A. Crooker/ Optical studies of spin polarized 2DEG in modulation-doped (Zn,Mn)Se/(Zn,Be)Se quantum wells in high magnetic fields ... 217

R.I. Dzhioev, V.L. Korenev, M.V. Lazarev, V.F. Sapega, R. Nötzel and K. Ploog/ Fine structure of excitons in quantum wires ... 233

P. Lavallard, G. Lamouche, and S.V. Goupalov/ Electromagnetic field resonance in semiconductor nanocrystals ... 239

S.M. Ryabchenko, O.V. Terletskij, Yu.G. Semenov, and F.V. Kirichenko/ Co-manifestation of interfaces asymmetry and magnetic field influence on luminescence polarization anisotropy of [100]-oriented QW with semimagnetic barriers ... 247

V.R. Shaginyan/ Model of strongly correlated 2D Fermi liquids based on fermion-condensation quantum phase transition ... 259

R.A. Sergeev and R.A. Suris/ The heavy-hole $X^+$ trion in double quantum wells ... 279

Contributors ... 289

Index ... 291

# Preface

This volume of NATO Science Series presents the Proceedings of the NATO Advances Research Workshop "Optical Properties of 2D systems with interacting electrons" which was held at the Educational Center of the Ioffe Physico-Technical Institute in the period of June 13–16, 2002. More than 40 scientists from 14 countries participated and gave 24 lectures and 16 poster presentations. The main goal of the Workshop was to bring together leading scientists working in the field of optical properties of correlated electron systems in semiconductor nanostructures and to organize a vital and informal discussions of newest results in the field. The location and the format of the Workshop provided an atmosphere of friendly and fruitful communications.

It is important to note that the Workshop continued a series of meetings concerning the same topic: Warsaw 1999, Würzburg 2000 (NATO ARW), and Berlin 2001.

The subject agenda of the ARW was focused on the following topics:

- Evolution of optical spectra from the excitonic peaks to the Fermi-edge singularity
- Negatively and positively charged excitons
- Reconstruction of one-particle and collective excitation spectra with increase of the electron density (combined exciton-cyclotron resonance and shake-up processes)
- Effect of spatial inhomogeneity on the carrier interaction in nanostructures
- Spin-sensitive interaction and spin-spin interaction in confined systems
- Many-particle effects in semimagnetic semiconductor heterostructures.

A wide spectrum of the optical methods for the analysis of many-body phenomena in semiconductor heterostructures had been presented at the ARW:

- Photoluminescence spectroscopy
- Raman spectroscopy
- Reflectivity
- Near-field spectroscopy
- Magneto-optics in magnetic fields up to 50 T
- Pico- and femto-second time-resolved spectroscopy,
- Polarized light spectroscopy (including Faraday rotation effect in transmission)
- Optically detected resonance spectroscopy under far-infrared and microwave radiation.

The combination of optical and transport measurements was reported in some presented papers. The technique let extract information on extended electronic states (mainly contributing in the transport phenomena) and on localized states (manifesting themselves in the light emission) in the same sample.

About one third of the presentations had been given by theorists, which mirrors a ratio 2:1 between the experimental and theoretical studies in the field.

One of the central problems discussed was the scenario of the evolution of the energy spectrum of the electronic system with increasing electron density. The regimes of low and moderate electron density were of special interest. The optical spectra transforms from the resonance corresponding to the excitation of the trion (three-particle complex) to the so-called "Fermi edge singularity". It had been agreed that the existing concept of the resonance in the optical spectra, which is commonly assigned to the Fermi edge singularity, should be revised, especially in light of new experimental data for the systems based on II–VI semiconductors with a strong Coulomb interaction.

A considerable progress has been shown in the field of spin-dependent phenomena for systems of interacting carriers. Spin dynamics of trions has been examined in great details for two-dimensional systems (quantum wells) and zero-dimensional systems (quantum dots). A new extension of this field has been achieved by using diluted magnetic semiconductors, e.g. CdMnTe and ZnMnSe, where strong exchange interaction of carrier spins with the localized magnetic ions results in a full polarization of the electronic system already at very low magnetic fields. Ferromagnetic phase formation, which is caused by collective effect of the system of free carriers of high density on the system of magnetic ions, ha been reported for CdMnTe quantum wells with 2D hole gas.

The Workshop topic: "the properties of many-electron confined systems" is very wide and of great interest. This is why we are sure that this workshop held in St Petersburg, Russia is not last in this field.

We strongly thank NATO Science Program for the financial support. We would like to thank Dr. Mikhail Mizerov, Dr. Boris Egorov and Mrs. Elena Solovyova who carried out the grate organization work to make the workshop successful. The preparation of Proceedings was performed by Mr. Nikita Vsesvetskii and Mrs. Elena Solovyova whose work is greatly appreciated. Finally, we thank the Educational Center staff involved for very valuable assistance.

Winter 2002 
Wolfgang Ossau
Robert Suris

# CYCLOTRON RESONANCE OF COMPOSITE FERMIONS

I.V. KUKUSHKIN†, J.H. SMET, K. VON KLITZING
*Max-Planck-Institut für Festkörperforschung,*
*D-70569 Stuttgart, Germany,*
† *Institute of Solid State Physics, RAS,*
*Chernogolovka, 142432 Russia*

W. WEGSCHEIDER‡
*Walter Schottky Institut, Technische Universität München,*
*D-85748 Garching, Germany,*
‡ *Institut für Experimentelle und Angewandte Physik,*
*Universität Regensburg,*
*D-93040 Regensburg, Germany*

Abstract. The introduction of suitable fictitious entities occasionally permits to cast otherwise difficult strongly interacting many-body systems in a single particle form. We can then take the customary physical approach, using concepts and representations which formerly could only be applied to systems with weak interactions, and yet still capture the essential physics. A most notable recent example occurs in the conduction properties of a two dimensional electron system (2DES), when exposed to a strong perpendicular magnetic field $B$. They are governed by electron-electron interactions, that bring about the fractional quantum Hall effect (FQHE) [1]. Composite fermions, that do not experience the external magnetic field but a drastically reduced effective magnetic field $B^*$, were identified as apposite quasi-particles that simplify our understanding of the FQHE [2, 3]. They precess, like electrons, along circular cyclotron orbits, with a diameter determined by $B^*$ rather than $B$ [4–10]. The frequency of their cyclotron motion remained hitherto enigmatic, since the effective mass is no longer related to the band mass of the original electrons and is entirely generated from electron-electron interactions. Here, we demonstrate the enhanced absorption of a microwave field that resonates with the frequency of their circular motion. From this cyclotron resonance, we derive a composite fermion effective mass that varies from 0.7 to 1.2 times the electron mass in vacuum as their density is tuned from $0.6 \times 10^{11}/cm^2$ to $1.2 \times 10^{11}/cm^2$.

Key words: two-dimensional electrons, fractional quantum Hall effect, composite fermions, microwave cyclotron resonance, dimensional plasma resonance, Kohn theorem

Composite Fermions (CFs) are electrons dressed with two flux quanta (more generally an even number), that point opposite to the external $B$-field [2–4]. This attachment is a natural way to minimize the energy of the 2DES, since the associated vortex expels other electrons from its neighbourhood and decreases the repulsive interaction between the 2D-electrons. If two flux quanta penetrate the sample per electron, i.e. if the filling factor $\nu$ equals 1/2, the external field is effectively compensated and a metallic state of these compound particles emerges [4]. This state can be characterized by a Fermi wavevector and energy. A deviation from exact half filling results in the appearance of a non-zero effective field $B^*$, that quantizes the CF-motion and discretizes their energy spectrum into Landau levels. In this framework, the FQHE [1] is a manifestation of this Landau quantization and is equivalent to the integer quantum Hall effect of CFs. A variety of experimental observations [5–10] can be understood in semiclassical terms of nearly independent CFs.

Since the kinetic energy of electrons is entirely quenched in the course of applying a $B$-field, the CF cyclotron mass is not a renormalized version of the electron conduction band mass, but must be generated entirely from electron-electron interactions [4]. The search of the CF cyclotron resonance requires substantial sophistication over conventional methods used to detect the electron cyclotron resonance, since Kohn's theorem [11] must be outwitted. It states that in a translationally invariant system radiation can only couple to the center-of-mass coordinate and can not excite other internal degrees of freedom. Phenomena originating from electron-electron interactions will thus not be reflected in the absorption spectrum. An elegant way to bypass this theorem is to impose a periodic density modulation to break translational invariance. The non-zero wavevectors defined by the appropriately chosen modulation may then offer access to the cyclotron transitions of CFs, even though they are likely to remain very weak. Therefore, the development of an optical detection scheme, that boosts the sensitivity to resonant microwave absorption by up to two orders of magnitude in comparison with traditional techniques, was a prerequisite for our studies. Furthermore, we exploited to our benefit the accidental discovery that microwaves, already incident on the sample, set up a periodic modulation through the excitation of surface acoustic waves (SAW).

The 2DES, patterned into a disk with 1 mm diameter [12], is placed near the end of a 16- or 8-mm short-circuited waveguide in the electric field maximum of the microwave excitation inside a $He^3$-cryostat. At a fixed $B$-field, luminescence spectra *with* and *without* microwave excitation were recorded consecutively. The differential luminescence spectrum is obtained when subtracting both these spectra. To improve signal-to-noise ratio, the same procedure was repeated $N$ times ($N = 2-20$). Subsequently, we integrated the absolute value of the averaged differential spectrum over the entire spectral range and hereafter refer to the value of this integral as the microwave absorption amplitude. The same procedure is then repeated for different values of $B$. To establish trustworthiness in this unconventional scheme, we apply it in Fig. 1 to the well-known case of the electron cyclotron resonance $\omega_{cr} = eB/m^*$, with $m^*$ the effective mass of GaAs ($0.067m_0$). Due to its limited size, the sample also supports a dimensional plasma mode with a frequency $\omega_p$, that depends on both the density $n_{2D}$ and diameter $d$ of the sample, according to $\omega_p^2 = 3\pi^2 n_{2D} e^2/(2m^* \varepsilon_{\text{eff}} d)$. The plasma and cyclotron mode hybridize and the

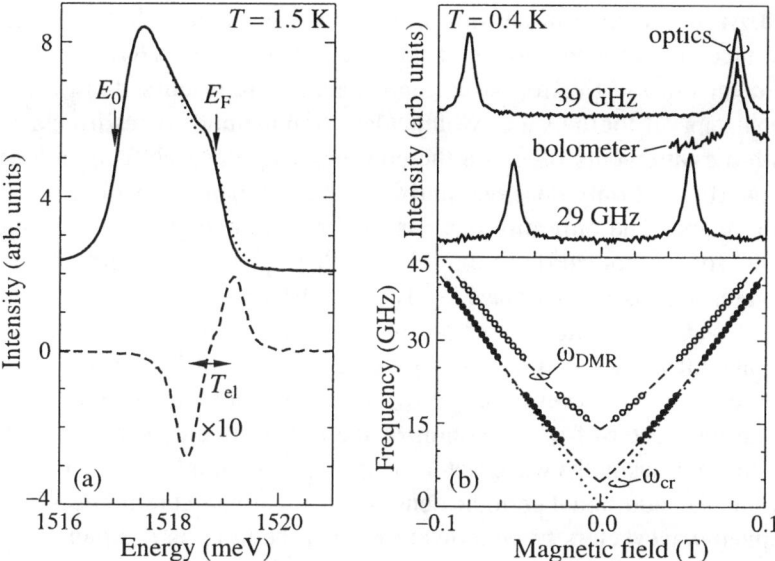

*Figure 1.* Illustration of the optical scheme to detect resonant microwave absorption for the electron cyclotron-magnetoplasmon hybrid mode at low $B$-fields. GaAs/Al$_{0.3}$Ga$_{0.7}$As heterostructures, containing a single 30 nm wide quantum well, served the investigation. The embedded 2DESs have densities and electron transport mobilities between $0.6–1.5 \times 10^{11}$/cm$^2$ and $3–5 \times 10^6$ cm$^2$/Vs respectively. (a) Luminescence spectrum in the presence of (dotted line) and without (solid line) a 50 $\mu$W microwave excitation of 18 GHz obtained on a disk-shaped 2DES with a diameter of 1 mm and carrier density $n_s = 5.8 \times 10^{10}$/cm$^2$ at a magnetic field $B = 22$ mT. The spectra were recorded by using a CCD-camera, a double-grating spectrometer that provides a spectral resolution of 0.03 meV, and a stabilized semiconductor laser operating at a wavelength of 750 nm and approximately 100 $\mu$W of cw-power. In the vicinity of the Fermi energy $E_F$ the spectrum is affected significantly under resonant microwave excitation due to heating. The dashed line represents the differential luminescence spectrum. The width of the differential spectrum reflects the increased electron temperature $T_e$. The integration of its absolute value across the entire spectral range yields the microwave absorption amplitude. (b) Top panel: The microwave absorption amplitude at 29 GHz and 39 GHz as a function of $B$-field by recording differential luminescence spectra as in (a) for 1 mT field increments at $n_s = 1.09 \times 10^{11}$/cm$^2$. The peaks, symmetrically arranged around zero field, are identified as the dimensional magnetoplasma-cyclotron hybrid mode. The inset shows a conventional bolometer measurement. Bottom panel: resonance position for $n_s = 1.09 \times 10^{11}$ (open circles) and $1.1 \times 10^{10}$/cm$^2$ (solid circles) as a function of incident microwave frequency. The intervals 10–20 GHz and 27–40 GHz were covered. The dashed lines represents the theoretical dependence of the hybrid dimensional magnetoplasma-cyclotron resonance. The dotted line corresponds to the cyclotron mode only.

resulting resonance frequency of the upper dimensional magnetoplasma-cyclotron mode $\omega_{\text{DMR}}$ equals $\omega_{\text{cr}}/2 + [\omega_p^2 + (\omega_{\text{cr}}/2)^2]^{1/2}$ [12, 13]. The optical method indeed recovers this mode. A comparison with the theoretical expression for $\omega_{\text{DMR}}$ yields excellent agreement. No fitting is required, since the density can be independently extracted from the luminescence at higher $B$-fields, where Landau levels can be resolved. At sufficiently low density, the influence of $\omega_p$ on the hybrid mode drops and one recovers at large enough $B$ the anticipated $\omega_{cr} = eB/m^*$-dependence. Further details of the electron cyclotron resonance are discussed elsewhere [12]. Additional support for the validity of the detection method comes from a comparison with measurements based on the conventional approach using a bolometer (inset Fig. 1b). Not only does one find the same resonance position, but also the same line shape. The only difference is the improved signal to noise ratio (30–100 times) for the optical detection scheme, that enables to observe the electron cyclotron resonance at microwave levels below 10 nW.

Disorder and the finite dimensions of the sample in principle suffice to break translational invariance as attested by the interaction of the cyclotron and dimensional plasma mode. However, they provide access to internal degrees of freedom other than the center-of-mass motion of the electrons either at poorly defined wavevectors or too small a wavevector for appropriate sample sizes. Therefore, the imposition of an additional periodic density modulation, that introduces larger and well-defined wavevectors to circumvent Kohn's theorem, is desirable. Transport experiments in the Hall bar geometry disclosed that additional processing is not required, since the microwaves, already incident on the sample, concomitantly induces a periodic modulation at sufficiently high power. A clear signature is the appearance of commensurability oscillations in the magneto-resistance due to the interplay between the $B$-dependent cyclotron radius of electrons and the length scale imposed by the modulation [14]. Examples are displayed in Fig. 2 and resemble the data in Ref. [15], where the modulation is produced with the help of SAW-transducers. Here, the following scenario is conceivable. Owing to the piezoelectric properties of the $Al_xGa_{1-x}As$-crystal, the radiation is partly transformed into SAW with opposite momentum, so that both energy and momentum are conserved. Reflection from cleaved boundaries of the sample then produces a standing wave with a periodicity determined by the sound wavelength. Photoexcitation creates a very poorly conducting parallel 3D-layer in the Si-doped portion of the AlGaAs-barrier and may enhance the influence of the standing acoustic waves. Carriers are collected in the nodes and affect the local density of the 2DES. The involvement of sound waves can be deduced from transport data, since from the minima we expect the modulation period to be approximately 200 and 250 nm for frequencies of 17 and 12 GHz respectively. The ratio of this period to the sound wavelength at these frequencies is 1.12 and 1.15.

Figure 3(a) depicts the $\mu$w-absorption amplitude up to high $B$-fields. Apart from the strong dimensional magnetoplasma-cyclotron resonance signal at low $B$-field discussed above, several peaks, that scale with a variation of the density, emerge near filling 1, 1/2 and 1/3. Those peak positions associated with $\nu = 1$ and 1/3 remain fixed when tuning the microwave frequency and are ascribed to heating induced by non-resonant absorption of microwave power. In contrast, the weak maxima surrounding filling 1/2 readily respond to a change in frequency as

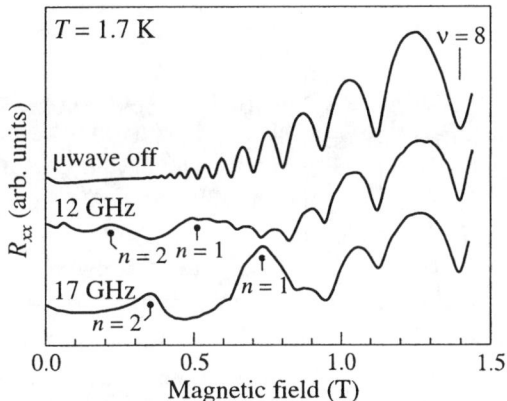

*Figure 2.* Magnetotransport data without (top curve) and under 100 $\mu$W of microwave radiation at 12 (middle curve) and 17 GHz (bottom curve). Curves are offset for clarity. Besides the well-known Shubnikov–de Haas oscillations, additional magnetoresistance oscillations appear under microwave radiation. They are commonly observed in 2DESs on which a static periodic modulation of the density has been imposed. There is no sign of such commensurability effects in the corresponding optical experiments. This can ultimately be traced back to the fact that, contrary to optical quantities, transport is also sensitive to semi-classical phenomena unrelated to changes in the density of states.

illustrated in Fig. 3(b). They are symmetrically arranged around half filling and their splitting grows with frequency. The $B$-dependence is summarized in Fig. 3(c) for two densities. To underline the symmetry, $B^*$ was chosen as the abscissa. The linear relationship between frequency and field extrapolates to zero at vanishing $B^*$. We do not expect a deviation at small $B^*$ due to a plasma-like contribution as in Fig. 1(c). Excitations for the 1/3, 2/5, 3/7 and other fractional quantum Hall states exhibit in numerical simulations no magnetoplasmon-like linear contribution to the dispersion at small values of $k$ [16]. We conclude that the resonance in Fig. 3 is the long searched for cyclotron resonance of CFs. Geometric resonances (GR), as they occur in transport at low fields due to the density modulation (Fig. 2), are excluded as an alternative interpretation for the observed features on the following grounds:

I. In the optical data, only the electron cyclotron resonance peak is observed. Contrary to optical quantities, transport is also sensitive to semi-classical phenomena unrelated to changes in the density of states.

II. Even if the 2DES condenses in a FQHE-state and the chemical potential is located in a gap, the resonance peaks surrounding $\nu = 1/2$ occur (Fig. 3c). Commensurability effects are not observable in this regime.

III. The observation of GRs requires that the density modulation is temporally static on the time scale with which CFs execute their cyclotron orbit. For electrons at low fields this condition is met and accordingly *transport* displays GRs. For the anticipated enhanced mass of CFs, this condition is violated.

IV. Analogous resonance peaks were also detected for the higher order CFs around

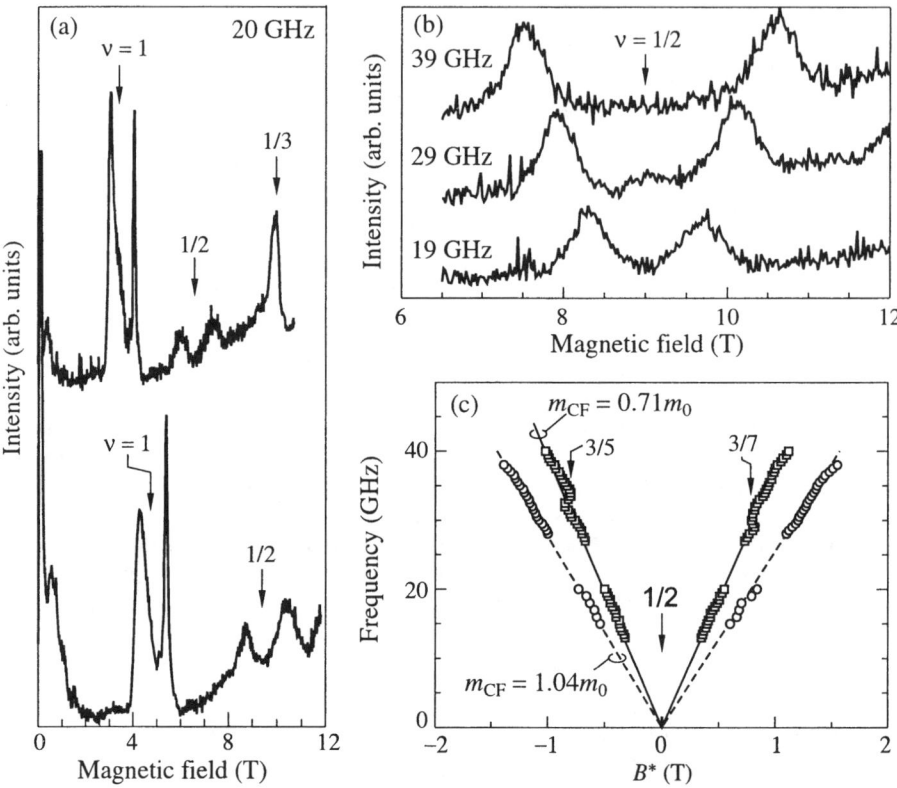

*Figure 3.* Microwave absorption amplitude and peak position for high $B$-fields in the samples with a disk shaped mesa at $T = 0.4$ K. (a) Microwave absorption amplitude at high magnetic fields for $n_s = 0.81 \times 10^{11}$ and $1.15 \times 10^{11}/\text{cm}^2$ and frequency of 20 GHz. The response near $\nu = 1$ and 1/3 does not shift with frequency. (b) Microwave absorption amplitude in the vicinity of $\nu = 1/2$ (20 mT step size) at three different frequencies and $n_s = 1.09 \times 10^{11}/\text{cm}^2$. The peak values are nearly two orders of magnitude weaker and considerably wider (about 30–50 times) than those due to the electron cyclotron resonance. (c) Position of the CF cyclotron mode as a function of the effective magnetic field $B^* = B - 2n_s \Phi_0$ ($\Phi_0$ is the elementary flux quantum) for $n_s = 1.09 \times 10^{11}/\text{cm}^2$ (circles) and $n_s = 0.59 \times 10^{11}/\text{cm}^2$ (squares). The CF effective mass equals $1.04m_0$ and $0.71m_0$ respectively. The resonances remain visible even if the 2DES condenses in the gapfull fractional quantum Hall states at filling factors 3/5 and 3/7.

$\nu = 1/4$. Since this CF metallic state is characterized by the same Fermi wavevector GRs would show up at the same distance from $\nu = 1/4$ as they do at $\nu = 1/2$. The observed peaks are located at different positions rendering a commensurability picture untenable (see below Fig. 4b).

In contrast to electron cyclotron resonance, the intensity of the CF cyclotron resonance is a strong non-linear function of $\mu$w-power (Fig. 4a). Moreover, its observability only at high power correlates with the first appearance of commensurability oscillations. The drop in intensity at even higher power is most probably

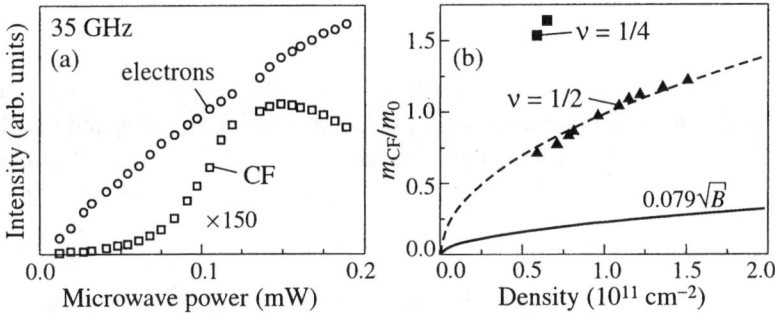

*Figure 4.* Cyclotron resonance amplitude as a function of incident microwave power and cyclotron mass of CFs at $\nu = 1/2$ and $1/4$ for various carrier densities. (a) Incident microwave power dependence of the amplitude at the cyclotron resonance (circles for electrons, squares for CFs). (b) Dependence of the CF effective mass near $\nu = 1/2$ on the carrier density $n_s$ (solid triangles). The mass for $\nu < 1/2$ is systematically a few percent heavier than for $\nu > 1/2$. The dashed line is a square root fit to the data. The solid curve is the prediction from theory reported in Ref. [17]. Analogous resonance peaks have also been detected around $\nu = 1/4$. The corresponding effective mass values are indicated as solid squares for two different densities. The large discrepancy in the extracted values of the effective masses for $\nu = 1/2$ and $\nu = 1/4$ allows to exclude that the resonant microwave absorption peaks originate from a commensurability effect.

due to heating. The intensity diminishes to zero at temperatures above 0.7 K, whereas the electron cyclotron resonance persists up to $T > 2$ K. The slope of the CF cyclotron frequency as a function of $B^*$ in Fig. 3(c) defines the cyclotron mass $m_{cr}^{cf}$. This mass is set by the electron-electron interaction scale, so that a square root behaviour on density or $B$-field is forecasted from a straightforward dimensional analysis [4]. Numerical calculations predict $m_{cr}^{cf}/m_0 = 0.079(B[T])^{1/2}$ for an ideal 2DES, not including Landau level mixing or finite width contributions [17]. The data, shown in Fig. 4(b), confirm qualitatively the strong enhancement in comparison with the electron mass (more than 10 times), however a fit to the square root dependence requires a prefactor that is four times larger. Previously reported mass values based on activation energy gap measurements [18, 19] must be distinguished from the cyclotron mass. The former corresponds to the limit of infinite momentum, whereas here $k$ approaches zero. Moreover, activation gaps can only be extracted at well-developed fractional quantum Hall states and their accurate determination suffers from disorder induced broadening. These and other limitations have been discussed in Ref. [19] for example. The technique discussed here can be performed at arbitrary filling factors.

In summary, the fortuitous breaking of translational invariance induced by the microwave irradiation combined with the virtues of an optical detection scheme for resonant absorption has enabled to unveil the cyclotron resonance frequency of CFs and to measure the corresponding cyclotron mass.

## Acknowledgements

Partial support by the Volkswagen Stiftung, the Russian Fund of Fundamental Research, INTAS, the German Ministry of Science and Education and the German Physical Society is gratefully acknowledged.

## References

1. Das Sarma, S., Pinczuk, A. (eds.) *Perspectives on quantum Hall effects* (Wiley, New York, 1996).
2. Jain, J.K. The composite fermion: a quantum particle and its quantum fluids. *Physics Today*, 39-45 (April 2000).
3. Jain, J.K. Composite fermion approach for the fractional quantum Hall effect. *Phys. Rev. Lett.* **63**, 199-202 (1989).
4. Halperin, B.I., Lee, P.A., Read, N. Theory of the half-filled Landau level. *Phys. Rev. B* **47**, 7312-7343 (1993).
5. Heinonen, O. (ed.) *Composite Fermions* (World Scientific Publishing, 1998).
6. Du, R.R., Stormer, H.L., Tsui, D.C., Pfeiffer, L.N., West, K.W. Experimental evidence for new particles in the fractional quantum Hall effect. *Phys. Rev. Lett.* **70**, 2944-2947 (1993).
7. Du, R.R. *et al.* Fractional quantum Hall effect around $\nu = 3/2$: composite fermions with a spin. *Phys. Rev. Lett.* **75**, 3926-3929 (1995).
8. Willett, R.L., Ruel, R.R., West, K.W., Pfeiffer, L.N. Experimental demonstration of a Fermi surface at one-half filling of the lowest Landau level. *Phys. Rev. Lett.* **71**, 3846-3849 (1993).
9. Goldman, V.J., Su, B., Jain, J.K. Detection of composite fermions by magnetic focusing. *Phys. Rev. Lett.* **72**, 2065-2068 (1994).
10. Smet, J.H., Weiss, D., Blick, R.H., Lütjering, G., von Klitzing, K. Magnetic focusing of composite fermions through arrays of cavities. *Phys. Rev. Lett.* **77**, 2272-2275 (1996).
11. Kohn, W. Cyclotron resonance and de Haas-van Alphen oscillations of an interacting electron gas. *Phys. Rev.* **123**, 1242-1244 (1961).
12. Gubarev, S.I. *et al.* Screening of excitonic states by low-density 2D charge carriers in GaAs/AlGaAs quantum wells. *JETP Lett.* **72**, 324-328 (2000).
13. Allen, S.J., Störmer, H.L., Hwang, J.C.M. Dimensional resonance of the two-dimensional electron gas in selectively doped GaAs/AlGaAs heterostructures. *Phys. Rev. B* **28**, 4875-4877 (1983).
14. Gerhardts, R.R., Weiss, D., von Klitzing, K. Novel magnetoresistance oscillations in a periodically modulated two-dimensional electron gas. *Phys. Rev. Lett.* **62**, 1173-1176 (1989).
15. Shilton, J.M. *et al.* Effect of spatial dispersion on acoustoelectric current in a high-mobility two-dimensional electron gas. *Phys. Rev. B* **51**, 14770-14773 (1995).
16. Jain, J.K.,Kamilla, R.K., Composite Fermions: particles of the lowest Landau level. *Composite Fermions*, 1-80 (Ed. O. Heinonen, World Scientific Publishing, 1998)
17. Park, K., Jain, J.K., Phase diagram of the spin polarization of composite fermions and a new effective mass. *Phys. Rev. Lett.* **80**, 4237-4240 (1998).
18. Du, R.R., Stormer, H.L., Tsui, D.C., Pfeiffer, L.N., West, K.W., *Phys. Rev. Lett.* **70**, 2944-2947 (1993).
19. Willett, R.L. Composite Fermions – experimental findings. *Composite Fermions*, 349-431 (Ed. O. Heinonen, World Scientific Publishing, 1998).

# COLLECTIVE BEHAVIOR OF INTERWELL EXCITONS IN GaAs/AlGaAs DOUBLE QUANTUM WELLS

A.V. LARIONOV AND V.B. TIMOFEEV
*Institute of Solid State Physics, RAS,
142432 Chernogolovka, Russia*

Key words: interwell excitons, double quantum wells, resonance excitation, coherent properties of collective phase

In recent years, the exciton Bose–Einstein condensation (BEC) has extensively been searched for in two-dimensional (2D) systems based on semiconductor heterostructures with spatially separated electrons and holes [1–9]. An interest in such 2D systems was stimulated by theoretical studies carried out as early as mid-70s. However, it should be recalled that an ideal unconfined 2D system cannot undergo BEC at nonzero temperatures because number of states diverged when the chemical potential $\mu \to 0$. It is pertinent to recall that according Bogolubov–Hoenberg theorem ideal 2D system cannot have a nonzero order parameter because it is destroyed by fluctuations. This theorem applies to both superfluid and superconductivity in ideal 2D system. Nevertheless, the BEC can occur at nonzero temperatures in quasi-2D systems and 2D systems with lateral confinement. The critical temperature of BEC in spatially confined 2D system, where the number of electronic states is finite and spectrum is discrete, is equal to $T_c = 2\pi \hbar N / m \log(N \cdot S)$, i.e., logarithmically decreases with increasing the area $S$ filled with a 2D Bose gas ($N$—density, $m$—mass of bose-particle). Finally, 2D semiconductor systems are obviously quasi-two-dimensional because the ratio of Coulomb energy to the size-quantization energy is not a small parameter in real cases. Besides, spatial confinement always arises in these systems because of the effect of random potential fluctuations.

Among the quasi-two-dimensional objects based on semiconductor heterostructures, double quantum wells (DQWs) are of special interest because they provide spatial separation of photoexcited electrons and holes in neighboring quantum wells [1]. In DQW with tilted bands due to bias, excitons can be excited whose electron and hole occur in adjacent quantum wells separated by a penetrable tunneling barrier. These excitons are called spatially indirect or interwell excitons (IEs) and differ from the direct intrawell (D) excitons whose electron and hole are in the same QW. Using GaAs/AlGaAs heterostructures it was realized the situation when excitons were indirect both spatially and in momentum spaces [2].

In contrast to intrawell excitons, interwell excitons are long-lived because the wave functions of their electron and hole in heterostructure growth direction only slightly overlap through a tunneling barrier. Therefore, the interwell excitons can easily be accumulated and a gas of these excitons can be cooled down to rather low temperatures. Due to broken inversion symmetry, IEs have a constant dipole moment even in the ground state. The dipole-dipole repulsion prevents these excitons from binding into interwell excitonic molecules.

Various possible scenarios of collective behavior in a dense system of spatially separated electrons and holes were theoretically considered [7–9]. In particular, Lozovik and Berman [9] have claimed that, despite the dipole-dipole repulsion between interwell excitons, a liquid dielectric phase of these excitons may become a metastable state of the e–h system at certain critical values of the dipole moment, density, and temperatures of IEs. Earlier, Xuejun Zhu et al. [7] pointed out that the condensed dielectric exciton phase (analogue of Bose-condensate) can arise only in presence of lateral confinement (spontaneous or artificially prepared) in the quantum well plane. In the presence of this confinement and its attendant compression, it is easier to accumulate interwell excitons up to the critical densities that are sufficient for the effects of collective exciton interaction to appear. The role of the exciton spin degrees of freedom in Bose condensation was also discussed in [8].

In the present study we have investigated the photoluminescence of interwell excitons in biased DQWs with a thin (4 monolayers) AlAs barrier separating QWs under cw and pulsed resonant tunable laser excitations. In studied structures because of used stop growth technique at heterojunction boundaries the long-range random potential fluctuations are presented. It is naturally expected that at fairly low temperatures interwell excitons will be located in these random large-scale lateral potential domains. It is interesting to know whether a system of interwell excitons will demonstrate critical behavior with increasing density and at low enough temperatures.

We investigated GaAs/AlGaAs n-i-n heterostructures with a GaAs/AlAs/GaAs DQWs and narrow AlAs tunnel barrier between wells (the width of GaAs QWs was about 120 Å). The entire structure was grown by MBE on [001]-oriented n-type doped GaAs substrate having a Si dopant concentration of $10^{18}$ cm$^{-3}$. Firstly, a 0.5 $\mu$m thick Si-doped GaAs buffer layer was grown on the substrate, followed by a 0.15 $\mu$m AlGaAs isolating layer ($x = 0.33$) and GaAs/AlAs/GaAs DQWs.

The heterojunction of each GaAs QW with the isolating AlGaAs layer was also separated by a narrow (4 monolayer) barrier. AlAs barriers were grown using a growth interruption technique creating the lateral fluctuations of the AlAs barrier widths of a large-scale. The DQWs were followed by a 0.15 $\mu$m thick isolating AlGaAs layer then a 0.1 $\mu$m thick Si-doped ($10^{18}$ cm$^{-3}$) GaAs layer. The upper part of the structure was covered with a 100 Å GaAs cap layer. Specially prepared mesas have dimensions of $1 \times 1$ mm$^2$ and were fabricated on the as-grown structure by a lithographic technique. Metal contacts of Au + Ge + Pt alloy were deposited as a frame on the upper part of the mesa and also the doped buffer layer. Besides, we have used n-i-n structures with metal masks (Al layer with thickness 0.12 nm) on the top of structure with lithographycally prepared windows (0.5–1 $\mu$m in diameter).

In this talk we are focused on some recent experiments where the main information on the interwell excitons properties is gained by analyzing the photoluminescence spectra measured under conditions of continuous or pulsed laser excitation and varying excitation intensity, temperature, and character of polarization of resonance optical pumping and magnetic field.

**1.** The luminescence spectra, shown for intrawell (D) and interwell (I) excitons in Fig. 1, were measured with resonance excitation and various applied bias. The optical transitions are schematically shown in insert to Fig. 1. In the region of intrawell PL at zero bias, two lines, 1s HH of free exciton on a heavy hole and of bound exciton, are observed. At negative biases, the spectra exhibit the interwell radiative recombination (I-line), which almost linearly shifts down in energy as the applied voltage increases, in accordance with Stark shift ($eF\Delta z$, where $F$ is the electric field) of spatially quantized levels in QWs. In this case, only the PL line of the charged exciton complex remains in the intrawell PL spectrum. At large negative biases $U > -0.4$ V (when the condition $eFz > E_D - E_I$ is fulfilled, $E_D$, $E_I$ — binding energies of intrawell and interwell excitons), the spectra are dominated only by IE PL, whereas the photoluminescence of the direct intrawell excitons and exciton complexes is considerably less intense. One can see that in considered case PL quantum efficiency is very high and nonradiative transitions can be ignored. This conclusion is based on the fact that, as the applied bias increases, the IEs lifetimes change by several order of the magnitude, while the PL intensity does not change significantly.

The IE PL line at low temperatures ($T \leq 2$ K) and a weak pumping is very broad (FWHM $\approx 4-5$ meV), and the line shape is asymmetric with an extended long-wave length tail and a rather sharp violet offset (see Fig. 1 and 2). These features of PL line of IEs are due to the strong localization of interwell excitons on the random potential fluctuations. In this case, the line width reflects the statistical distribution of random potential amplitudes and pumping is so weak that the average density of spatially separated electrons and holes is $n_{e-h} < 10^9$ cm$^{-2}$. At these densities, the statistically average IEs filling of the random potential domains with the linear scales $l < 1$ $\mu$m in QWs plane is less than unity and the inhomogeneous width of PL spectrum is large enough (see Fig. 2).

The intensity, shape, and width of IE PL line change drastically with increasing intensity of resonance excitation of the direct 1s HH intrawell excitons by circularly polarized light (see Fig. 2). As the pumping increases, IE line narrows down to

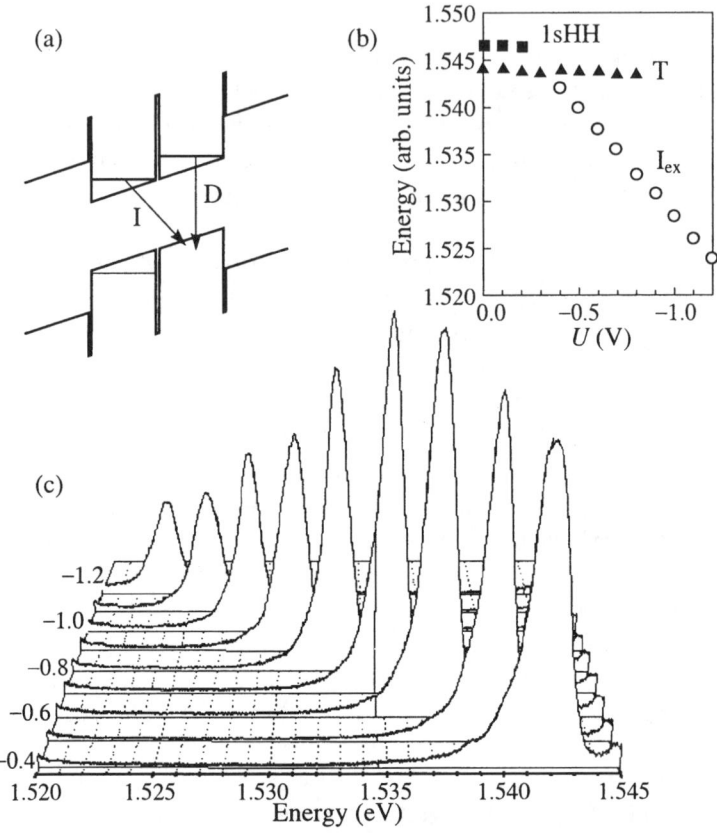

*Figure 1.* (a) Schematic diagram of optical transitions; (b) spectral positions of lines of the direct exciton, excitonic complex (T), and interwell exciton as a function of the electric displacement; and (c) the behavior of luminescence spectra of interwell excitons as a function of the applied voltage (numbers on the right of the spectra correspond to the electric voltage in volts) at $T = 2$ K.

1 meV, i.e. by almost a factor of four. In this case, the maximum intensity increases superlinearly and the line contour becomes almost symmetric. In the region of strong narrowing, the IEs line shifts to lower energy by a value of up to 1.5 meV as pumping increases. Only at rather strong pumping $P > 6$ W/cm$^2$, this line shifts to higher energy and broadens. The high energy line shift testifies to the screening of the applied electric field when the IE density becomes large. Therefore, by using the Gauss formula, one can obtain upper estimation for the IE density from spectral line shift. This estimation for IE density gives $n = 3 \times 10^{10}$ cm$^{-2}$ when the line width is minimal (around 1 meV). A considerable narrowing of the IE PL line is observed for the negative voltages in the range $-0.5$ to $-1.2$ V. At large negative biases, similar narrowing of the IE PL line occurs at appreciably smaller pumping values.

Strong narrowing of the IE PL line at low temperatures suggests that, as the

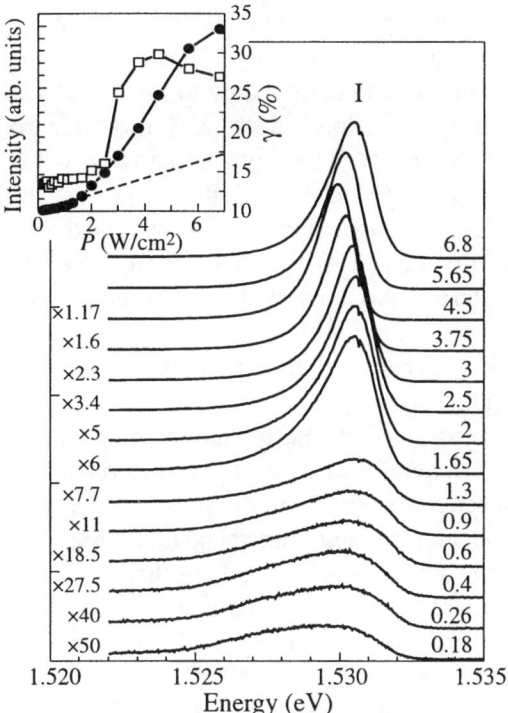

*Figure 2.* Photoluminescence spectra of interwell excitons (line I) at various powers of the resonance excitation of the direct 1sHH excitons as a function of the excitation by circularly polarized light ($\sigma^+$), applied displacement $-1$ V, and $T = 2$ K. Numbers on the right of the spectra correspond to pumping intensities in W/cm$^2$, those on the left correspond to the scale factors in respect to the two upper spectra. The inset presents the line intensity of interwell excitons (round symbols, left scale) and its degree of circular polarization (square symbols, right scale) as functions of the power density. The dashed line is an extrapolation of the linear dependence of the intensity.

excitation intensity increases, IEs first fill the localized states caused by random potential fluctuations. Each such state can be occupied by only one interwell exciton because of strong dipole-dipole repulsion between excitons. After the localized states are filled and upon further increase in pumping, IEs find themselves above the percolation threshold (or mobility threshold associated with strong localization effect) and becomes delocalized. These are precisely the delocalized IEs for which the narrow PL line shifts to lower energies with increasing IE density (see Fig. 2). This conclusion indicates that, despite the dipole-dipole repulsion, the ground state energy of interacting IEs decreases with increasing exciton density. This behavior is signature of a degenerate boson system with increasing bose-particles density at sufficiently low temperature.

The degree of circular polarization of the interwell exciton PL exhibits interesting behavior in the region where its intensity increases superlinearly with increasing resonant excitation power (see Fig. 2). In our experiments using circularly polarized light, intrawell, completely spin-aligned 1sHH excitons were created for which the heavy-hole angular momentum is $J_h = +3/2$ and the electron spin $S_z = -1/2$. As a result of carriers tunneling and binding to form interwell excitons, and also as a result of spin-lattice relaxation and strong spin-orbit interaction for the holes, the spin memory of the interwell excitons is partially lost but still remains appreciable and is almost 5–10% at low excitation power, although the interwell excitons are mainly localized under these conditions and the corresponding PL line is inhomogeneously broadened. At fixed excitation power the degree of circular polarization of the interwell excitons PL decreases monotonically with increasing bias voltage. As the power density of the resonant photoexcitation increases when the IEs PL-line exhibits substantial narrowing, the degree of circular polarization of the corresponding line abruptly several-fold increases. Assuming that the rate of spin relaxation varies little with increasing pumping, this increase in the degree of circular polarization is naturally attributed to a reduction in the lifetime of interwell exciton. This is deduced from a simple rate expression linking the degree of circular polarization with the lifetime and spin relaxation time: $\gamma = \gamma_0/(1 + \tau_d/\tau_s)$. Using this expression we can easily conclude that observed significant growth of the degree of circular polarization of the interwell excitons PL with increasing excitation power is a consequence of at least fivefold increase in their radiative annihilation.

When direct 1sHH excitons were excited resonantly by linearly polarized light, as the pump power increased we observed a threshold increase in the linear polarization of IE narrow PL line (due to alignment of interwell excitons) in the region of superlinearly increasing intensity. Under low excitation densities, when the interwell excitons were strongly localized due to fluctuation of random potential, their spectrum remained weakly polarized. Experiments results clearly show an abrupt increase in the linear polarization of the interwell excitons PL in a narrow range of pumping. The alignment of the lateral dipole momentum of the interwell excitons reached a maximum and then decreased on further increasing of excitation power, when the screening of the applied electric voltage became appreciable.

Thus, as the concentration of interwell excitons increases, the intensity of the corresponding PL-line increases superlinearly and the line exhibits strong narrowing while its degree of polarization increases, which indirectly indicates that the radiative lifetime of the interwell excitons is reduced. These effects are very sensitive to temperature. We observed that when the temperature increased above critical values at high constant pump power, the line width of the interwell excitons abruptly increased and the degree of circular polarization and the PL line width dropped to its previous level. The temperature behavior of the degree of circular polarization and the PL line width of the interwell excitons are illustrated in Fig. 3. It can be seen that the critical temperature at which such dramatic spectral changes occurred in this case is $T_c \leq 6$ K ($\Delta T = \pm 1$ K).

2. Now we shall discuss the time evolution of the luminescence spectra of

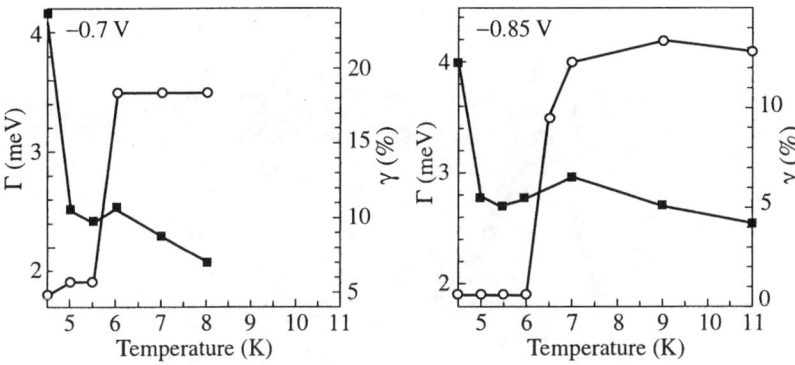

*Figure 3.* Temperature dependence of the luminescence line width of interwell excitons Γ (round symbols, left scale) and its degree of circular polarization γ (square symbols, right scale) for electric displacements of −0.7 and −0.85 V.

intrawell and interwell excitons and also kinetics of the intensities of the corresponding PL spectra under conditions of pulsed resonance photoexcitation using a tunable picosecond laser. Figure 4 demonstrates the time evolution of PL spectra under pulse excitation measured with various time delays relative to the exciting laser pulse at $T = 1.8$ K and the applied voltage $U = +0.5$ V. Time-resolved spectra were detected under excitation of circularly polarized light ($\sigma^+$ components, full curves) resonantly with 1s HH intrawell excitons and were analyzed in $\sigma^+ - \sigma^-$ circular polarizations. The excitation power density was selected in such a way that, on integrating over all laser pulses, the PL line width was a minimum at the highest degree of circular polarization. This condition was fulfilled at the peak power density $P \approx 30$ kWt/cm$^2$. An estimation of the IE concentration in this case gives the value $n \approx 3 \times 10^{10}$ cm$^{-2}$. It is under these conditions, as previously shown, the collective properties of IEs are most pronounced.

At zero delays, IE PL line is strongly polarized (more than 70%) on the high energy side and has a width of about 3 meV. As the time delay increases, the line intensity increases superlinearly, and the line narrows and somewhat displaces toward the long-wavelength part of the spectrum. At delays of 5–6 ns, its width is a minimum and comprises about 1 meV. The maximum intensity of IE PL line is reached at delays of about 3 ns. This time is necessary for the formation of IE upon the resonance tunneling of electrons and holes to the neighboring QWs and their relaxation in the energy to the equilibrium values of density and temperature. This behavior of the IE PL line is also demonstrated by PL decay curves (see inset in Fig. 4). It is evident that it takes about 2.5–3 ns after the arrival of a laser pulse to attain the maximum intensity of the PL line. In this case, the maximum PL intensity in the other ($\sigma^-$) polarization is additionally displaced with respect to the beginning of the laser pulse action by approximately 1 ns. At the same time, the maximum degree of circular polarization corresponds to the beginning of the laser pulse action.

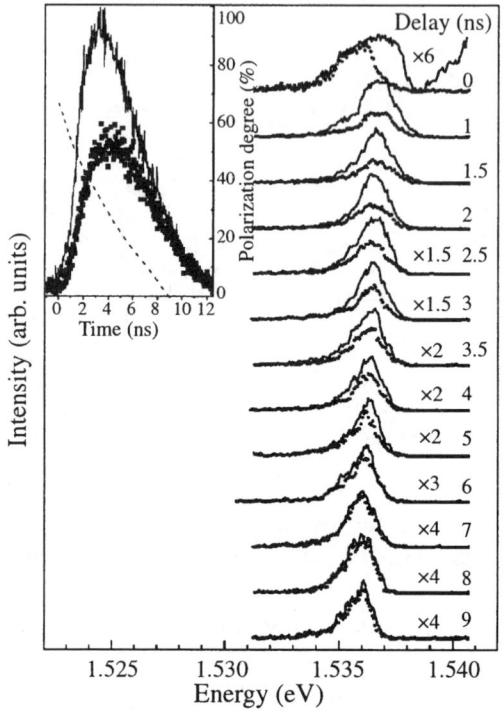

*Figure 4.* Time-delayed IE PL spectra (numbers to the right from spectra correspond to the time delay relative to the laser pulse in ns) under the resonance excitation of direct 1sHH exciton by circularly polarized light ($\sigma^+$, full curves) integrated over 1 ns and recorded at an applied bias of +0.5 V, $T = 2$ K and $P = 30$ kW/cm$^2$. The inset displays IE PL decay curves ($\sigma^+$, full curves, and $\sigma^-$, square symbols) for detection at the spectral position 1.5365 eV. The dashed curve (the scale on the right) gives the degree of circular polarization.

The time evolution of the degree of circular polarization reflects the process of IE spin polarization. It is evident in Fig. 4 that this process is different along the PL line contour. In the first nanosecond, just after arrival the laser pulse, the IE PL line is strongly polarized only on the high energy side of the spectrum, next the degree of polarization remains equal along the entire line contour for approximately 3 ns, and then only the high energy part of the spectrum remains polarized.

Temperature is an important factor that, along with the excitation power density, affects all the kinetic events in which interwell excitons participate. Figure 5 displays the FWHM of the PL line and the degree of circular polarization as function of temperature. The high sensitivity of the PL line width to variation of temperature is pronounced well already from 2 to 4.5 K. There is no such a strong narrowing of the line with increasing delay in the time dependencies at $T = 4.5$ K, and the dependence itself is not nonmonotonic, as it is at $T = 2$ K. The effect of line narrowing at $T = 10$ K is virtually absent. Simultaneously, the behavior the degree of circular polarization also changes significantly. The initial

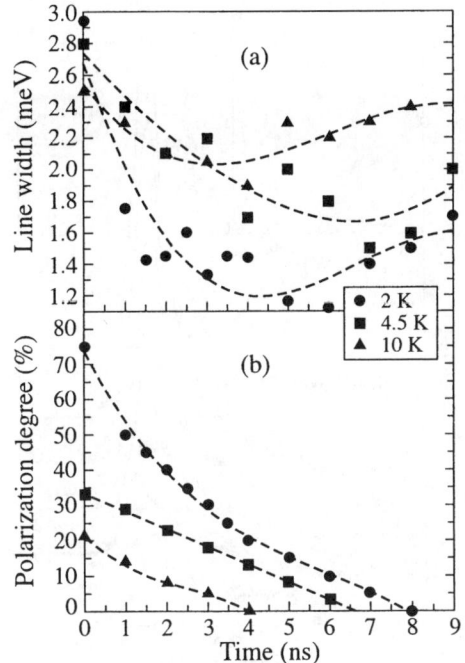

*Figure 5.* (a) Temperature dependence of the IE PL line width as a function of time delay at an applied bias of +0.5 V, $T = 2$ K and $P = 30$ kW/cm$^2$. Dashed curves are given for convenience. (b) Temperature dependence of the degree of circular polarization of the IE PL line under excitation by circularly polarized light ($\sigma^+$) as a function of time delay for the spectral position 1.5365 eV. Dashed curves are given for convenience.

degree of circular polarization at $T = 4.5$ K is only about 30% (more than 70% at $T = 2$ K); it drops monotonically, and becomes equal to zero at delays of more than 5 ns. At $T = 10$ K, the maximum degree of circular polarization is less than 20%, and the line polarization vanishes after 3 ns.

3. All the outlined above experimental results confirm the previously made suggestion about collective nature of the interwell excitons of high enough density at low temperatures [7]. Qualitatively, the origination of the collective exciton phase looks as follows. At $T \leq 2K$, as the density of the optical excitation power increases, IEs fill the random potential relief in the QW plane, which arises from residual impurities, defects, and other structure imperfections. This is manifested as a narrowing of the PL line with an increase in pumping, which ceases to reflect the statistical distribution of fluctuation amplitudes of chaotic potential. In our opinion, the sharp narrowing of the PL line and the superlinear rise in its intensity cannot be associated with only the attainment of the percolation threshold by IEs, because this event is very sensitive to temperature, though it has no sharp temperature boundary. An essential amendment was made in [7], whose authors claimed that interwell excitons condensation (analog of BEC) in real systems can

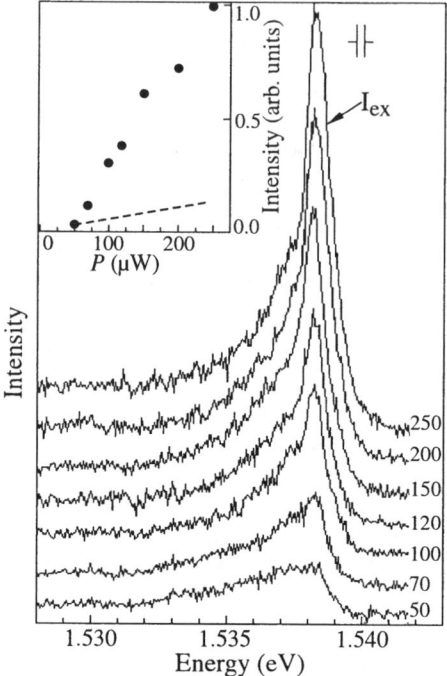

*Figure 6.* Photoluminescence spectra of interwell excitons (line $I_{ex}$) at various powers of the resonance excitation of the direct 1sHH exciton, applied displacement 0.3 V, and $T = 1.51$ K. Numbers on the right of the spectra correspond to pumping intensities in $\mu$W. The inset presents the line intensity of interwell excitons (solid symbols) as functions of the power density. The dashed line is an extrapolation of the linear dependence of the intensity.

most probably occurs in regions with lateral confinement (in QW plane). In studied structures, where epitaxial growth interruption technique has been used (in our case, growth interruption at heterojunction boundaries reached 2 min), large-scale fluctuations of the widths of QWs and barriers arise in plane of heteroboundaries (the geometrical size of fluctuations in the direction of epitaxial growth is of the order of one monolayer). The characteristic lateral linear scales of such fluctuations can reach up to 1 $\mu$m [10].

Because of such fluctuations, lateral wells or domains arise in QWs themselves. As was judged from the characteristic doublet structure in PLE spectra of interwell excitons measured in our samples, the potential depth of such domains reaches around 1.5 meV. Interwell excitons can accumulate in these domains, because the boundaries of lateral domains prevent IEs from spreading randomly in QW plane. We believe that in these domains where the density of excitons and their temperature reach critical values the IEs demonstrate a collective behavior. Thus, as the exciton density increases, the chaotic fluctuations connected with defects become shielded to certain extent. With a further increase in density exceeding percolation threshold,

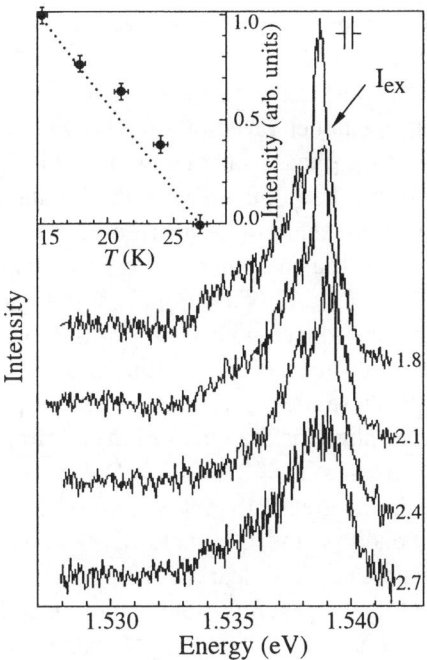

*Figure 7.* Photoluminescence spectra of interwell excitons (line $I_{ex}$) at various temperatures at the resonance excitation of the direct 1sHH exciton, applied displacement 0.3 V, and $P = 150~\mu W$. Numbers on the right of the spectra correspond to temperature in K. The inset presents the line intensity of interwell excitons (solid symbols) as functions of the temperature. The dashed line is an extrapolation of the linear dependence of the intensity $I_T \propto (1 - T/T_c)$.


the IEs become delocalized within macroscopically large domains; however, the dimensions of the domains in which excitons are accumulated spatially confine their motion. IEs are composite bosons, therefore excitons can condense upon reaching the critical concentration and temperature values. Assuming the translational exciton mass $m_{ex} = 0.25 m_0$ and the domain size $0.5~\mu m^2$ we obtain critical temperature $T_c \approx 3$ K for exciton concentration $n_{ex} \approx 3 \times 10^{10}$ cm$^{-2}$. This is very close to the values observed experimentally. It should also be noted that the measurements under our experimental conditions are carried out simultaneously with several tens of domains with regard to the fact that the smallest geometrical size of excitation spot on a sample from which PL spectra are detected is about 20 $\mu$m.

Considering the dispersion of the lateral sizes of domains and the integration of PL spectra from domains differing in lateral size, we are not surprised that we do not observe a sharp threshold of critical behavior in temperature in the experiments above described. For the same reasons, the smallest observed PL line width (about 1 meV) is inhomogeneous in fact, because domains differing in size contribute to the line width. This consideration is confirmed by specially performed experiments on DQWs samples covered by metal masks with lithographycally prepared windows with a minimum size less than 0.5 $\mu$m. Figure 6 and 7 illustrate behavior of IE PL spectra just detected from a window with diameter less than 1 $\mu$m on the sample surface under variation of excitation power and temperature, respectively. One can see significant narrowing of IE PL line width on excitation power growth (FWHM less than 350 $\mu$eV) with simultaneous low energy shift of spectral line, besides more sharp offset is seen in the temperature dependence of the line width. Therefore, the sharp narrowing of the interwell exciton PL line observed experimentally at $T < T_c$ ($T_c < 3.4$ K) and the low energy shift of this line (about 0.5 meV) in accordance with filling of the lowest energy state in the domain are a serious manifestation of Bose properties of interwell excitons.

4. The condensed IE phase must exhibit coherent properties. This means that IEs must possess the same phase on the scales of the de Broglie wavelength, which is close to linear domain sizes (at $T = 2$ K $\lambda_{dB} = h/\sqrt{\pi m k T} = 1.5 \times 10^3$ Å) and is more than order of magnitude greater than the exciton Bohr radius. In our opinion, this phase coherence of excitons must, in its turn, affect the radiative annihilation rate, and this rate must increase. It is clearly evident from the kinetic of PL spectra that the lifetime of the collective exciton state is more than an order of magnitude shorter than the PL time decay of localized IEs. Thus, the increase in the radiative decay rate of IEs and the resulting increase in the degree of circular polarization of PL are particular manifestation of the coherence of the collective exciton state.

## References

1. Fukuzawa, T., Mendez, E.E., and Hong, J.M. (1990) Phase transition of an exciton system in GaAs coupled quantum wells, *Phys. Rev. Lett.* **64**, 3066.
2. Butov, L.V. (1996) Anomalous transport and luminescence of indirect excitons in coupled quantum wells, *Proc. 23d Int. Conf. on Physics of Semiconductors*, Ed. M. Schefler and R. Zimmermann, Berlin.
3. Krivolapchuk, V.V., Moskalenko, E.S., Zhmodikov, A.L. *et al.* (1999) Collective properties of spatially indirect excitons in asymmetric double quantum wells, *Solid State Commun.* **111**, 49.
4. Larionov, A.V., Timofeev, V.B., Hvam, J.M. *et al.* (2000) Interwell excitons in GaAs/AlGaAs double quantum wells and their collective properties, *JETP* **90**, 1093.
5. Butov, L.V., Ivanov, A.L., Imamoglu, A.E., *et al.* (2001) Stimulated scattering of indirect excitons in coupled quantum wells: signature of degenerate Bose-gas of excitons, *Phys. Rev. Lett.* **86**, 5608.
6. Larionov, A.V., Timofeev, V.B., Hvam, J.M. *et al.* (2002) Collective state of interwell excitons under pulse resonant photoexcitation, *JETP Lett.* **75**, 233.
7. Xuejun, Zhu, Littlewood, P.L., Hybersten, M.S., and Rice, T. (1995) Exciton condensate in semiconductor quantum well structures, *Phys. Rev. Lett.* **74**, 1633.
8. Fernandez-Rossier, J. and Tejedor, C. (1997) Spin degree of freedom in two dimensional exciton condensate, *Phys. Rev. Lett.* **78**, 4809.
9. Lozovik, Yu.E. and Berman, O.L. (1997) Phase transitions in system of spatially separated electrons and holes, *JETP* **84**, 1027.
10. Brown, S.W., Kennedy, T.A., Gammon, D. *et al.* (1996) Spectrally resolved overhauser shifts in single quantum dots, *Phys. Rev.* **B 54**, R17339.

# INTERNAL TRANSITIONS OF NEGATIVELY CHARGED EXCITONS AND MANY-ELECTRON EFFECTS IN GaAs QUANTUM WELLS

B.D. MCCOMBE, A.B. DZYUBENKO, H.A. NICKEL, T. YEO,
C.J. MEINING, T. SANDER, A. PETROU
*Department of Physics, University at Buffalo,
SUNY, Buffalo, NY 14260, USA*

A.YU. SIVACHENKO
*Department of Physics, University of Utah,
Salt Lake City, UT 84112, USA*

Abstract. States of charged magnetoexcitons in quasi-two-dimensional systems are investigated. The effect of excess electrons on internal transitions of negatively charged excitons $X^-$ in GaAs quantum wells is studied experimentally by optically detected resonance (ODR) spectroscopy and theoretically. An experimentally observed blue shift with excess electron density is explained in terms of collective excitations – magnetoplasmon bound to a mobile valence band hole. A possibility to observe photoluminescence of dark $X^-$ states in angle-resolved experiments is also discussed.

Key words: charged magnetoexciton, optically detected resonance, quantum well, internal transitions

## 1. Introduction

Since the initial observations of negatively charged excitons ($X^-$) or trions in CdTe [1] and GaAs [2] quantum wells (QWs), there has been considerable experimental (e.g., [3–5] and references therein) and theoretical (e.g., [6–12] and references therein) work on the bound states of both negatively (two electrons and one hole) and positively (two valence band holes and one electron) charged electron-hole complexes in quasi-two-dimensional (2D) semiconductor systems. This work has almost exclusively focused on the interband optical and magneto-optical properties and binding energies of "bright" and "dark" $X^-$ states [9, 10, 13–15]. The observation of both spin singlet and triplet bound states has been reported by many groups. Confinement enhances the binding of the charged complexes significantly [16, 17].

The negatively charged exciton is often considered to be the semiconductor analog of the negatively charged hydrogen ion ($H^-$) of atomic physics, which provides a central example of the role of electron–electron interaction in determining the energy eigenstates in a Coulomb potential. In addition, however, these charged complexes in QW-structures offer opportunities for exploring new physics relative to $H^-$: 1) the positively charged particles in semiconductors (holes) have a mass comparable to the electron mass; 2) the magnetic field energy can be dominant at laboratory magnetic fields; 3) the system is quasi-2D; and 4) in QW systems the density of excess electrons in the wells can be controlled and varied (at low excess electron densities, $X^-$ is the *stable ground state* of the photoexcited carrier system). The electron-electron interaction leads to large corrections to the single particle states, which are directly measurable by optical experiments. The manner in which the electron-hole system in quasi-2D structures evolves with excess electron concentration from isolated neutral excitons to a few-hole-many-electron plasma, and the effects of a magnetic field on this evolution have been studied extensively by interband emission and absorption (see, e.g., [2, 4, 5, 13–15, 18–22]).

At zero field the $X^-$-feature at low electron densities evolves into the Fermi-edge singularity of the electron-hole plasma with increasing density; the density at which the crossover takes place depends on the inherent disorder in the sample [2, 4, 5]. For samples in which the electrons and holes are confined in the same spatial region (along the growth direction) with increasing magnetic field the magneto-PL or magnetoabsorption changes rather abruptly at Landau level (LL) filling factor $\nu = 2$ from Landau-level to Landau-level-like transitions that are linear in B ($\nu > 2$) to exciton-like behavior ($\nu < 2$) irrespective of the electron density [4, 21]. The nature of the "exciton-like" states at high fields and at high electron densities is still not clear [4, 13–15] and recent work on the ZnSe quantum well system [18], which has very large Coulomb energy relative to Fermi energies has served to bring this lack of understanding into sharper focus. A full appreciation of this complex many-body system depends on first understanding the behavior of neutral and negatively charged exciton systems in the presence of a low density of excess electrons. Internal exciton transitions (IETs) probe directly the ground and excited states and yield important insight into their properties [22, 23]. The IETs thus offer an additional tool for probing the excitonic state in the dilute situation and its evolution with excess electron density and magnetic field.

The $X^-$-complex, although superficially similar to its close relative, the negatively charged donor ($D^-$) [24] differs in one very important respect: $D^-$ is a localized (fixed) positive charge binding two electron, while in ideal systems the positive charge in $X^-$ is free. A symmetry associated with the resulting free center of mass motion (magnetic translational invariance) leads to a new, exact electric-dipole selection rule that imposes severe limitations on interband photoluminescence (PL) transitions of isolated $X^-$ complexes [9]. It also *prohibits* certain bound-to-bound internal intraband transitions between families of $X^-$ states belonging to different Landau levels (LLs) [9]. In particular, the $X^-$-singlet and -triplet bound-to-bound transitions that dominate the spectra of $D^-$ [24] are *strictly forbidden* for $X^-$; the selection rule permits only *bound-to-continuum* transitions to the first electron LL that gain strength with increasing magnetic field [9].

The present experimental observation and studies of internal transitions of $X^-$ in GaAs QW structures as functions of excess electron density and magnetic field provide new insight and a deeper understanding of this complex system. At low excess electron densities, intraband triplet bound-to-bound transitions are absent and both singlet and triplet internal transition features appear as bands, with positions in excellent quantitative agreement with numerical calculations for the bound-to-continuum transitions. At higher excess electron densities $\nu > 2$ internal transitions are not observed, while for $\nu < 2$ internal transitions *are* observed, but they are blue-shifted in energy from their low-density counterparts. This work thus provides clear experimental verification of the predicted consequences of the magnetic translational symmetry [9] for charged electron-hole complexes, and also shows that the feature identified as $X^-$ in intraband measurements in high density samples represents the collective response of a few-hole/many-electron system — a magnetoplasmon bound to a mobile valence band hole [25, 26]. In the following sections we present an outline of the relevant theory of the internal transitions of negatively charged excitons, the extensions to the case of excess electrons and the resulting blue shift of the internal transitions, and some recent work on the angular dependence of dark states in magneto photoluminescence. We follow this by a brief description of the experimental technique and samples. In Section 4 we present results of ODR experiments, analysis and comparison with theory, initially on negatively charged excitons followed by the effects of excess electrons in the quantum wells. Finally, we summarize and conclude.

## 2. Theory

The theory of charged quasi-2D excitons in magnetic fields was developed in a number of papers [6–12] (see also [17, 25] and references cited therein). Wojs and Hawrylak [6] showed that in strong magnetic fields there exists a bound triplet $X^-$ state in a laterally confined quantum dot. This triplet state turns out to be the only bound $X^-$ state that exists in strictly-2D systems in the zero LL in the limit of strong magnetic fields; there are no bound singlet $X^-$ states in this limit [7, 8]. At zero magnetic field, on the other hand, only the singlet bound $X^-$ state exists in 2D or in 3D systems (see, e.g., [16, 17]). This means that (at least) in 2D systems the e-e repulsion overcomes the e-h attraction with increasing magnetic field so that the triplet electron configuration becomes energetically favorable. In quasi-2D systems, therefore, a singlet-triplet crossing should occur at finite magnetic fields. The values of the magnetic field at which the crossing occurs turn out to be very sensitive to the particulars of the systems such as the electron and hole effective masses, the width and the depth of the QW, and to the approximations made [8, 12, 17].

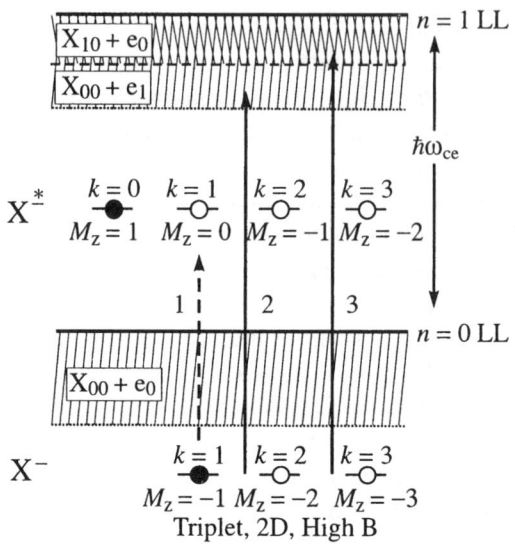

*Figure 1.* Schematic energy level diagram of the triplet X⁻ state in two lowest electron Landau levels in the strictly-2D system in the limit of strong magnetic fields. After Ref. [9].

The coupling of the center-of-mass and internal motions in magnetic fields plays a qualitatively important role for charged electron-hole (e-h) complexes [9, 25, 27]. It leads, in particular, to stringent limitations on allowed optical transitions of charged excitons in magnetic fields that follow from the exact selection rules. In order to illustrate this, we consider the eigenspectra of the X⁻ triplet states in the strictly-2D high-field limit (Fig. 1). The X⁻ eigenstates in addition to the spin quantum numbers are labeled by the following exact orbital quantum numbers: the total angular momentum projection, $M_z$, and by the oscillator quantum number $k = 0, 1, \ldots$. The latter characterizes the center of the cyclotron motion of the charged complex as a whole, and incorporates the Landau degeneracy of energy. The degenerate states that are characterized by different quantum numbers $k = 0, 1, \ldots$ comprise a *family* of states that starts with a Parent State that has $k = 0$ (roughly speaking, rotates about the origin) and a particular value of $M_z$, which can only be determined by solving the Shroedinger equation. Figure 1 shows the bound triplet states in the two lowest electron LLs and illustrates this classification of states. Importantly, in each family of states the quantum numbers $k$ and $M_z$ are *not independent*: raising $k$ leads to a corresponding lowering of $M_z$ (see Fig. 1). The same classification holds also for unbound charged three-particle states X + e⁻ (neutral magnetoexciton X plus one electron in a scattering state) that form the continua below the free LLs. A more detailed theoretical description and the formalism that allows one to separate one degree of freedom while keeping both axial symmetry (rotations about the **B**-axis) and magnetic translations intact can be found in Refs. [9, 27].

It was shown [9] that the exact optical selection rules for the dipole-allowed intraband internal $X^-$ transitions in the $\sigma^+$ polarization are $\Delta M_z = +1$ and $\Delta k = 0$. Physically, the latter selection rule means that the absorption or emission of the photon – in the dipole approximation – cannot displace the center of the cyclotron motion. Because the quantum numbers $M_z$ and $k$ in each family of states are not independent, the combination of the selection rules prohibits the bound-to-bound transition to the first electron LL (dashed line in Fig. 1). As a result, only the bound-to-continuum photoionizing transitions $T_1$ and $T_2$ to the edges of the two overlapping magnetoexciton (MX) continua in the first electron LL are allowed and gain strength with increasing magnetic field. As detailed theoretical calculations show [23, 26], the same qualitative features hold also for the singlet $X^-$ state that exists in quasi-2D QWs at finite magnetic fields. A comparison between the theory predictions for internal $X^-$ singlet and triplet transitions and the experimental results is given below in Sect. 5.

## 3. How Dark Are "Dark" $X^-$ States?

Magnetic translational invariance has important consequences for magneto-photoluminescence of charged excitons $X^- \to e^- +$ photon. The combination of the exact selection rules for the envelope functions in this case is $\Delta M_z = 0$ and $\Delta k = 0$ and leads to a very simple but powerful result: For a family of $X^-$ states for which the Parent State has the total angular momentum projection $M_z$, the electron in the final state can only belong to the LL with the quantum number $n = M_z$ [9]. This means, in particular, that the triplet $X^-$ state with $M_z = -1$ is dark in PL. This result is a manifestation of the axial and translational symmetries, therefore it holds as long as these symmetries are preserved in the system; for example, even at finite magnetic fields, in quasi-2D QWs with a complex valence band structure, when warping is neglected, *etc.* (cf. with the work in the spherical geometry [10]).

Somewhat surprisingly, there is another symmetry – with a more narrow range of applicability – that also governs interband optical transitions of charged magnetoexcitons. Both symmetries predict, in particular, that the ground $X^-$ triplet state must be dark in photoluminescence. This other so-called "hidden symmetry" [28–31] is applicable to 2D symmetric electron-hole systems in the limit of strong magnetic fields, when no LL mixing is allowed. The requirement on the interaction potentials is that all are identical up to the sign: $U_{ee} = U_{hh} = -U_{eh}$. In such systems the neutral magnetoexcitons (MXs) of zero total momentum $\mathbf{K} = \mathbf{0}$ form an ideal gas of non-interacting composite Bosons. This was first shown diagrammatically by Lerner and Lozovik [28] and later was established by direct quantum-mechanical means in a number of papers [29–31]. A straightforward way is to consider the exact quantum equation of motion [29] $[H_{\text{int}}, Q_0] = E_0 Q_0$, which involves the interaction Hamiltonian of the two-component e–h system in the extreme magnetic quantum limit, $H_{\text{int}}$, and the annihilation operator of a 2D MX of zero center-of mass momentum, $Q_0 = \sum_m a_m b_m$. Here $a_m (b_m)$ is the electron (hole) annihilation operator in zero LL with the oscillator quantum number $m$, $E_0 = \sqrt{\pi/2} e^2/\varepsilon l_B$ is the isolated MX binding energy, and $l_B = (\hbar c/eB)^{1/2}$ is the magnetic length. Physically the above result means that the particles $Q_0$ form an *ideal gas* of composite Bosons that do not interact with each other or with other particles (electrons, holes, other excitons $Q_K$) present in the system.

The relevance of this result for optics of two-dimensional systems (both two-component electron-hole and one-component electron systems) in strong magnetic fields can be understood from considerations of the luminescence operator that describes annihilation of an e-h pair, $L_{PL} = p_{cv} \int d\mathbf{r} \, \Psi_e(\mathbf{r}) \Psi_h(\mathbf{r})$. Here $p_{cv}$ is the interband dipole transition matrix element. After projecting the operator $L_{PL}$ onto zero LLs one obtains an interesting and important result $L_{PL} = p_{cv} Q_0$ that shows that the photoluminescence operator $L_{PL}$, up to some factor, coincides with the annihilation operator of the 2D neutral MX $Q_0$. Taking into account that the $\mathbf{K} = \mathbf{0}$ 2D MXs form an ideal gas in symmetric electron-hole systems [28, 29], we arrive to the well-known result [30,31] that the PL emission from such systems consists of a single line at the energy of an *isolated* 2D neutral MX. Formally, this follows from the fact that the matrix elements of $L_{PL}$ are non-zero only between the exact many-body states of the Hamiltonian $H_{int}$ that differ in energy exactly by the isolated MX binding energy $E_0$. Physically this result means that the PL is not sensitive to many-body correlations in strictly-2D symmetric systems in the strong magnetic field limit. In particular, when applied to the $X^-$ triplet state that has energy $-1.043 E_0$, which is lower than that of the neutral MX $-E_0$, the hidden symmetry prohibits [7] PL of the $X^-$ triplet.

The above results hold, however, only for emission of light *perpendicular* to the QW. Importantly, because the translational invariance is broken along the growth direction in quasi-2D systems, all excitons within the light cone, i.e., with momenta

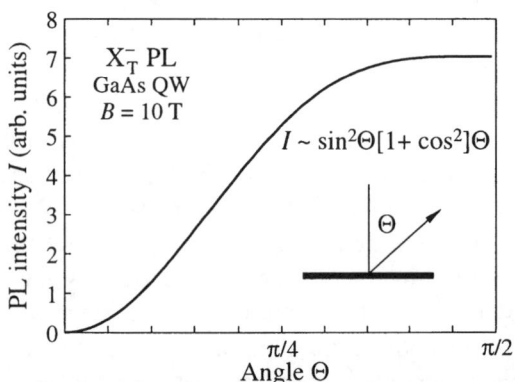

*Figure 2.* Angular dependence of the quadrupole emission from the "dark" triplet $X^-$ in 2D system in the limit of strong magnetic fields. The inset shows the considered geometry.

$|\mathbf{K}| < K_0 \approx nE_{gap}/\hbar c$, are optically active, here $n$ is the index of refraction; for GaAs $K_0 \approx 2.7 \times 10^5$ cm$^{-1}$. Optically active excitons with $\mathbf{K} \neq 0$ emit light at finite angles relative to the normal of the 2D structure (see inset to Fig. 3). In the strong magnetic field limit, the PL operator describing such emission is $L_{PL}(\mathbf{q}) = p_{cv} \int d\mathbf{r} e^{i\mathbf{q}\cdot\mathbf{r}} \Psi_e(\mathbf{r}) \Psi_h(\mathbf{r})$, where $\mathbf{q} = (\mathbf{K}, q_z)$ is the momentum of the 3D photon. Projecting this operator onto zero LLs we obtain $L_{PL}(\mathbf{q}) = p_{cv} Q_\mathbf{K}$ which means that now the operators of magnetoexcitons with finite in-plane momenta $\mathbf{K}$ are involved. Since finite momenta excitons $Q_\mathbf{K}$ do not form an ideal gas, the angle-resolved PL should therefore be sensitive to many-body correlations.

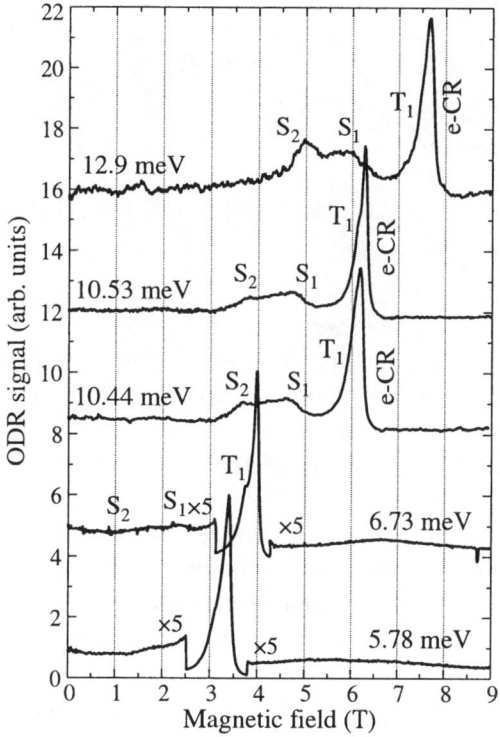

*Figure 3.* ODR spectra for sample 1 at five far infrared laser photon energies. The features labeled $S_1$, $S_2$, and $T_1$ are the singlet and triplet transitions of $X^-$ discussed in the text. Data were obtained by tracking the neutral exciton.

Here, as an important example, we consider the angle-resolved emission of the "dark" $M_z = -1$ $X^-$ triplet state in the strictly-2D high-magnetic field limit. This state can participate in the *quadrupole* emission of light, when the photon is allowed to carry away the angular momentum $m_z=-1$. The angle dependence of the intensity of the quadrupole emission is given by $I_T(\Theta) \propto (K_0 a)^2 (1 + \cos^2\Theta) \sin^2\Theta$. Note that this holds in the absence of scattering (by disorder, 2DEG) in which case the emission is zero in the direction perpendicular to the QW: $I_T(\Theta = 0) = 0$ (see Fig. 2). Higher multipole emission channels (corresponding to higher photon magnetic quantum numbers $m_z$ and/or breaking of the $\Delta k=0$ selection rule) are also open but are suppressed as higher powers of the dimensionless parameter $(K_0 a)^2$, where $a$ is the effective size of the system. The $X^-$ triplet effective size turns out to be rather large $a \approx 4.1 l_B$ so that at $B = 10$ T the parameter $(K_0 a)^2 \approx 0.8$ is not small which means that the quadrupole $X^-$ triplet emission is not expected to be very weak – in comparison with that of the "bright" singlet $X^-$ state – and perhaps can be observed experimentally. Note that the "bright" singlet $X^-$ state has $M_z = 0$ and can participate in the usual dipole emission characterized by the PL angular dependence $I_S(\Theta) \propto (1 + \cos^2 \Theta)$. The singlet PL intensity $I_S(\Theta)$ does not contain the factor $(K_0 a)^2$ and has its maximum at $\Theta = 0$. In general, dark $X^-$ states that have $M_z < 0$ and exist in higher LLs [27], become optically-active in higher-multipole emission channels: The lowest multiplicity of allowed PL from an $X^-$ state with some fixed value of $M_z < 0$ is $2|M_z| + 2$.

## 4. Experiment and Samples

Several GaAs/AlGaAs multiple-quantum-well (MQW) samples were investigated by optically detected resonance (ODR) spectroscopy. ODR spectroscopy is a highly sensitive technique in which the photoluminescence (PL) signal excited by a visible or near infrared laser is monitored, typically as a function of magnetic field, while the sample is simultaneously illuminated with a far infrared (FIR) laser beam. Resonant absorption of the FIR beam modulates the intensity of the PL. A particular PL feature is tracked by the spectrometer window as the magnetic field is scanned; changes in the strength of the PL that are synchronous with the chopped FIR beam are detected by a Si photodiode or a photomultiplier tube. Very weak absorption features can be detected under the proper conditions. In the present experiments electrons, holes and excitons are continuously excited with the 632.8 nm line of a He-Ne laser coupled to the sample (at low temperature in the Faraday geometry at the field center of a 15/17 T superconducting magnet) via an optical fiber; PL was collected with a second fiber [32].

Results from five MBE-grown GaAs/ $Al_{0.3}Ga_{0.7}As$ MQW samples are presented in the following sections. Samples 1 and 2 are nominally undoped with 20 nm GaAs wells and $Al_{0.3}Ga_{0.7}As$ barriers: sample 1 – 20 wells with 60 nm barriers; sample 2 – 10 wells with 20 nm barriers. Sample 3 also has 20 nm wells but is modulation $\delta$-doped with Si in the centers of the 40 nm barriers at $2 \times 10^{10}$ cm$^{-2}$, and there are 40 repetitions. Samples 4 and 5 have 24 nm wells and are Si doped in the central third of the barriers at sheet densities of $8 \times 10^{10}$ cm$^{-2}$ and $2.8 \times 10^{11}$ cm$^{-2}$, respectively: sample 4 (5) has 20 (10) wells with 48 (24) nm barriers.

## 5. Results and Discussion

### 5.1. ISOLATED $X^-$

Raw ODR data from sample 1 at several FIR laser photon energies taken while tracking the neutral heavy-hole exciton are shown in Fig. 3. Very similar, but negative-going features are observed when tracking $X^-$. A sharp, strong electron cyclotron resonance (e-CR) and several weaker features are apparent. Three features are attributed to $X^-$ internal transitions ($S_1$, $S_2$ and $T_1$). The shoulder on the low-field side of e-CR, and most apparent at FIR photon energies of 10.53 and 6.73 meV, is ascribed to the dominant triplet ionizing transition $T_1$.

The peak of this band occurs for transitions terminating near the edge of the continuum; so the observed peak is shifted to higher energies than e-CR (lower fields) by an amount equal to the small triplet binding energy; and there is a tail to higher energies (lower fields). The labeling for the singlet transitions $S_1$ and $S_2$ is analogous to that for the triplet in Fig. 1. The singlet binding energy at finite fields and finite confinement is significantly larger than that of the triplet; so the singlet features occur at substantially lower fields.

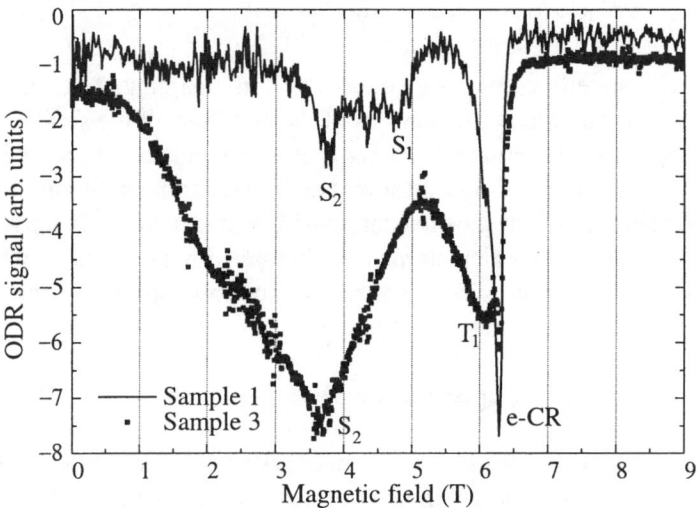

*Figure 4.* Comparison of 4.5 K ODR spectra for samples 1 (solid line) and 3 (dots) at a FIR laser photon energy of 10.44 meV. Data were obtained by tracking the $X^-$ PL.

Figure 4 shows a comparison of the ODR spectra for samples 1 and 3 taken at a FIR photon energy of 10.44 meV while tracking the $X^-$ PL. For sample 3 (with a low density of excess electrons in the wells from the modulation doping) there is a dramatic increase in the relative strength of the features associated with the $X^-$ internal transitions relative to that of e-CR. The weaker $S_1$ line in this sample is masked by the tail of the triplet transition(s). However, studies of the temperature dependence (up to 20 K) of the ODR spectra for this sample reveal the $S_1$ band clearly at about 10 K. There is also evidence of the $T_2$ band in these data. This comparison lends support to the assignment of these features to internal transitions of $X^-$.

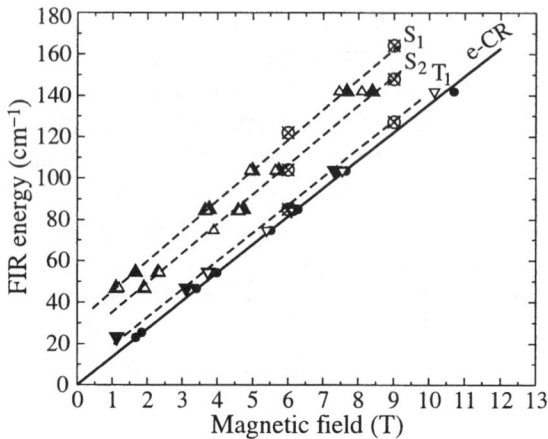

*Figure 5.* Summary plot of internal singlet ($S_1$, $S_2$), and triplet ($T_1$) transitions and comparison with numerical calculations (crossed circles). After Ref. [26].

The most compelling evidence supporting this assignment, however, is provided by detailed numerical calculations of the positions of the singlet and triplet bands for 20 nm wells. Figure 5 shows a summary of the energy positions of these features for samples 1 and 2 vs. magnetic field compared with quantitative numerical calculations (see [26] for details) for a 20 nm well. There is excellent agreement between these calculations and the experimental results, which lends additional credence and provides quantitative support for the assignment of the observed features.

5.2. EFFECTS OF EXCESS ELECTRONS

Figure 6 shows magneto-photoluminescence (MPL) data for sample 5. Overlaying the contour plot are data points indicating peak positions that were obtained by peak-fitting the individual MPL spectra. The solid lines indicating LL-to-LL-like recombination are guides to the eye. Note the sharp demarcation between the linear behavior and the "exciton-like" behavior (nearly quadratic in magnetic field) near filling factor $\nu = 2$. This is characteristic of samples having electrons and holes localized in the same region along the growth direction. To explore further the effects of large densities of excess electrons and the difference in behavior of the magneto-PL at fields above and below $\nu = 2$, we have investigated two modulation-doped samples (samples 4 and 5) by ODR at a series of FIR photon energies such that the internal transitions occur at fields corresponding to $\nu = 2$.

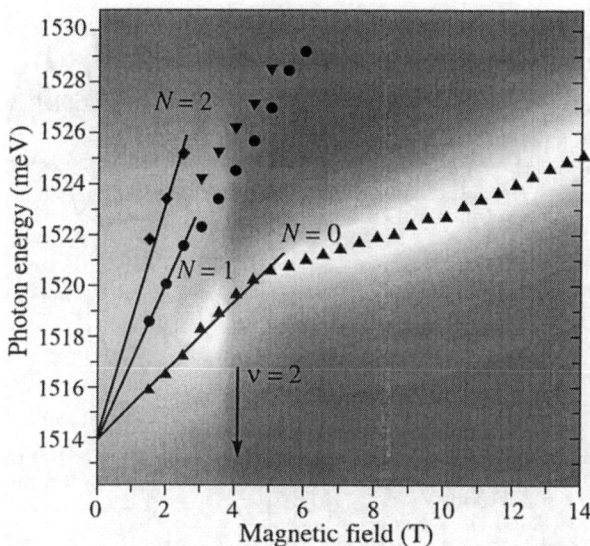

*Figure 6.* Magneto-PL for a MQW sample having nominal electron density in the wells of $2.8 \times 10^{11}$ cm$^{-2}$ (actual density $2.4 \times 10^{11}$ cm$^{-2}$). The straight lines are guides to the eye. The field position of filling factor $\nu = 2$ for a density of $2.4 \times 10^{11}$ cm$^{-2}$ is indicated.

Figure 7(a) shows ODR spectra for sample 4 for several FIR laser photon energies. These data were obtained by tracking the main PL feature with magnetic field. The magnetic field corresponding to filling factor $\nu = 2$ occurs at $B = 1.66$ T for this sample. The sharp feature present in all traces is electron cyclotron resonance (e-CR). Note that no feature is observed in any scan at a higher magnetic field than that of e-CR, in agreement with the selection rule [9, 26], prohibiting bound-to-bound IETs of $X^-$. Two other lines are observed in the ODR spectra for high FIR photon energies. These features, which are blue-shifted from those observed in the undoped GaAs samples, are attributed to bands of ionizing singlet- (vertical arrows) and triplet-like (diagonal arrows) internal transitions of $X^-$. The resonant fields for both singlet- and triplet-like $X^-$ transitions occur for $\nu < 2$ (*above* the field corresponding to $\nu = 2$). At lower FIR photon energies, the triplet feature remains strong, but the singlet feature weakens and is not observable at 6.73 meV (the predicted field position corresponds to $\nu = 2$). For the heavily doped sample 5 these features become generally weaker and are further blue-shifted as shown in Fig. 7(b). Note that when e-CR occurs at fields at or below that corresponding to $\nu = 2$ the line becomes symmetric and no evidence of the triplet-like or singlet-like transitions is seen in either sample.

Figure 8 summarizes the measured positions of the singlet (sample 1) and singlet-like (samples 4 and 5) $S_2$ transitions and illustrates the observed blue shift. This demonstrates clearly that the transitions in the presence of many electrons are collective and the transition energies are clearly influenced substantially by the interaction.

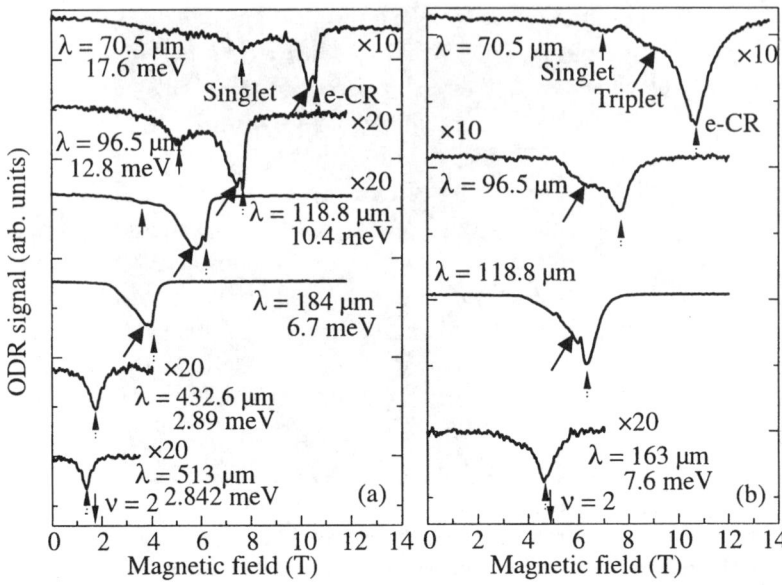

*Figure 7.* ODR scans for two modulation-doped MQW samples. (a) sample 4: $8 \times 10^{10}$ cm$^{-2}$, and (b) sample 5: $2.8 \times 10^{11}$ cm$^{-2}$ (nominal). Arrows indicate the singlet-like (↑) and triplet-like (gray arrows) transitions (after Nickel *et al.* [26]).

In the presence of a large density of excess electrons, the picture of a screened $X^-$ state is no longer applicable. The surprising robustness of the charged trions $X^-$

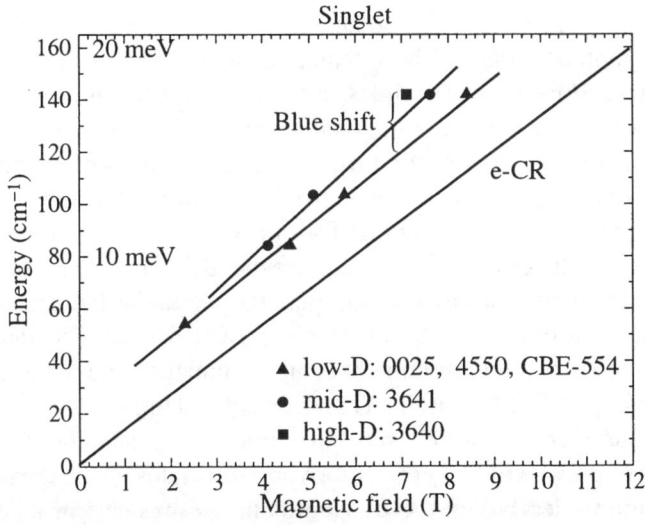

*Figure 8.* Summary of the positions of the singlet and singlet-like $S_2$ transitions for several samples showing the blue shift.

in the magneto-PL spectra for $\nu < 2$ (as if they are embedded in the sea of electrons and barely feel them) can be understood to some extent by "hidden symmetry" arguments [21]. However, as we see from the above, no such exact compensation of the e-e and e-h interactions occurs in intraband transitions [22, 26]. This can be qualitatively understood by considering intraband excitations from the ground state of the 2D system with integer filling of electron LLs $\nu = 1, 2$ containing a low-density gas of mobile valence band holes. The collective excitations correspond to transitions to the correlated e-h final state that can be considered as a *positive* trion $X^+$ that contains the conduction band hole in an otherwise filled zero LL. Similar ideas have been used earlier [33, 34] for a description of the "$D^-$"-transitions [24] in the presence of excess electrons and (for work on interband PL from the 2DEG in high fields (see, e.g, [19, 20, 21]). Our calculations for 2D systems with integer filling factors $\nu = 1, 2$ show that in the region of energies *larger* than $\hbar\omega_{ce}$ there is one prominent absorption peak. It corresponds to a magnetoplasmon bound to the mobile hole. These excitations resemble the magnetoplasma modes bound to the fixed donor ion $D^+$ in the presence of many electrons; but the magnetic translational invariance is broken for the latter. Energies of bound collective modes experience discontinuities at integer $\nu$ and increase with increasing $\nu$, reflecting the enhanced contribution of the exchange-correlation effects and explaining the blue shift. As seen in Fig. 9 the measured blue shifts for the singlet $X^-$ with the mobile hole are, larger than those for the $D^-$ for fields above 5 T. This is in qualitative agreement with theory for the blue shift of the $X^-$ triplet internal transition (see Fig. 9). The

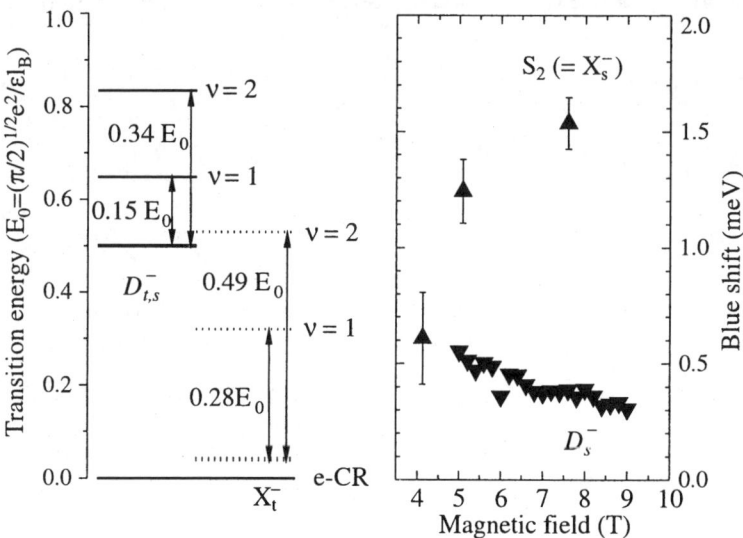

*Figure 9.* Comparison of the blue shift for $X^-$ and $D^-$. Left panel — theory for 2D high field limit. Right panel — experiment. The data were taken for the singlet transition in both cases on modulation-doped samples with the same nominal electron density and approximately the same well-widths (20 nm for the $D^-$ and 24 nm for the $X^-$).

larger blue shift for the $X^-$ results from the diminished negative contribution of the Coulomb e-h interaction to the final state energy for the mobile hole.

## 6. Summary and Conclusions

We have observed singlet and triplet $X^-$ internal transitions in undoped GaAs quantum wells. Results support predictions based on magnetic translational invariance; the positions of the observed ODR peaks are in excellent agreement with detailed numerical calculations. Studies of the dependence of the maxima of the observed bands on electron filling factor $\nu = 1$ have shown: (1) there are no observable singlet- or triplet-like bands appearing at fields corresponding to filling factors $\nu > 2$, and (2) when observed for $\nu < 2$, the bands exhibit a systematic blue shift relative to the corresponding bands in nominally undoped samples. This has been qualitatively explained theoretically in terms of a collective response of the system, a FIR-active magnetoplasmon bound to a mobile hole. We conclude that "$X^-$-like" PL features seen in QWs with excess electrons at $\nu < 2$ represent a collective response of the electron-hole system, which approaches the isolated $X^-$ only for $\nu \ll 1$.

A possibility of probing few- and many-body correlations in 2D systems in high magnetic fields by angle-resolved PL has been indicated. It has been shown, in particular, that the "dark" $X^-$ states become PL-active in the quadrupole emission, which is sensitive to disorder and $X^-$ scattering. This deserves, in our opinion, further theoretical and experimental investigation.

## Acknowledgements

This work was supported in part by NSF grant DMR-0203560.

## References

1. Kheng, G.K, Cox, R.T., Daubigne, Y.M., Bassani, F., Saminadayar, K., and Tatarenko, S. (1993) Observation of negatively charged excitons $X^-$ in semiconductor quantum wells, *Phys. Rev. Lett.* **71**,1752–1755.
2. Finkelstein, G.H. Shtrikman, H., and Bar-Joseph, I. (1995) Optical spectroscopy of a two-dimensional electron gas near the metal-insulator transition, *Phys. Rev. Lett.* **74**, 976–979.
3. Shields, A.J., Pepper, M., Simmons, M.Y., and Ritchie, D.A. (1995) Spin-triplet negatively charged excitons in GaAs quantum wells, *Phys. Rev. B* **52**, 7841–7844.
4. Gekhtman, D., Cohen, E., Ron, A., and Pfeiffer, L.N. (1996) Charged and neutral exciton phase formation in the magnetically quantized two-dimensional electron gas, *Phys. Rev. B* **54**, 10320–10323.
5. Huard, V., Cox, R.T., Saminadayar, K., Arnoult, A., and Tatarenko, S. (2000) Bound states in optical absorption of semiconductor quantum wells containing a two-dimensional electron gas, *Phys. Rev. Lett.* **84**, 187–190.

6. Wojs, A. and Hawrylak, P. (1995) Negatively charged magnetoexcitons in quantum dots, *Phys. Rev. B* **51**, 10880–10885.
7. Palacios, J.J., Yoshioka, D., and MacDonald, A.H. (1996) Long-lived charged multiple-exciton complexes in strong magnetic fields, *Phys. Rev. B* **54**, R2296–R2299.
8. Whittaker, D.M. and Shields, A.J. (1997) Theory of $X^-$ at high magnetic fields, *Phys. Rev. B* **56**, 15185–15194.
9. Dzyubenko, A.B. and Sivachenko, A.Yu. (2000) Charged magnetoexcitons in two-dimensions: Magnetic translations and families of dark states, *Phys. Rev. Lett.* **84**, 4429–4431.
10. Wójs, A., Quinn, J.J., and Hawrylak, P.(2000) Charged excitons in a dilute two-dimensional electron gas in a high magnetic field, *Phys. Rev. B* **62**, 4630–4637.
11. Esser, A., Runge, E., Zimmermann, R., and Langbein, W. (2000) Photoluminescence and radiative lifetime of trions in GaAs quantum wells, *Phys. Rev. B* **62**, 8232–8239.
12. Riva, C., Peeters, F.M., and Varga, K. (2001) Magnetic field dependence of the energy of negatively charged excitons in semiconductor quantum wells, *Phys. Rev. B* **63**, 115302.
13. Yusa, G., Shtrikman, H., and Bar-Joseph, I. (2001) Charged Excitons in the Fractional Quantum Hall Regime, *Phys. Rev. Lett.* **87**, 216402.
14. Schüller, C., Broocks, K.-B., Heyn, Ch., and Heitmann, D. (2002) Oscillator strengths of dark charged excitons at low electron filling factors, *Phys. Rev. B* **65**, 081301.
15. Sanvitto, D., Whittaker, D.M., Shields, A.J., Simmons, M.Y., Ritchie, D.A., and Pepper, M. (2002) Dynamic of spin triplet and singlet trions in a GaAs quantum well, *Phys. Stat. Sol. A* **190**, 809–812.
16. Stebe, B. and Ainane, A. (1989) Ground state energy and optical absorption of excitonic trions in two-dimensional, *Superlat. Microstruct.* **5**, 545–548.
17. Peeters, F.M., Riva, C., and Varga, K. (2001) Trions in quantum wells, *Physica B* **300**, 139–155.
18. Gurevich, A.S., Astakhov, G.V., Suris, R.A., Kochereshko, V.P., Yakovlev, D.R., Ossau, W., Crooker, S.A., and Karzewski, G. (2002) Filling-factor dependence of magneto-luminescence in II-VI QWs with 2DEG, to be published.
19. Cooper, N.R. and Chklovskii, D.B. (1997) Theory of photoluminescence of the $\nu = 1$ quantum Hall state:Excitons, spin waves, and spin textures, *Phys. Rev. B* **55**, 2436–2455.
20. Hawrylak, P. and Potemski, M. (1997) Theory of photoluminescence from an interacting two-dimensional electron gas in strong magnetic fields, *Phys. Rev. B.* **56**, 12386–12394.
21. Rashba, E.I. and Sturge, M.D. (2001) Hidden symmetry and magnetospectroscopy of quantum wells near filling factor $\nu = 2$, *Phys. Rev. B* **63**, 045305.
22. Nickel, H. A., Yeo, T., Meining, C.J., Yakovlev, D.R., Furis, M., Dzyubenko, A.B., McCombe, B.D., and Petrou, A. (2002) Interaction of an electron gas with photo-excited electron-hole pairs in modulation-doped GaAs and CdTe quantum wells, *Physica E* **12**, 499–502.
23. Dzyubenko, A.B., Sivachenko, A.Yu., Nickel, H.A., Yeo, T., Kioseoglou, G., McCombe, B.D., and Petrou, A. (2000) Internal transitions of two-dimensional charged magneto-excitons $X^-$: Theory and experiment, *Physica E* **6**, 156–160.
24. Cheng, J.P. Wang, Y.J., McCombe, B.D., and Schaff, W. (1993) Many-electron effects on quasi-2D shallow donor impurity states in high magnetic, *Phys. Rev. Lett.* **70**, 489–492.
25. Dzyubenko, A.B, Nickel, H.A., Yeo, T., McCombe, B.D., and Petrou, A. (2001) Charged Magnetoexcitons in Two Dimensions: Isolated $X^-$ and Many-Electron Effects, *Phys. Stat. Sol. B* **227**, 365–379.
26. Nickel, H.A., Yeo, T.M., Dzyubenko, A.B., McCombe, B.D., Petrou, A., Sivachenko, A.Yu., Schaff, W., and Umansky, A. (2002) Internal transitions of negatively charged magneto-excitons and many body effects in a two-dimensional electron gas, *Phys. Rev. Lett.* **88**, 056801.
27. Dzyubenko, A.B. (2002) Charged hydrogenic problem in a magnetic field: Noncommutative translations, unitary transformations, and coherent states, *Phys. Rev. B* **65**, 035318.

28. Lerner, I.V. and Lozovik, Yu.E. (1981) Two-dimensional electron-hole system in a strong magnetic field as an almost ideal exciton gas, *JETP* **53**, 763–772.
29. Dzyubenko, A.B. and Lozovik, Yu.E. (1984) Quasi-2D electron-hole pair condensate in strong magnetic fields, *Sov. Phys. – Solid State* **26**, 938.
30. MacDonald, A.H. and Rezayi, E.H. (1990) Fractional quantum Hall effect in a two-dimensional electron-hole fluid, *Phys. Rev. B* **42**, 3224–3227.
31. Apalkov, V.M. and Rashba, E.I. (1992) Interaction of excitons with an incompressible quantum liquid, *Phys. Rev. B* **46**, 1628–1638.
32. Kono, J., Lee, S.T., Salib, M.S., Herold, G.S., Petrou, A., and McCombe, B.D. (1995) Optically detected far-infrared resonances in doped GaAs quantum wells, *Phys. Rev. B* **52**, R8654–R8657.
33. Dzyubenko, A.B. and Lozovik, Yu.E. (1993) Localized magnetoplasma and spin excitations in 2D electron-hole system in a high magnetic field, *JETP* **77**, 617–624.
34. Hawrylak, P. (1994) Many-electron effects on donor states in a 2D electron gas in a strong magnetic, *Phys. Rev. Lett.* **72**, 2943–2946.

# POSITIVELY AND NEGATIVELY CHARGED TRIONS IN ZnSe-BASED QUANTUM WELLS

W. OSSAU, G.V. ASTAKHOV[†], D.R. YAKOVLEV[†],
W. FASCHINGER
*Physikalisches Institut der Universität Würzburg,
97074 Würzburg, Germany*
[†] *Ioffe Physico-Technical Institute, Russian Academy of Sciences,
194017, St Petersburg, Russia*

V.P. KOCHERESHKO
*Ioffe Physico-Technical Institute, Russian Academy of Sciences,
194017, St Petersburg, Russia*

J. PULS, F. HENNEBERGER
*Humboldt-Universität zu Berlin, Institut für Physik,
10115 Berlin, Germany*

S.A. CROOKER, Q. MCCULLOCH
*National High Magnetic Field Laboratory,
Los Alamos, New Mexico 87545, USA*

A. WAAG
*Abteilung Halbleiterphysik, Universität Ulm,
89081 Ulm, Germany*

Abstract. Excitons and charged excitons (trions) are investigated in ZnSe-based quantum well structures with (Zn,Be,Mg)Se and (Zn,Mg)(S,Se) barriers by means of magneto-optical spectroscopy. Binding energies of negatively ($X^-$) and positively ($X^+$) charged excitons are measured as functions of quantum well width, free carrier density and in external magnetic fields up to 47 T. The binding energy of $X^-$ shows a strong increase from 1.4 to 8.9 meV with decreasing quantum well width from 19.0 to 2.9 nm. The binding energies of $X^+$ are about 25% smaller than the $X^-$ binding energies in the same structures. The magnetic field behavior of $X^-$ and $X^+$ binding energies differ qualitatively. With growing magnetic field strength, $X^-$ increases its binding energy by 35–150%, while for $X^+$ the binding energy decreases by 25%.

Key words: charged exciton, trion, modulation doped QWs, 2DEG, ZnSe

## 1. Introduction

Charged excitons (or trions) are exciton complexes consisting of three particles. Two electrons and one hole form a negatively charged exciton $X^-$. Two holes and one electron can be organized in a positively charged exciton $X^+$. Trion complexes in bulk semiconductors, i.e. in three dimensions, are fragile, but become stable in low-dimensional systems. That is why the theoretical prediction of Lampert from 1958 [1] was followed by a confident experimental observation of trions only in 1993 for the quasi-two dimensional electronic system in CdTe/(Cd,Zn)Te quantum wells (QW's) [2]. Since, positively- and negatively charged excitons have been studied experimentally in III–V heterostructures based on GaAs and in II–VI quantum well structures based on CdTe, (Cd,Mn)Te, ZnSe and (Zn,Mn)Se (see e.g. Refs. [3–6] and references therein).

II–VI semiconductors are very suitable for trion studies due to their strong Coulombic interaction compared with III–V materials. E.g. exciton binding energies (exciton Rydberg) in GaAs, CdTe and ZnSe are 4.2, 10 and 20 meV, respectively. Among these materials ZnSe has the strongest Coulombic interaction. However, after the first report of a $X^-$ observation in $Zn_{0.9}Cd_{0.1}Se/ZnSe$ QW's in 1994 [7], detailed investigations were started from 1998 only, when high-quality ZnSe-based structures with binary quantum well layers were fabricated [8–10]. At present rather detailed experimental information on trions in ZnSe QW's is available [5, 8, 11–16].

Theoretical results for this material system are limited to the calculation of the trion binding energy *vs* well width [17] and its variation in high magnetic fields [12]. The agreement with experimental data was rather qualitative — one of the reasons for this is the uncertainty of the parameters used for the calculations.

We present a detailed study of trion binding energies in ZnSe-based structures as a function of quantum well width and applied magnetic field. Parameters of exciton and trion states were determined by means of magneto-optical experiments, calculated on the base of a variational approach, and evaluated from the best fit of the experimental dependencies. The paper is organized as follows: Sec. 2 details the structures, while exciton parameters (measured and calculated) are discussed in Sec. 3. In Sec. 4 results on the binding energies of trions are collected and discussed. Finally, in Sec. 5 the results of the modification of the singlet trion state in high magnetic fields and with increasing carrier density are presented.

In this paper we deal with positively- ($X^+$) and negatively ($X^-$) charged excitons. We will label them in this way when the difference in the charge structure of trions is important. The term "trion" (T) will be used as a general definition for both positively- and negatively charged excitons.

## 2. Experimental Details

### 2.1. QUANTUM WELL STRUCTURES

ZnSe-based quantum well heterostructures with a binary QW material were grown by molecular-beam epitaxy on (100)-oriented GaAs substrates. Studied structures contain single quantum wells, with thickness varying from 2.9 to 19.0 nm. Schemes for the structure designs are presented in Fig. 1. Different barrier materials were used, namely (Zn,Be,Mg)Se, (Zn,Mg)(S,Se) and (Zn,Be)Se. For each barrier material different types of structures has been assigned. Parameters of the barrier materials were chosen with an aim to make them lattice-matched to GaAs substrates, which allows growing QW's of very high structural quality.

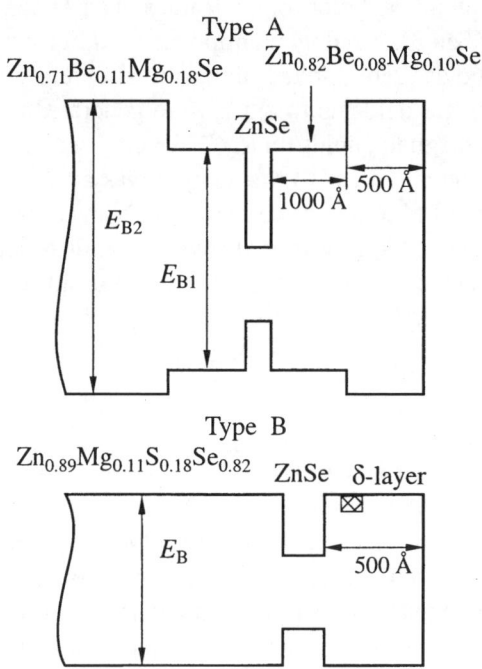

Figure 1. Schematics of the studied structures. Type A is ZnSe/$Zn_{0.82}Be_{0.08}Mg_{0.10}Se$ single QW surrounded by additional $Zn_{0.71}Be_{0.11}Mg_{0.18}Se$ barrier. Type B is ZnSe/$Zn_{0.89}Mg_{0.11}S_{0.18}Se_{0.82}$ single QW with modulation doping.

Most of the structures used in this study were nominally undoped. The background carrier density in them was tuned by additional above-barrier illumination. The range of tuning depends on the QW width, allowing in the widest QW to vary electron density from $5 \times 10^9$ to $10^{11}$ cm$^{-2}$. Details of the illumination technique will be presented in Sec. 4.2. In the sample zq1038 free electrons in the QW were provided by $n$-type doping with a 3.0 nm-thick, Cl doped layer (donor concentration of $5 \times 10^{17}$ cm$^{-3}$) separated from the QW by a 10.0 nm-thick spacer. The sample zq1113 was $p$-type doped with nitrogen (RF plasma cell at a power of 350 W and a background pressure of $5 \times 10^{-6}$ Torr). In this sample, symmetric doping was achieved by uniform doping of the barriers excluding 3.0 nm-thick spacer layers. The concentration of the two-dimensional hole gas (2DHG) in the QW of this sample is about $n_h \approx 3 \times 10^{10}$ cm$^{-2}$ and was insensitive to additional illumination.

2.2. EXPERIMENTAL METHODS

Photoluminescence (PL), PL excitation, reflectivity (R) and spin-flip Raman scattering (SFRS) spectroscopies were exploited for experimental study of trion parameters. Optical spectra were detected at a low temperature $T = 1.6$ K. Different CW lasers were used for photoexcitation, e.g. UV lines of an Ar-ion laser, a He-Cd laser and a dye laser (Stylben 3). A halogen lamp was used in reflectivity experiments. External magnetic fields were applied along the structure growth axis (Faraday geometry). DC magnetic fields up to 7.5 T were generated by a superconducting solenoid and pulsed magnetic fields up to 47 T were used. In the case of DC field experiments direct optical access to the sample was available via windows. For pulsed field experiments fiber optics were used. In both cases circular polarization degree of emitted/reflected light was analyzed. A complete set of field-dependent PL spectra was collected during each magnetic field pulse (for details see Ref. [18]). Experiments in a capacitor-driven 50 T mid-pulse magnet ($\sim$ 400 ms decay) were performed at the National High Magnetic Field Laboratory (Los Alamos, USA).

## 3. Properties of Confined Excitons

In the studies of trions, similar to excitons, the Rydberg energy of the exciton in bulk semiconductor is often chosen as a characteristic energy to parameterize the problem. In the case of quantum confined heterostructures it is also instructive to compare the binding energies of trion states with the binding energies of confined excitons. We will follow this tradition in our investigations of charged excitons in ZnSe-based QW's. Published information on the properties of excitons in ZnSe-based QW's with binary well material is rather limited. Therefore, we forestall the results on trions in this section where the exciton parameters for the ZnSe-based QW's will be evaluated from optical and magneto-optical experiments and variational calculations.

3.1. THEORETICAL MODEL FOR MAGNETO-EXCITONS

All the calculations of exciton states presented in this paper where made within a parabolic approximation, i.e. the admixture of the light-hole states and all effects of nonparabolicity are neglected. In the frame of this approach we have calculated exciton energies for the studied QW structures *vs* QW width, exciton binding energies and modification of these parameters in external magnetic fields up to 50 T. Details of the calculations are published elsewhere [19] and some of the results are included in Figs. 4 and 5.

### 3.2. OPTICAL SPECTRA OF EXCITONS

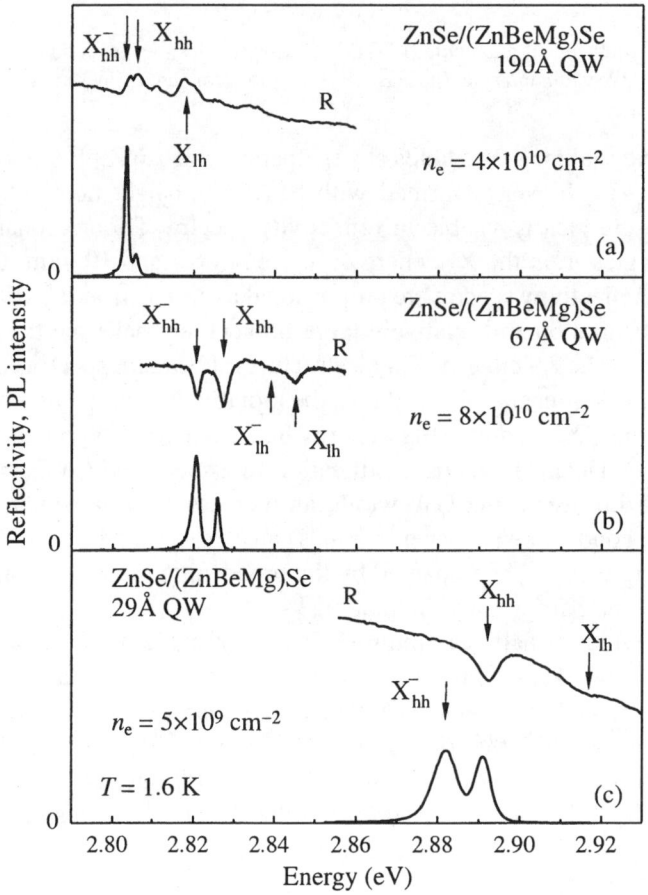

*Figure 2.* Reflectivity and photoluminescence spectra taken from ZnSe/Zn$_{0.82}$Be$_{0.08}$Mg$_{0.10}$Se (Type A) structures with different QW widths of 19.0 nm (a), 6.7 nm (b) and 2.9 nm (c). Arrows indicate the heavy-hole exciton ($X_{hh}$) the light-hole exciton $X_{lh}$, and the negatively charged excitons ($X_{hh}^-$ and $X_{lh}^-$). Electron concentration $n_e$ in the QW is given in the figure.

Figure 2 displays typical optical spectra for three ZnSe/Zn$_{0.82}$Be$_{0.08}$Mg$_{0.10}$Se single QW's, which covers the whole range of the studied QW widths $L_z$ from 2.9 to 19.0 nm. Photoluminescence and reflectivity spectra were measured in the

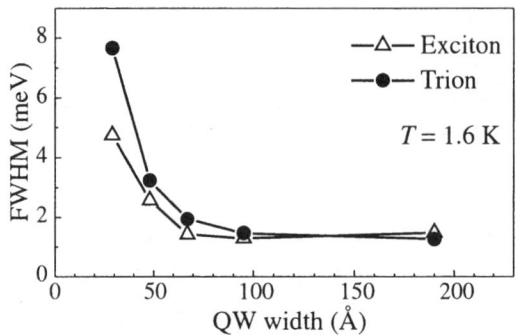

*Figure 3.* Full width at half maximum (FWHM) of trion (circles)- and exciton (triangles) PL lines as a function of QW width in ZnSe/Zn$_{0.82}$Be$_{0.08}$Mg$_{0.10}$Se structures. Lines are guides to the eye.

absence of external magnetic fields at a temperature of 1.6 K. Exciton resonances corresponding to the states formed with heavy- and light holes ($X_{hh}$ and $X_{lh}$, respectively) are clearly visible in reflectivity spectra. Trion resonances shifted to lower energies from the $X_{hh}$ energy are seen in 6.7 and 19.0 nm QW's. Their intensities in reflectivity spectra are proportional to the electrondensities [16]. Low electron concentration and relatively large broadening make the trion resonance unresolvable for the 2.9 nm QW. For all structures shown, PL spectra consist of two lines, where the low-energy line is due to the radiative recombination of negatively charged excitons ($X_{hh}^-$) and the high-energy line is due to recombination of neutral excitons ($X_{hh}$). Details of their identification in ZnSe-based QW's can be found in Ref. [5]. With decreasing QW width, an increase of confinement energies of carriers in the conduction and valence bands causes a shift of the exciton transitions to higher energies. It is accompanied by the broadening of exciton transitions due to QW width and barrier alloy fluctuations.

The full width at half maximum (FWHM) of exciton and trion PL lines is plotted in Fig. 3. Exciton linewidth for $L_z > 6.7$ nm is smaller than 1.2 meV, which evidence the high structural and optical quality of the studied samples. For this range of QW width the trion linewidth roughly coincides with the excitonic one. It is interesting that for $L_z < 10.0$ nm, exciton PL lines are narrower than the trion lines. The difference achieved 60% in the 2.9 nm QW. Possible reasons for this will be discussed below in Sec. 5.2.

3.3. EXCITONS IN HIGH MAGNETIC FIELDS

Application of external magnetic fields allows evaluating important exciton parameters such as the in-plane reduced effective mass $\mu$, and the $g$-factor that characterizes the spin splitting of excitons due to the Zeeman effect. These parameters are important for understanding and calculating the spin- and energy structure of trions and excitons. Experiments were performed in pulsed magnetic fields up to 47 Tesla. Application of high magnetic fields was required to induce sufficient

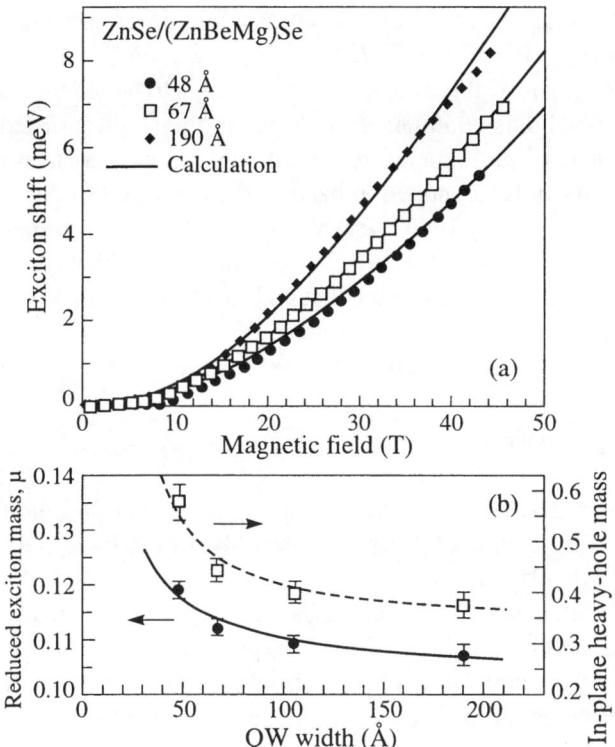

*Figure 4.* (a) Exciton energy vs magnetic field strength for different type A QW's: 19.0 nm (diamonds), 6.7 nm (squares) and 4.8 nm (circles). Center-of-gravity of the exciton spin doublet in PL spectra is taken for the exciton energy. $T = 1.6$ K. Lines show results of model calculations. (b) Reduced mass of exciton $\mu$ (left axis) and in-plane heavy-hole mass $m_{hh}^{xy}$ (right axis) vs QW width. Symbols are experimental data. A solid line is an interpolation of data points by the hyperbolic function $\mu = (0.103 + 0.07/L_z[\text{nm}])m_0$. The dashed line is a result of the equation $m_{hh}^{xy} = m_e/(m_e/\mu - 1)$.

energy shift of strongly bound excitons in ZnSe-based QW's. Photoluminescence spectra were measured in two circular polarizations corresponding to the two spin states of the optically active excitons. Evolution of PL spectra with increasing magnetic fields is discussed in detail in Refs. [8, 12]. Here we concentrate on the energy shift of exciton with increasing magnetic fields only.

To avoid spin splittings the center-of-gravity of the exciton spin doublet was evaluated and plotted as a function of magnetic field strength in Fig. 4(a) for QW's of different widths. Characteristic diamagnetic shift of excitons is seen for all samples. Exciton shift increases in wider QWs, which coincides with decreasing binding energy of excitons.

Solid lines in Fig. 4(a) show the best fit of experimental data in the frame of the model described shortly in Sec. 3.1. The exciton reduced mass $\mu$ is the only free parameter in the fit. Determined values of $\mu$ are also plotted in Fig. 4(b) vs $L_z$. The reduced mass increases for thinner QW's with a functional dependence that can be interpolated as $\mu = (0.103 + 0.07/L_z[\text{nm}])m_0$ (see solid line in Fig. 4(b)). Taking the value of the electron effective mass $m_e = 0.15m_0$ to be independent of the well width (which is valid as confinement energies in the studied structures are small and do not exceed 60 meV, therefore, one should not expect strong contributions of nonparabolicity in the conduction band to the electron effective mass), the in-plane values for the heavy-hole effective mass $m_{hh}^{xy}$ are determined by means of the relationship $m_{hh}^{xy} = \mu m_e/(m_e - \mu)$. These values are displayed in Fig. 4(b) by open squares (right axis). One can see that $m_{hh}^{xy}$ increases significantly from $0.37m_0$ in a 19.0 nm QW to 0.58 $m_0$ in a 4.8 nm QW. This fact should be accounted in model calculations of trions binding energies *vs* QW width (e.g. [17]).

### 3.4. CONFINED EXCITONS

QW width dependencies for exciton energy $E_X$ and exciton binding energy $E_B^X$ were calculated by means of the model described in Sec. 3.1. Results of these calculations are displayed in Fig. 5.

In Fig. 5(a) calculated exciton energies for different types of structures (which differ by barrier heights and band offsets) are plotted by lines. We use these dependencies and experimental values of the exciton energies to determine QW width in the studied structures. In Fig. 5(b) binding energies of 1s and 2s exciton states are plotted as a function of $L_z$ for ZnSe/Zn$_{0.82}$Be$_{0.08}$Mg$_{0.10}$Se structures. One can see that in ZnSe QW's, the exciton binding energy $E_B^X$ (i.e. the binding energy of 1s state) has its maximum for QW's with $L_z \approx 2.0$ nm. This value is 40 meV, which is about twice as large as the bulk exciton Rydberg $R = 20$ meV. The value indicates that in the ZnSe QW's, the exciton is quasi-two-dimensional, as its binding energy is considerably smaller than the binding energy of the 2D limit, $4R = 80$ meV.

## 4. Charged Excitons

### 4.1. IDENTIFICATION OF NEGATIVELY- AND POSITIVELY CHARGED EXCITONS

Charged exciton states in optical spectra can be identified by their specific polarization properties in external magnetic fields. Analysis of the circular polarization degree of photoluminescence is rather complicated. In addition to the spin polarization of the free carriers, the spin-dependent trion formation and spin relaxation of trions are involved [13]. However, polarization properties of the trion states in reflectivity, absorption or transmission spectra allow to distinguish trions from excitons and positively- and negatively charged excitons from each other. Here we present in short principles of the identification, and further details can be found in Refs. [5, 16].

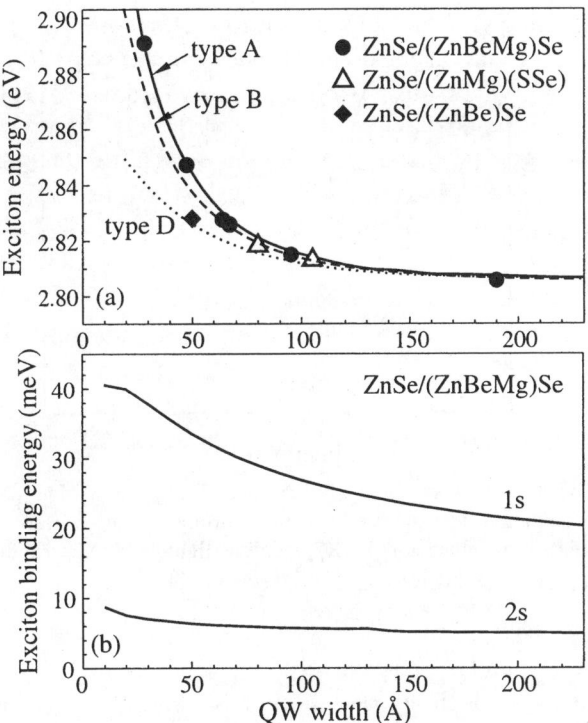

*Figure 5.* (a) Lines are exciton energies calculated for different types of structures as a function of QW width. Symbols present experimental data. The nominal (i.e. technological) values of QW width were slightly corrected to put experimental points at calculated dependencies. (b) The calculated exciton binding energy (1s- and 2s states) as a function of QW width for type A ZnSe/Zn$_{0.82}$Be$_{0.08}$Mg$_{0.10}$Se structures.

The polarization degree of the trion resonance in reflectivity spectra mirrors the polarization of free carriers in QW's which is caused by thermal distribution of the carriers among the Zeeman sublevels. This is due to the singlet spin structure of the trion ground state, i.e. spins of two carriers with the same charges (electrons in X$^-$ and holes in X$^+$) in the trion complex are oriented antiparallel. A triplet trion state with parallel orientation of these spins is unbound at zero magnetic field and becomes bound in high fields only [12]. When free carriers are fully polarized by magnetic field the trions can be excited optically for one circular polarization only. In case of negatively charged excitons in ZnSe QW's with a positive electron $g$ factor it is $\sigma^-$ polarization.

In Fig. 6 typical reflectivity spectra containing strongly polarized resonances of negatively charged excitons associated with heavy-hole excitons ($X_{hh}^-$) and light-hole excitons ($X_{lh}^-$) are given. Results are shown for a 6.7 nm ZnSe/Zn$_{0.82}$Be$_{0.08}$Mg$_{0.10}$Se QW and a magnetic field of 4 T. In accordance with the selection rules discussed above, $X_{hh}^-$ and $X_{lh}^-$ resonances show up in different

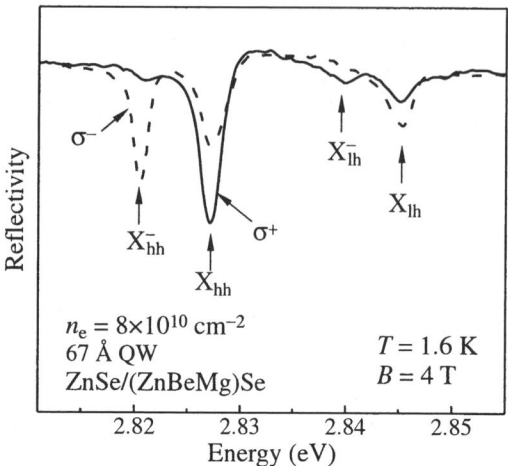

*Figure 6.* Reflectivity spectra taken from a 6.7 nm ZnSe/Zn$_{0.82}$Be$_{0.08}$Mg$_{0.10}$Se QW detected in different circular polarization at a magnetic field of 4 T. Strongly polarized resonances of negatively charged excitons related to the heavy-hole ($X^-_{hh}$) and the light-hole ($X^-_{lh}$) excitons are labeled by arrows. Electron concentration in QW is $n_e = 8 \times 10^{10}$ cm$^{-2}$.

polarizations, i.e. $\sigma^-$ and $\sigma^+$ respectively.

Examples of the opposite polarization of $X^-$ and $X^+$ in ZnSe QW's can be found in Ref. [5] and in Fig. 8 in the next section, where a recharging effect of the QW by above-barrier illumination is discussed.

### 4.2. OPTICAL TUNING OF CARRIER DENSITY IN QW'S

Optical tuning of the carrier density in QW's is a very reliable method that is widely exploited for trion studies [20–23]. Different structure designs have been suggested for this purpose. The principle of the method is based on the spatial separation of electron-hole pairs photogenerated by photons with energies exceeding the barrier band gap. Depending on the structure design, one type of carrier is captured by the surface states, trapped centers in barriers or additional quantum well. The other type of carrier is collected into the quantum well where it is involved in the trion formation. Free carrier concentration in the QW is tuned by the intensity of the above-barrier illumination. However, the dependence of the concentration on the illumination intensity can be very nonlinear with a pronounced saturation at higher intensities. The optical method can also used be for the fine-tuning of carrier densities in structures with modulation doping and/or under applied gate-voltages.

We will show here that optical tuning is very effective for the types A and B of ZnSe-based heterostructures investigated in this paper. Let us start with the type B structure zq1038, where the 8.0 nm QW is separated from the surface by a 60.0 nm Zn$_{0.89}$Mg$_{0.11}$S$_{0.18}$Se$_{0.82}$ barrier. Reflectivity spectra measured under different illumination intensities are presented in Fig. 7. Laser light with $\hbar\omega_L = 3.5$ eV was

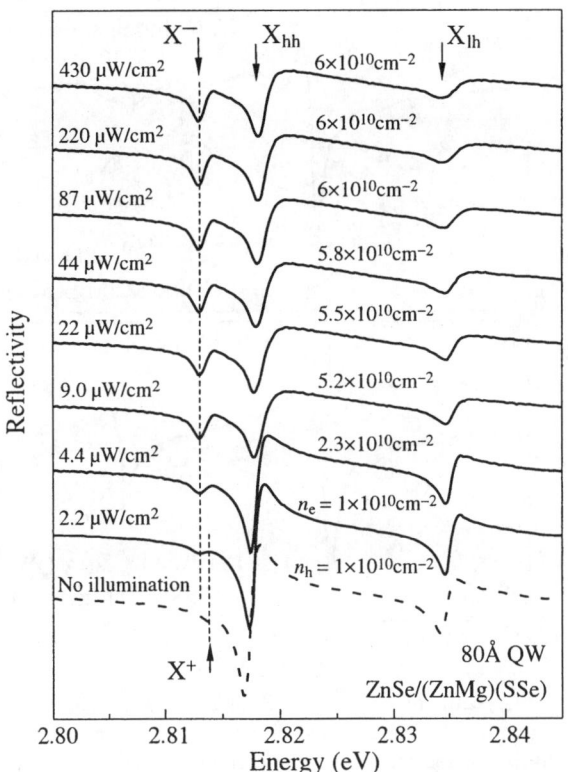

*Figure 7.* Tuning of electron (hole) gas concentration by additional above-barrier illumination. The figure presents reflectivity spectra of an 8.0 nm ZnSe/Zn$_{0.89}$Mg$_{0.11}$S$_{0.18}$Se$_{0.82}$ QW vs illumination intensity of Ar-ion laser (3.5 eV). The laser power is given in the figure. $B = 0$ T, $T = 1.6$ K.

used for illumination and the high-energy part of the halogen lamp spectrum was cut by a 420 nm edge filter. Without laser illumination strong exciton resonances $X_{hh}$ and $X_{lh}$ dominate the reflectivity spectrum shown by dashed line. Only a weak $X^+$ resonance is detectable 3.3 meV to low energies from the $X_{hh}$ one. With increasing illumination power the $X^+$ resonance vanishes and a new $X^-$ resonance appears. Its energy distance from the exciton energy is 4.4 meV. We note here that the same method has been applied in Ref. [22] for GaAs-based QW's. It is elegant and very convincing as it allows measuring parameters of $X^+$ and $X^-$ resonances in the same structure, thus avoiding technological and growth uncertainties.

Identification of charged exciton resonances were based on their polarization properties in external magnetic fields (see Fig. 8). We conclude from the data of Fig. 7 that without illumination all donor-electrons from the modulation-doped layer are either captured by charged surface states or remain on donors. In this condition the QW contains a very diluted hole gas with $n_h = 1 \times 10^{10}$ cm$^{-2}$. This value was determined from the oscillator strength of the $X^+$ transition, which for low carrier densities is linearly proportional to the carrier concentration (see

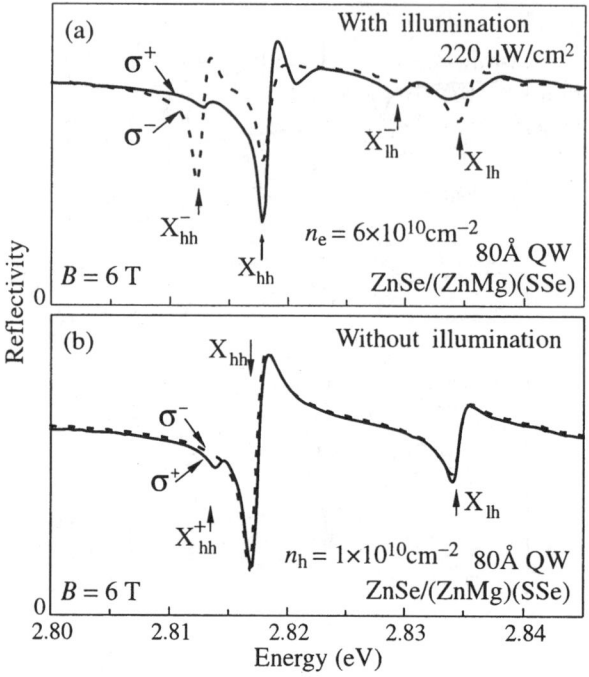

*Figure 8.* Reflectivity spectra of an 8.0 nm ZnSe/Zn$_{0.89}$Mg$_{0.11}$S$_{0.18}$Se$_{0.82}$ QW: (a) With above-barrier illumination detected at different circular polarization at a magnetic field of 6 T. Arrows indicate negatively charged exciton X$^-$ and excitons formed with heavy-hole and light-hole. (b) Without above-barrier illumination at a magnetic field of 6 T. Arrows indicate positively charged exciton X$^+$ and excitons formed with heavy-hole and light-hole.

details in Refs. [16, 24]). Laser illumination redistributes the carrier location in the structure by supplying the QW with electrons. An increase of the electron density saturates at $n_e = 6 \times 10^{10}$ cm$^{-2}$ (for detailed behavior see [19]), which is still much lower than the concentration of donors in the modulation doped layer $n_d = 3 \times 10^{11}$ cm$^{-2}$ evaluated from the technological calibration. We suggest that the reason is a relatively small conduction band offset in this structure ($\Delta E_C = 100$ meV).

In the type A structure cb1172, a 19.0 nm QW is separated from the surface by two barriers of different heights. Note that this structure is nominally undoped. Instead of the laser we use for illumination the light of the halogen lamp selected by edge filters. Reflectivity spectra are shown in Fig. 12. Only exciton transitions are visible in the spectrum measured with a 420 nm filter, i.e. when photocarriers are excited only in the ZnSe layer of the 19.0 nm-thick QW. A threshold-like increase of the electron density in the QW starts when the energy of illumination light exceeds the band gap of the highest barrier (3.21 eV). The electron density in this structure is varied from $5 \times 10^9$ to $9 \times 10^{10}$ cm$^{-2}$. $n_e$ was evaluated from the analysis of the polarization degree of the trion line. The procedure has been suggested in Ref. [16] and detailed later in Ref. [24]. It is based on the fitting of the magnetic-field-induced polarization of trion resonance in the frame of the approach accounting for the Fermi-Dirac statistics of the electron gas. The Fermi energy that is determined from the best fit of experimental data points is directly linked to the electron density. From the threshold-like effect of the illumination on the electron density in the QW we conclude that a recharging of surface states (namely a capture of photo-holes by the surface states) is the main mechanism for the carrier separation that supplies the QW with free electrons.

## 4.3. EXCITON-TRION ENERGY SEPARATION. EFFECT OF THE FERMI ENERGY

Now we turn our attention to the binding energy of the trions $E_B^T$, defined as the energy required to dissociate an isolated trion into a neutral exciton and an electron (for) $X^-$ or a hole (for $X^+$). In the limit of a very diluted carrier gas, $E_B^T$ is given by the energy difference between the exciton and trion lines (i.e. energy separation between bound and unbound states) $\Delta_{XT} = E_X - E_T$. A deviation from this "bare" value of $E_B^T$ takes place with increasing carrier density. In Fig. 10 we show the exciton-trion separation $\Delta_{XT}$ as a function of electron density $n_e$ for two QW's. In the case of an 8.0 nm ZnSe/Zn$_{0.89}$Mg$_{0.11}$S$_{0.18}$Se$_{0.82}$ QW (squares) the electron concentration was varied by modulation doping. For a 19.0 nm ZnSe/Zn$_{0.82}$Be$_{0.08}$Mg$_{0.10}$Se QW (circles) the electron density was tuned via additional illumination (see Sec. 4.2). For both cases the exciton-trion energy separation increases remarkably with the electron density. Solid lines in Fig. 10 have a slope of the Fermi energy increasing with growing electron density $E_F = \pi\hbar^2 n_e/m_e$. In ZnSe QW's with $m_e = 0.15m_0$ $E_F$[meV] $= 1.53 \times 10^{-11} n_e$ [cm$^{-2}$]. Comparing solid lines and data points in the figure one can establish that the concentration dependence of the exciton-trion energy separation is approximately given by the Fermi energy: $\Delta_{XT} = E_B^T + E_F$. This result has been also reported recently for CdTe-based QW's [25].

Such a behavior of $\Delta_{XT}(n_e)$ does not correspond to a real increase of the trion binding energy and can be quantitatively explained in terms of exciton-trion repulsion due to their mixing [26]. This mixing is provided by mutual transformation of exciton and trion states via exchange of additional electron. The detailed investigation of mixed exciton-trion states will be published elsewhere.

To obtain the $E_B^T$ value we extrapolate experimental dependencies $\Delta_{XT}(n_e)$ to the limit $n_e \to 0$ (see Fig. 10) getting the "bare" binding energy of trion $E_B^T$. We performed this procedure for all studied structures in order to receive information on binding energies of "isolated" trions that can be directly compared with theoretical calculations.

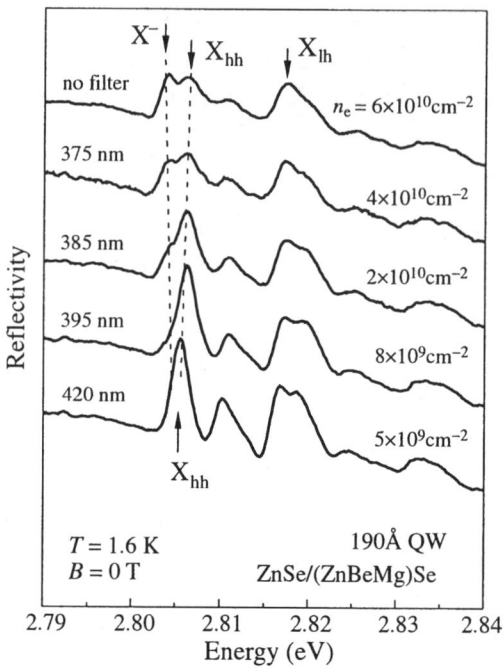

*Figure 9.* Tuning of electron gas concentration by an additional above-barrier illumination. The figure presents reflectivity spectra taken from a 19.0 nm ZnSe/Zn$_{0.82}$Be$_{0.08}$Mg$_{0.10}$Se QW as a function of above-barrier illumination. The illumination is provided by a white-light source together with longpass optical filters having different absorption edges. The values of the filter absorption edge are given in the figure. $B = 0$ T, $T = 1.6$ K.

### 4.4. BINDING ENERGY OF TRIONS

Binding energies of trions $E_B^T$ determined for the low carrier density regime are displayed in Fig. 11(a) as a function of QW width. Solid symbols correspond to $X_{hh}^-$, open symbols show $X_{hh}^-$ and crosses are used for $X_{hh}^+$. We discuss first the data for the negatively charged excitons related to heavy-hole excitons. The line in Fig. 11(a) is an interpolation made for $X_{hh}^-$ data points (solid symbols). The trion binding energy increases strongly from 1.4 meV in a 19.0 nm QW up to 8.9 meV in a 2.9 nm QW. The increase for $E_B^T$ is 6.4 times while the exciton binding energy increases only twice (see Fig. 5(b)). Stronger sensitivity of the trion binding energy to confinement conditions is due to the lager extension of the trion wave function and to the strong effect of reduction of dimensionality on the trion stability [27]. Theoretical calculations show that trion states are very weakly bound in three-dimensional systems, which hinders their experimental observation in bulk semiconductors. Reduction of dimensionality from 3D to 2D is a crucial factor for increasing trion stability, and the trion binding energy grows by a factor of ten [28]. We believe that increase of $E_B^T$ shown in Fig. 11(a) is dominated by

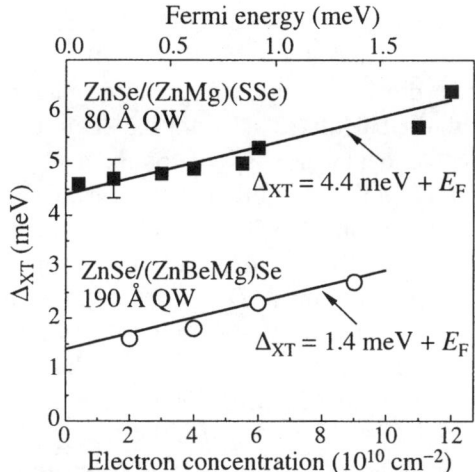

*Figure 10.* Exciton-trion separation as a function of 2DEG density for a 19.0 nm ZnSe/Zn$_{0.82}$Be$_{0.08}$Mg$_{0.10}$Se QW (circles) and an 8.0 nm ZnSe/Zn$_{0.89}$Mg$_{0.11}$S$_{0.18}$Se$_{0.82}$QW (squares). Lines are the sum of a trion binding energy in corresponding QW and the Fermi energy of a 2DEG.

localization of carrier wave functions along the structure growth axis, i.e. by the increasingly two-dimensional character of the carrier wave functions. Contribution of the in-plane localization of trions is minor except perhaps for the very narrow 2.9 nm QW. This conclusion is based on comparing the exciton linewidths (see Fig. 3), which gives us characteristic energies for the in-plane exciton localization, with the trion binding energies. Exciton linewidth is weakly sensitive to the QW width and is below 2 meV for the range 50–190 Å. It increases to 5.3 meV in the very narrow QW, but even in this case it stays smaller than the trion binding energy of 8.9 meV.

A detailed comparison of the trion and exciton modifications with decreasing QW width is given in Fig. 11(b), where the ratio of the trion and the exciton binding energies $E_B^T/E_B^X$ is presented. For the 19.0 nm QW this ratio $E_B^T/E_B^X = 0.065$. It increases linearly with decreasing QW width achieving a value of 0.235 in the 2.9 nm QW. Theoretical calculations of this ratio performed for the two-dimensional limit give a value $E_B^T/E_B^X \approx 0.12$, which is rather insensitive to the ratio of electron and hole effective masses [28, 29]. The experimental value for the 2.9 nm QW exceeds the theoretical limit by a factor of two. We explain this by the fact that our experimental situation corresponds to the quasi-2D case rather than to strictly 2D one. This is confirmed by the moderate increase of the exciton binding energy, which is twice as large as the bulk Rydberg in narrow QW's and, respectively, twice as small as the 2D limit of four Rydbergs. A dimensional transition for a Coulombic state in QW structures is determined by a ratio of the Bohr radius of the states to the QW width. Obviously, for trions with larger Bohr radius this transition will happen in wider QW's than for excitons, whose wave function is more compact. Thus at a given QW width, excitons and trions have different degrees of two-dimensionality, which causes a larger measured value of $E_B^T/E_B^X$ compared with the calculated value for the 2D limit.

It is interesting to note that the strength of confinement potentials in our structures plays a minor effect on the trion binding energies. Data points in Fig. 11(a) for structures with different materials with $\Delta E_g$ value varied from 200 to 250 meV follow the same dependence. Only a small deviation from this dependence was found for the type D structure with $\Delta E_g = 70$ meV. For very shallow 7.0 nm ZnSe/Zn(S,Se) QW's with $\Delta E_g = 25-35$ meV a trion binding energy of 2.7–2.9 meV has been reported [11]. This is consistent with our data from Fig. 11(a) and evidences that decreasing the electron confinement leads to smaller binding energies for $X^-$.

We are aware of one paper only where the binding energies of $X_{hh}^-$ were calculated for ZnSe-based QW's [17]. The quantitative agreement was not satisfactory and authors suggested that the polaron effect, which in ZnSe QW's could give an additional 1.3–2.6 meV contribution to the trion binding energy, should be considered. For our structures we found a relatively strong dependence of the exciton reduced mass on the QW width (see Fig. 4b). We believe that incorporating this factor into calculations will increase their reliability and coincidence with experiment. A. Esser has performed calculation for an 8.0 nm QW (zq1038) with our new parameters and got a value of 4.2 meV for $X_{hh}^-$ which is in good agreement with our experimental value of 4.4 meV even whithout introduction of the polaronic correction [30].

Binding energies of trions based on the light-hole excitons (open symbols in Fig. 11a) are 20–30% smaller than the $X_{hh}^-$ binding energies. To the best of our knowledge no detailed investigation of $X_{lh}^-$ states has been reported so far. Trions associated with light-hole excitons were observed in PL excitation spectra of GaAs-based QW's [31] and in the reflectivity spectra of monomolecular CdTe islands [32]. In both cases the $X_{lh}^-$ binding energy was very close to that of $X_{hh}^-$. Numerical calculations performed for GaAs QW's give, for example, a 40% difference in favor of $X_{hh}^-$ in an 10.0 nm GaAs/Ga$_{0.85}$Al$_{0.15}$As QW [27]. However, the model used in Ref. [27] has not accounted for the modification of the in-plane effective mass in the valence band which is essential for the quantitative comparison with experimental results.

It is interesting that positively charged excitons show binding energies reduced by about 25% compared with their negatively charged partners. E.g. in an 8.0 nm QW of type B (zq1038), binding energies for negatively- and positively charged excitons are 4.4 and 3.3 meV, respectively. We are very confident of this result, as it has been measured in the same structure (see Figs. 7 and 8) where the type of free carriers occupying the QW was reversed by the illumination. Calculations performed for the 3D and 2D limits give the $X^+$ binding energy larger than the

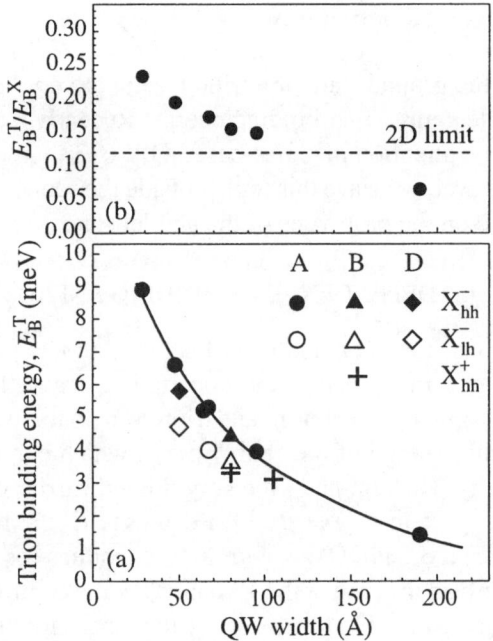

*Figure 11.* (a) Trion binding energy as a function of QW width for ZnSe/Zn$_{0.82}$Be$_{0.08}$Mg$_{0.10}$Se (type A), ZnSe/Zn$_{0.89}$Mg$_{0.11}$S$_{0.18}$Se$_{0.82}$ (type B) and ZnSe/Zn$_{0.96}$Be$_{0.04}$Se (type D) structures. Solid symbols show data for negatively charged excitons formed with heavy-hole ($X_{hh}^-$), open symbols are for negatively charged excitons formed with light-hole ($X_{lh}^-$), and crosses are for positively charged excitons formed with heavy-hole ($X^+$). Solid line is interpolation for $X_{hh}^-$ values. (b) Ratio of the $X_{hh}^-$ binding energy to the binding energy of quasi-two-dimensional exciton taken from Fig. 5(b). Theoretical value of $E_B^T/E_B^X = 0.12$ for a two-dimensional case is shown by a dashed line [28,29].

$X^-$ one [28, 29]; this result is explained qualitatively by the heavier effective mass of holes compared to electrons. However, in the quasi-two-dimensional case the situation can differ qualitatively. Recent calculations performed by A. Esser with parameters of the structure zq1038 give 4.2 and 4.0 meV binding energies for $X^-$ and $X^+$, respectively [30]. The smaller binding energy of the positively charged exciton is explained by the "effective" hole-hole Coulomb repulsion to be stronger in this QW than the electron-electron one. The calculations qualitatively reproduce experimental trends. Better quantitative agreement for $X^+$ state is still desired. Note that for GaAs QW's identical binding energies for $X^-$ and $X^+$ have been reported [22].

Comprehensive theoretical consideration for the trion binding energy data collected in Fig. 11(a) is still missing. We hope that these data and the set of exciton parameters given in the paper will encourage such activity.

## 5. Properties of Singlet Trion States

The singlet state is the ground state of a trion except at very high magnetic fields where the triplet state gains larger binding energy. Recently the triplet states have attracted considerable attention in GaAs-based QW's [33, 40] and in ZnSe-based structures [12]. However, we leave this topic outside the scope of the present paper and concentrate here on the properties of the singlet state.

### 5.1. MAGNETIC FIELD DEPENDENCE OF BINDING ENERGY

Energy distance between exciton and trion PL lines $\Delta_{XT}$ is plotted in Fig. 12 vs magnetic field strength. In order to avoid uncertainties caused by spin splittings, data for the center-of-gravity of exciton- and trion spin doublets are given. Our task is to study the binding energy of the "bare" trions which exhibit no contribution from the Fermi energy. The regime of the very diluted carrier gas is fulfilled for a 10.5 nm QW with $X^+$, where $\Delta_{XT}$ at $B = 0$ T equals to $E_B^T$. Similar statements can be made for 4.8 nm and 6.7 nm QW's with $X^-$. $E_B^T$ values for QW's with $X^-$ are shown by arrows. Only in the case of the 19.0 nm QW the contribution of the Fermi energy to $\Delta_{XT}$ was considerable for the set of data measured in pulsed magnetic fields (shown by solid diamonds in Fig. 12). We have repeated measurements for this structure in DC magnetic fields $B < 8$ T keeping a low density of the 2DEG. Results are given by open diamonds. One can see that the difference between the two data sets vanishes with increasing magnetic fields and disappears for $B > 7$ T. This means that the contribution of $E_F$ decreases with growing magnetic fields, which can be explained by an increase of the density of states of the Landau levels. For the following discussion we consider the $E_B^T(B)$ dependence for a 19.0 nm QW consisting of open diamonds at low fields and of solid diamonds at high fields ($B > 8$ T).

Binding energies of $X^-$ in all studied QW's show a monotonic increase with growing magnetic fields and a tendency to saturation in high fields $B > 25$ T. The increase is stronger in wider QW's with smaller $E_B^T$, e.g. it amounts to 150% in a 19.0 nm QW and has only 35% in a 4.8 nm QW. Being more compact in narrow QW's the singlet state becomes less sensitive to compression by external magnetic fields. Qualitatively $E_B^T(B)$ dependencies for $X^-$ from Fig. 12 are consistent with theoretical predictions for the singlet state [12, 33, 35].

Magnetic field dependence of the positively charged exciton differs drastically from the $X^-$ behavior. Binding energy of $X^+$ shows no dependence on magnetic fields for $B < 6$ T and decreases by 25% at higher fields (see stars in Fig. 12). In the field range 26–32 T the $X^+$ singlet state line shows irregular behavior caused by its crossing with the triplet state. These results will be published elsewhere and here, for clarity, we do not show data points for this field range. Principally different behavior of $X^-$ and $X^+$ states in external magnetic fields has been established first for GaAs QW's [4]. Our results confirm this for ZnSe-based QW's. We are not

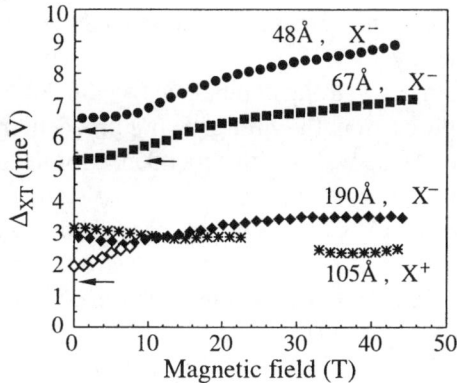

*Figure 12.* Exciton-trion separation as a function of magnetic field for QW's of different width. $X^-$ was measured in ZnSe/Zn$_{0.82}$Be$_{0.08}$Mg$_{0.10}$Se QW's: 19.0 nm (diamonds), 6.7 nm (squares), 4.8 nm (circles). Solid symbols correspond to the PL excited by laser with energy above barrier and open symbols are for below-barrier excitation. Arrows indicate "bare" trion binding energy. $X^+$ is taken for a 10.5 nm ZnSe/Zn$_{0.89}$Mg$_{0.11}$S$_{0.18}$Se$_{0.82}$ QW's (stars). $T = 1.6$ K.

aware of theoretical attempts to model $X^+$ behavior in magnetic fields. However, it is clear that the difference in magnetic field behavior of $X^+$ and $X^-$ binding energies is due to the very different structure of wave functions of these complexes (see discussion in Ref. [29]). $X^-$ is constructed of two light particles (electrons) rotated around one heavy particle (hole). This complex has one center and magnetic field will localize the electron wave functions around the hole, thus inducing an increase of the binding energy. In contrast $X^+$ has two heavy particles, i.e. two centers, and one light particle moving between two centers. In this case shrinking of electron wave function by magnetic fields hinders it from optimal adjustment for two centers, which results in decreasing binding energy of $X^+$ complex.

## 5.2. TRION LINEWIDTH

We discuss here the linewidth of exciton and trion luminescence lines presented in Fig. 3. It was found that for QW's thinner than 10.0 nm, the trion line is systematically broader than the exciton one. The difference in linewidths grows up to 60% in a 2.9 nm QW. At least two physical reasons for that can be suggested:
(i) PL linewidth is contributed by localization energies for carriers. In the case of excitons it is summed up from the electron and hole contributions, where the electron plays a dominant role. In case of trions, two electrons and one hole participate. Qualitatively it should result in larger broadening, but the quantitative approach to this problem does not seems to be very trivial, because it will depend strongly on the choice of a model for localizing potential.
(ii) Another reason is related to a certain freedom in the energy conservation law in case of the trion recombination. An electron, which is left after trion recombination, can have a finite kinetic energy. The energy of emitted photon will be reduced by

this amount. Respectively, the trion line will exhibit additional broadening due to the electron kinetic energy.

The first mechanism has a strong dependence on the QW width — its contribution should increase proportionally with growing broadening of the exciton line. However, the character of the well width dependence for the second mechanism is not very obvious for us.

The second mechanism has been studied theoretically and experimentally for GaAs-based QW's [68]. It was shown that it has a strong temperature dependence and at $T = 2$ K the additional broadening of the trion line is about 0.04 of the exciton binding energy. Applying this estimation to our QW's we get the contribution of the second mechanism of 0.8 meV for a 19.0 nm QW and of 1.5 meV for a 2.9 nm QW. In the narrow QW, exciton and trion linewidths are 4 and 6.5 meV, respectively, i.e. they differ by 2.5 meV. From this we suggest that both mechanisms have comparable contribution to the broadening of the trion emission line. Further experiments including the careful analysis of the temperature dependencies of the trion linewidth are required to separate the role of two mechanisms.

## 6. Conclusions

Negatively and positively charged excitons in ZnSe-based QW's were investigated in structures with various QW widths and free carrier densities. The binding energy of $X^-$ shows a strong dependence on the QW width, increasing from 1.4 to 8.9 meV as the well width decreases from 19.0 to 2.9 nm. This variation is 6.4 times while the neutral exciton binding energy increases only twice. The binding energy of $X^+$ is 25% smaller than that of $X^-$. This observation is in qualitative agreement with model calculations and is explained by stronger "effective" Coulomb repulsion in case of hole-hole interaction compared with electron-electron interaction. Qualitatively different behavior for $X^-$ and $X^+$ is found in external magnetic fields. $X^-$ increases its binding energy depending on the QW width by 35–150%, while in contrast $X^+$ shows a decrease of its binding energy by 25%. A detailed set of exciton parameters for the studied structures is collected in the paper. We hope that this will encourage theoretical efforts for better understanding the energy- and spin structure of trions.

## Acknowledgements

We acknowledge stimulating discussions with A.B. Dzyubenko and R.A. Suris. We are thankful to A. Esser for allowing us to use in this paper his unpublished results on calculation of trion binding energies. The work was supported in part by the Deutsche Forschungsgemeinschaft through Sonderforschungbereich 410 and grant Nos. Os98/6 and 436 RUS 113/557, as well as by grants of the Russian Foundation for Basic Research (Grants Nos. 00-02-04020 and 01-02-04010)

We feel deep sorrow at the death of our colleague W. Faschinger, who was an outstanding scientists and very kind person.

## References

1. Lampert, M.A. (1958) *Phys. Rev. Lett.* **1**, 450.
2. Kheng, K., Cox, R.T., Merle d'Aubigne, Y., Bassani, F., Saminadayar, K., and Tatarenko, S. (1993) *Phys. Rev. Lett.* **71**, 1752.
3. Cox, R.T., Huard, V., Kheng, K., Lovisa, S., Miller, R.B., Saminadayar, K., Arnoult, A., Cibert, J., Tatarenko, S., and Potemski, M. (1998) *Acta Phys. Pol. A* **94**, 99.
4. Glasberg, S., Finkelstein, G., Shtrikman, H., and Bar-Joseph, I. (1999) *Phys. Rev. B* **59**, R10425.
5. Astakhov, G.V., Yakovlev, D.R., Kochereshko, V.P., Ossau, W., Nürnberger, J., Faschinger, W., and Landwehr, G. (1999) *Phys. Rev. B* **60**, R8485.
6. Crooker, S.A., Johnston-Halperin, E., Awschalom, D.D., Knobel, R., and Samarth, N. (2000) *Phys. Rev. B* **61**, R16307.
7. Kheng, K., Cox, R.T., Kochereshko, V.P., Saminadayar, K., Tatarenko, S., Bassani, F., and Franciosi, A. (1994) *Superlatt. & Microstruct.* **15**, 253.
8. Lovisa, S., Cox, R.T., Baron, T., Keim, M., Waag, A., and Landwehr, G. (1998) *Appl. Phys. Lett.* **73**, 656.
9. Ivanov, V.Yu., Godlewski, M., Bergman, J.P., Monemar, B., Behringer, M., and Hommel, D. 1998 *Proc. $24^{th}$ Int. Conf. on the Physics of Semiconductors*, Jerusalem, Israel 1998, (World Scientific, Singapore 1999), Ed. by D.Gershoni, published on CD.
10. Homburg, O., Sebald, K., Michler, P., Gutowski, J., Wenisch, H., and Hommel, D. (2000) *Phys. Rev. B* **62**, 7413.
11. Yakovlev, D.R., Astakhov, G.V., Ossau, W., Crooker, S.A., Uchida, K., Miura, N., Waag, A., Gippius, N.A., Sivachenko, A.Yu., and Dzyubenko, A.B. (2001) *Phys. Stat. Sol. (b)* **227**, 353.
12. Yakovlev, D.R., Nickel, H.A., McCombe, B.D., Keller, A., Astakhov, G.V., Kochereshko, V.P., Ossau, W., Nürnberger, J., Faschinger, W., and Landwehr, G. (2000) *J. Cryst. Growth* **214-215**, 823.
13. Yakovlev, D.R., Puls, J., Mikhailov, G.V., Astakhov, G.V., Kochereshko, V.P., Ossau, W., Nürnberger, J., Faschinger, W., Henneberger, F., and Landwehr, G. (2000) *Phys. Stat. Sol. (a)* **178**, 501.
14. Wagner, H.P., Tranitz, H.-P., and Schuster, R. (1999) *Phys. Rev. B* **60**, 15542.
15. Astakhov, (2000), Kochereshko, V.P., Yakovlev, D.R., Ossau, W., Nürnberger, J., Faschinger, W., and Landwehr, G. (2000) *Phys. Rev. B* **62**, 10345.
16. Riva, C., Peeters, F.M., and Varga, K. (2000) *Phys. Rev. B* **61**, 13873.
17. Crooker, S.A., Rickel, D.G., Lyo, S.K., Samarth, N., and Awschalom, D.D. (1999) *Phys. Rev. B* **60**, R2173.
18. Astakhov, G.V., Yakovlev, D.R., Kochereshko, V.P., Ossau, W., Faschinger, W., Puls, J., Henneberger, F., Crooker, S.A., McCulloch, Q., Wolverson, D., Gippius, N.A., and Waag, A. (2002) *Phys. Rev. B* **65**, 165335.
19. Naumov, A., Mi, D., Sturge, M.D., Ge, W., Le Si Dang, Mariette, H., and Magnea, N. (1995) *J. Appl. Phys.* **78**, 1196.
20. Shields, A.J., Osborne, J.L., Simmons, M.Y., Pepper, M., and Ritchie, D.A. (1995) *Phys. Rev. B* **52**, R5523.
21. Glasberg, S., Finkelstein, G., Shtrikman, H., and Bar-Joseph, I. (1999) *Phys. Rev. B* **59**, R10425.
22. Kossacki, P., Cibert, J., Ferrand, D., Merle d'Aubigne, Y., Arnoult, A., Wasiela, A., Tatarenko, S., and Gaj, J.A. (1999) *Phys. Rev. B* **60**, 16018.

23. Astakhov, G.V., Kochereshko, V.P., Yakovlev, D.R., Ossau, W., Nürnberger, J., Faschinger, W., Landwehr, G., Wojtowicz, T., Karczewski, G., and Kossut, J. (2002) *Phys. Rev. B* **65**, 115310.
24. Huard, V., Cox, R.T., Saminadayar, K., Arnoult, A., and Tatarenko, S. (2000) *Phys. Rev. Lett.* **84**, 187.
25. Suris, R.A., Kochereshko, V.P., Astakhov, G.V., Yakovlev, D.R., Ossau, W., Nürnberger, J., Faschinger, W., Landwehr, G., Wojtowicz, T., Karczewski, G. and Kossut, J. (2001) *Phys. Stat. Sol. (b)* **227**, 343.
26. Stebe, B., Munschy, G., Stauffer, L., Dujardin, F., and Murat, J. (1997) *Phys. Rev. B* **56**, 12454.
27. Stebe, B. and Ainane, A. (1989) *Superlatt. & Microstruct.* **5**, 545.
28. Sergeev, R.A. and Suris, R.A. (2001) *Phys. Solid State* **43**, 746.
29. A.Esser, *private communication*
30. Finkelstein, G. and Bar-Joseph, I. (1995) *Nouvo Cimento D* **17**, 1239.
31. Taliercio, T., Lefebvre, P., Calvo, V., Scalbert, D., Magnea, N., Mathieu, H., and Allegre, J. (1998) *Phys. Rev. B* **58**, 15408.
32. Vanhoucke, T., Hayne, M., Henini, M., and Moshchalkov, V.V. (2001) *Phys. Rev. B* **63**, 125331.
33. Yusa, G., Shtrikman, H., and Bar-Joseph, I. 2001 arXiv: cond.mat/0103561, 27 March 2001.
34. Whittaker, D.M. and Shields, A.J. (1997) *Phys. Rev. B* **56**, 15185.

# NON-LINEAR EFFECTS ON THE SPIN DYNAMICS OF POLARITONS IN II–VI MICROCAVITIES

G. AICHMAYR[†], M.D. MARTIN[‡], L. VIÑA
Dept. Física de Materiales, Universidad Autónoma,
E-28049 Madrid, Spain,
[†] Infineon Technologies,
Königsbrücker St. 180 D-01099, Dresden, Germany
[‡] Dept. of Physics and Astronomy, University of Southampton,
SO17 1BJ, Southampton, UK

R. ANDRÉ
Lab. Spectrometrie Physique (CNRS), Univ. J. Fourier 1,
F-38402 Grenoble, France

V. CIULIN, J.D. GANIERE AND B. DEVEAUD
Physics Dept., Swiss Federal Institute of Technology,
CH-1015 Lausanne, Switzerland

Abstract. We have studied the polarization of the light emitted by a semiconductor microcavity as a function of the detuning between the cavity-mode and the exciton. Under high excitation conditions, when the cavity is in a non-linear regime, the emission originates from the cavity-like branch of the polaritons, i.e. the lower polariton branch (LPB) for negative detuning and the upper polaritons branch (UPB) for positive detuning. The time dependence of the polarization, which represents the spin dynamics of the polaritons, shows a very rich and novel behavior in this non-linear regime, as compared to that under low excitation conditions. In the latter case, the polarization decays exponentially to zero after a pulsed excitation, in a similar way to that known for bare excitons in quantum wells, while in the non-linear regime the polarization reaches its maximum at a finite time and furthermore, its sign is strongly dependent on the cavity-exciton detuning ($\delta = E_C - E_X$): it is positive for $\delta > 0$ and negative for $\delta < 0$. The negative polarization is directly related with an energy splitting between the $\sigma^+$- and $\sigma^-$-polarized components of the emission, which appears when the excitation density drives the cavity into the non-linear regime.

Key words: microcavities, polaritons, spin dynamics, stimulated emission, time-resolved optical spectroscopies

## 1. Introduction

Semiconductor microcavities have attracted increasing interest in the last decade because they allow a precise control of the radiation-matter interaction. This interaction is strongest when the characteristic frequencies of photons (radiation) and excitons (matter) are brought into resonance. Two different regimes can be established under this resonance condition: the strong and weak coupling regimes. The largest effort has been devoted to the study of the strong coupling regime (SCR), in which the eigenstates of the system are no longer pure exciton or photon but a superposition of both, known as cavity-polaritons [1]. The resonant frequencies of excitons and photons are split, leading to the so-called *Rabi splitting*, in analogy with atomic cavities [2]. On the contrary, in the weak coupling regime (WCR), the polaritons are bleached. A transition from the SCR to the WCR can occur, for example, when the quantum wells (QWs) are close to a node of the standing wave in the cavity or when the exciton population is very large [3]. The latter situation hindered the study of the non-linear properties of cavity polaritons. Only in the last years it has been possible to observe the polariton non-linear emission in both, III–V [4–6] and II–VI [7–10] semiconductor microcavities. In particular, in the work of Le Si Dang and coworkers, a transition to the non-linear regime for moderate excitation densities is observed, and only for much larger powers a transition to the WCR is found [7].

Other issue that has drawn a lot of attention in the non-linear SCR is the existence of a polariton-polariton scattering mechanism stimulated by the final state population. This mechanism will be active in a bosonic system, such as cavity polaritons, as soon as the final state population approaches unity. Clear experimental evidences of this stimulated scattering have been reported recently in the literature [5, 11]. In those experiments, the parametric scattering was enhanced by a convenient choice of the angle of incidence of the excitation beams. The result is a macroscopic polariton occupancy ("condensation") of the states at $K \sim 0$ and $K \sim 2k_{pump}$, where $k_{pump}$ is the incident pump wave vector.

Concomitantly, a rekindled interest on the carriers' spin in semiconductor structures has given rise to a new field, *spintronics*, which explores the possibility of designing new spin-based devices. The polarization of the emission is directly related with the spin of the elementary excitations of the system. The properties of spin polarized carriers in bulk semiconductors, their generation, characterization and the mechanisms responsible for the spin relaxation have been profusely studied in the past [12–15]. The spin relaxation, i.e. the change in the spin state, takes place through scattering with phonons or impurities, as a consequence of the change in momentum together with the spin-orbit coupling. This process for electrons in the conduction band of bulk semiconductors appears from the mixing with finite momentum states of the valence band: Elliott–Yafet (EY) [16, 17] and Dyakonov'–Perel' (DP) [18–20] mechanisms. In the case of a simultaneous existence of electrons and holes, after an optical excitation, this relaxation is due to the exchange interaction between both kinds of carriers, which is known as Bir–Aronov–Pikus (BAP) mechanism [21]. The spin relaxation processes of excitons, electrons and holes in low dimensional semiconductors have been also extensively investigated in the last decades both, experimentally [22–25] and theoretically [26–28]. In the particular case of cavity polaritons, due to the mixed photon-exciton character, significant changes on their spin dynamics with respect to bare quantum wells are expected. Nevertheless, the spin has been considered only very recently both under cw [29] and pulsed excitation [11, 30–32].

In this work, we review the role that the polariton spin plays in the stimulated scattering process, and illustrate the large values of the polarization degree of the emission in the SCR. We show that the sign of the polarization is controlled by the exciton-cavity detuning, through a breaking of the degeneracy of the spin-up and -down cavity-like states: the ground state of the system is spin-up (-down) at positive (negative) detunings. The rest of the manuscript is organized as follows: Section 2 gives the experimental details. The dynamics of the polariton spin and its detuning dependence is shown in Section 3. A short account of the angular dependence, with respect to the sample normal, of the polarization is given in Section 4. Finally, we summarize in Section 5.

## 2. Experimental Details

The experiments were performed either in a cold-finger or in an immersion cryostat, with the temperature kept at 5 K and 1.2 K, respectively. The samples were optically excited with light pulses from a Ti:Sapphire mode-locked laser pumped by an $Ar^+$-ion laser. The photoluminescence (PL) was time-resolved either in a standard up-conversion spectrometer (at the Universidad Autónoma de Madrid –UAM–) or using a streak-camera (at the Ecole Polytechnique Federale de Lausanne –EPFL–). The exciting light was circularly polarized by means of a $\lambda/4$ plate, and the PL was analyzed into its $\sigma^+$ and $\sigma^-$ components. The time resolution using the up-conversion set-up (streak-camera) was 1.2 ps (5 ps).

Two different microcavities were used in this study. Sample A is a $Cd_{0.40}Mg_{0.60}Te$ microcavity of $\lambda/2$ thickness, sandwiched between the top (bottom) distributed Bragg reflectors (DBRs). These mirrors are made of 17.5 (23) pairs, of alternating $\lambda/4$-thick layers of $Cd_{0.40}Mg_{0.60}Te/Cd_{0.75}Mn_{0.25}Te$. Two 90-Å thick CdTe quantum wells are placed in the centre of the cavity to obtain the optimum radiation-matter interaction, which leads to a Rabi splitting of 10 meV (see Fig. 1). Sample B has a $\lambda$ thickness, with the CdTe QWs placed, which leads to a Rabi splitting of 12 meV. The optical excitation was performed at the first minimum above the stop-band of the DBRs, $\sim$ 90 meV higher than the emission energy of the cavity-like polariton branch, to assure the same excitation conditions in all the experiments.

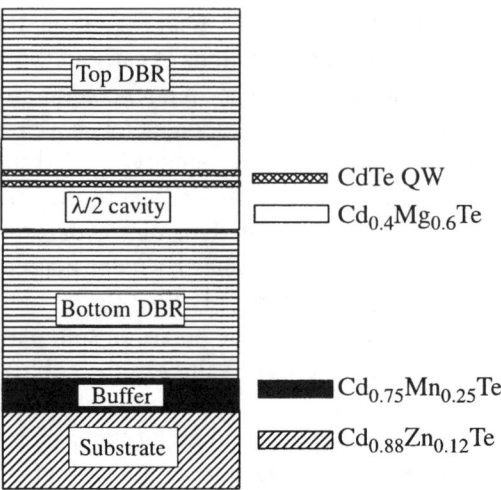

*Figure 1.* Scheme of the microcavity, sample A, used in the up-conversion experiments.

A slight wedge in the cavity thickness allows tuning the cavity and the exciton into resonance by moving the excitation spot across the wafer.

## 3. Detuning Dependence

The results presented in this section correspond to sample A and were performed with the up-conversion set-up at the UAM. We investigated the time evolution of the polariton PL as a function of the cavity-exciton detuning ($\delta = E_C - E_X$) and of the excitation power density. Let us start by a characterization of the sample. Figure 2 shows the energies of the upper- and the lower-polariton branches, labeled UPB and LPB, respectively, for different positions of the excitation beam on the sample surface. These energies are obtained from time-resolved photoluminescence spectra taken 10 ps after excitation with 2 W/cm$^2$. The normal mode splitting is clearly resolvable, obtaining a Rabi splitting of 10 meV, similar to that obtained in cw-reflectivity experiments. The insets in the figure depict the measured spectra at the positions marked by the arrows, demonstrating the resolution of both polariton branches at negative and positive detunings in time-resolved emission at short times after the excitation.

The time evolution, at resonance ($\delta = 0$ meV), is displayed in Fig. 3 for the lower polariton branch at two different excitation conditions. The rise of the PL in Fig. 3(a), corresponding to a power density of 1.5 W cm$^{-2}$, has several contributions: the photocreated electrons and holes bind into large **K** excitons in the first tens of picoseconds [33]; these excitons scatter into LPB and UPB states and continue reducing their energy and momentum via the emission of acoustic phonons. The PL reaches its maximum when the polaritons have reached **K** $\sim$ 0 states. This **K** $\sim$ 0 population exponentially disappears via radiative recombination and escape of the

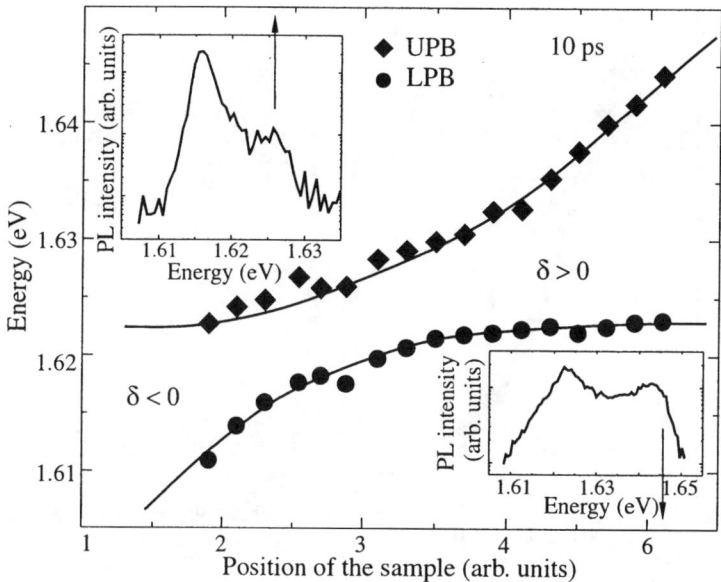

*Figure 2.* LPB (•) and UPB (◇) energies as a function of the position of the excitation spot on sample A, for a time delay of 10 ps. The lines are guides to the eye. The inset show representative spectra used to extract the peak positions at the sample positions marked by the arrows.

polaritons out of the cavity. The time dependence, with slow rise and decay times, is qualitatively similar for both polariton branches. An increase of the excitation density modifies strongly the dynamics of the light emission, as shown in the trace depicted in Fig. 3(b) for a density of 10 W cm$^{-2}$. At this density, the system enters in a non-linear regime, where the emission intensity increases exponentially with increasing the excitation density. The rise and decay dynamics become very fast and even the curvature of the time traces at very short times changes sign (see inset in the figure), a fact already reported in GaAs microcavities [31]. One should also mention that the structure observed in Fig. 3(b) at $t \sim 40$ ps is reproducible, and it will be discussed in Section 4.

The dynamics of the light emission in the linear-regime depends also of the detuning as demonstrated in Fig. 4. The circles (diamonds) represent the decay time of the PL for the LPB (UPB) as a function of the detuning, for a density of 1.5 W cm$^{-2}$. These times are similar at a negative detuning of $-5$ meV and increase for the LPB with increasing detuning, as its excitonic character grows. An opposite behavior is obtained for the UPB, where the decay times decrease with increasing detuning and become very fast ($\sim 10$ ps) at $\delta = +15$ meV (the inset in the figure depicts the corresponding time-traces).

The time evolution of the emission is presented in Fig. 5 for different detunings, where solid (dashed) lines show the co-(counter-) polarized components. The left panels correspond to the LP branch under low excitation conditions: the increase

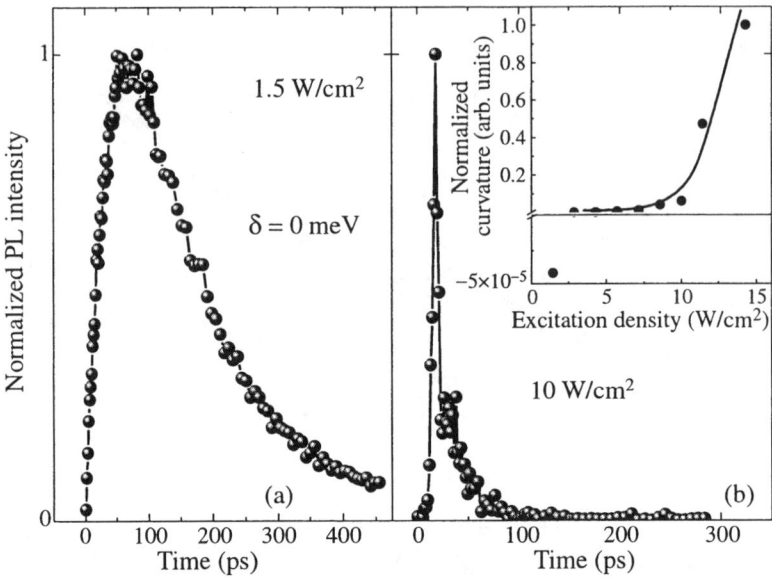

*Figure 3.* Time-evolution of the PL emitted by the LPB in sample A at resonance ($\delta = 0$ meV) for an excitation density of 1.5 W cm$^{-2}$ (a) and 10 W cm$^{-2}$ (b). The inset depicts the normalized curvature of the initial rise of the LPB, at 10 ps, as a function of the excitation density. The lines are guide to the eye.

in the decay rate is clearly seen going from positive ($\delta = +15$ meV) to negative ($\delta = -21$ meV) detunings, while the difference in intensity of both polarized components is very small and independent of $\delta$. However, this situation changes drastically when the system is brought into the non-linear emission regime, as can be observed in the right panels of Fig. 5, which correspond to an excitation power of 15 mW, and the emission stems from the branch with the largest photonic content. For $\delta = +15$ meV the co-polarized emission ($\sigma^+\sigma^+$) is considerably larger than the counter-polarized ($\sigma^+\sigma^-$), the difference becomes even larger and the decay rates also increase at $\delta = +8$ meV. However, the intensities become comparable close to resonance ($\delta = 0$ meV) and the counter-polarized emission becomes dominant for negative detunings. It is also worthwhile mentioning that the integrated emission intensity of this photon-like polariton branch displays an exponential growth with the excitation power. Similar exponential growths have been reported in the literature and have been attributed to bosonic final state stimulated scattering [5, 11].

The polarization degree of the PL is defined as $\wp = (I^+ - I^-)/(I^+ + I^-)$, where $I^{+/-}$ is the emission intensity of the $\sigma^{+/-}$ component of the PL. $\wp$ will be denoted in the following as polarization. The time evolution of the polarization is depicted in Fig. 6. The open points correspond to the low excitation regime, which shows a dynamic of the polarization independent of the detuning and very similar to

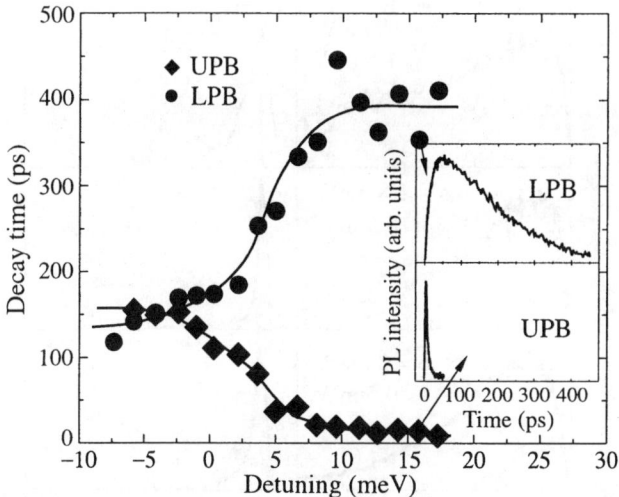

*Figure 4.* Decay times of the LPB (•) and UPB (◊) for a density of 1.5 W cm$^{-2}$ (sample A). The lines are a guide to the eye. The inset show the time traces for the points marked by the arrows ($\delta = +15$ meV).

that found for bare excitons in quantum wells. The traces with the full symbols are taken at an excitation density of 15 mW and demonstrate the very reach behavior of the polarization when the cavity is in the non-linear emission regime. At positive detunings, the polarization is positive and increases up to its maximum value, which can be as high as 90%, in 20 ps; subsequently the polarization decreases to zero. The rise of the polarization degree ($0 < t < 20$ ps) can be interpreted as follows: the initial $\sigma^+$-polarized pulse creates a larger +1 spin population. A fast scattering process will bring all the photocreated excitons to the cavity mode before any spin relaxation can occur, what will result in a larger +1 spin population of $\mathbf{K} \sim 0$ states at $t = 0$. This initial +1 spin population will act as a seed for the stimulated scattering process and therefore a large number of +1 spins will be transferred to $K \sim 0$ states. The radiative recombination of this population will result in a bigger $\sigma^+$-polarized emission, i.e. a very large positive polarization. After reaching the maximum $+1/-1$ spin population difference ($t \sim 20$ ps) the +1 spin population disappears very quickly through the $\sigma^+$-polarized stimulated emission process, taking the polarization to zero ($20 < t < 40$ ps). This initial rise of the polarization is unique to microcavities and contrasts with the behavior of ℘ for bare excitons that always decays with time monotonically to zero from its maximum initial value. The degree of polarization presents a very short time-interval of negative values, when $\delta$ is close to zero, followed by a maximum, before ℘ decreases steadily to zero. At negative detunings, ℘ becomes basically negative, reaching values as large as $-75\%$ for $\delta \sim -12$ meV. It even shows a plateau lasting $\sim 30$ ps at $\delta = -14$ meV, before the absolute values of ℘ begin to decrease when

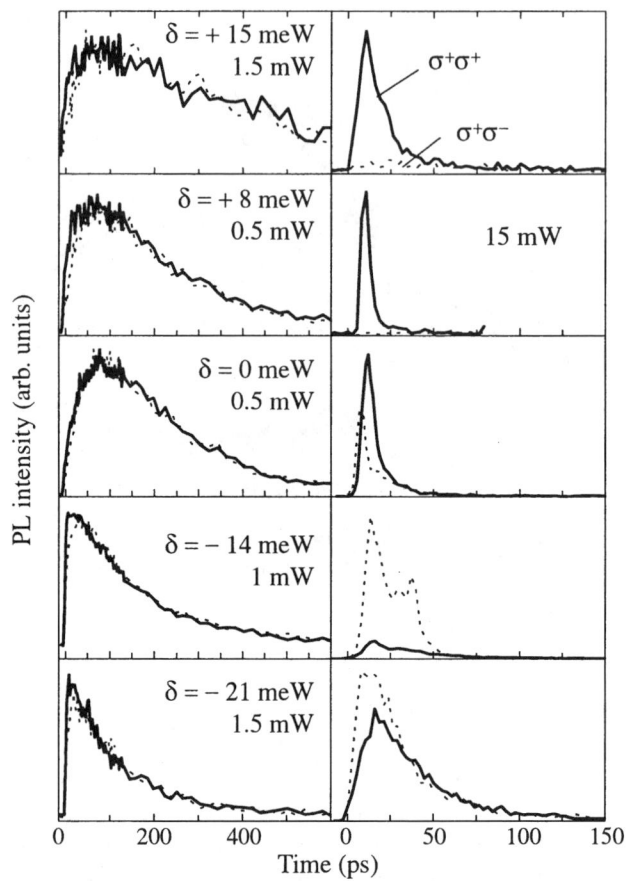

*Figure 5.* Time evolution of the PL from sample A. Solid (dashed) lines depict the $\sigma^+$ ($\sigma^-$) polarized component of the emission after excitation with a $\sigma^+$ pulse. Left panels correspond to low-excitation conditions and the emission arises from the LP branch. Right panels show the PL under high-excitation, which arises from the LP (UP) branch for negative (positive) detunings.

the detuning is further increased to negative values. This behavior is related to the spin-selective polariton-polariton stimulated scattering to the final state, which is strongly dependent of the exciton-cavity detuning.

In the case of $\delta < 0$, the initial $\sigma^+$-polarized pulse creates a larger +1 spin population, which is reflected at $\mathbf{K} \sim 0$ by the positive polarization degree at $t = 0$. The relaxation of the non-resonantly created polaritons to $\mathbf{K} \sim 0$ is governed by the final state stimulated scattering. Nevertheless, the scattering to the −1 spin states is more efficient than to +1 spin states. The accumulation of −1 spin polaritons results in a larger $\sigma^-$-polarized emission and therefore, a very large negative polarization. This negative polarization could be attributed to a resonant excitation of light hole excitons; however, this possibility can be discarded by energy arguments (the excitation energy is always at least 30 meV above the light-

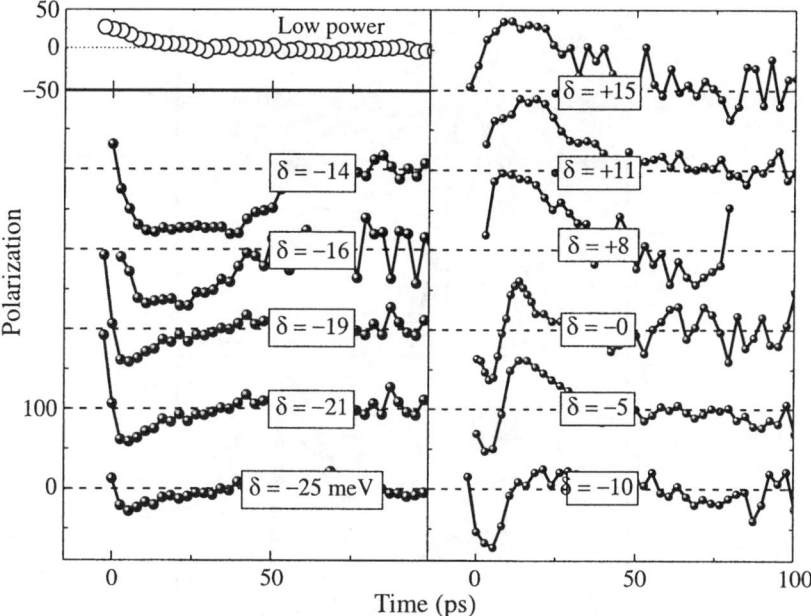

*Figure 6.* Time evolution of the polarization degree of the emission of the photonic branch of sample A. Open points correspond to a low excitation condition. Solid points depict the results in the case when the cavity is in the non-linear regime for an excitation power of 15 mW. The detuning for each curve is indicated in a box. Dashed lines indicate the zero of the degree of polarization for each trace.

hole exciton resonance) and furthermore, such a resonant excitation would lead to a negative initial polarization degree, in contrast with the experimental findings.

The different scattering efficiencies might be related with an energy splitting observed between the two circularly polarized component of the PL at very short times. This splitting is mirroring the different energies of the $+1$ and the $-1$ spin levels and is demonstrated in Figs. 7(a) and 7(b), for a delay time of 20 ps (which is the time when the emission reaches its maximum intensity) and an excitation density of 20 W/cm$^2$, for negative and positive detunings, respectively. This spin splitting increases with excitation power, saturating at $\sim 0.5$ meV for 20 W/cm$^2$. In the case of $\delta > 0$, the splitting is observed only for the UP branch. At negative (positive) detunings, the energy of the $-1$ $(+1)$ spin states at $\mathbf{K} \sim 0$ is smaller than that of the $+1$ $(-1)$ spin states; this fact, in addition with the different efficiencies of the $+1/-1$ stimulated scattering processes would account for the large $\sigma^-$ $(\sigma^+)$-polarized emission intensity and the observed negative (positive) polarization.

The physical origin of this energy splitting between the two spin states at $\mathbf{K} \sim 0$ still needs to be clarified, but it is likely to account for the reversal of the polarization degree of the PL with changing the exciton-cavity detuning. The splitting would be compatible with a decrease in the light-matter interaction strength for the majority

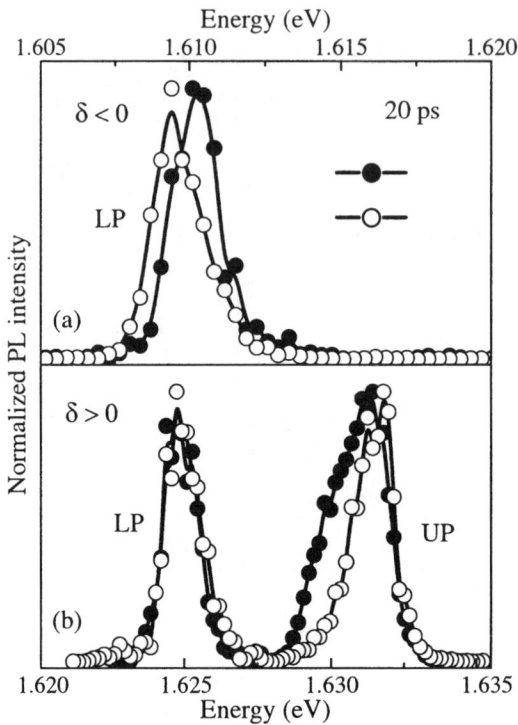

*Figure 7.* Spectra of sample A taken 20 ps after excitation with a $\sigma^+$-polarized pulse analyzed into its $\sigma^+$ ($\lambda$) and $\sigma^-$ (•) components for an excitation power of 20 mW. Panel (a) corresponds to a negative detuning of $-10$ meV, while (b) depicts the results for $\delta = +10$ meV.

polaritons (+1), which are initially created by the $\sigma^+$-polarized excitation, as compared to that of the minority (−1) polaritons. This would imply that for negative (positive) detuning, the +1 states would lie above (below) the −1, rendering a $\Delta < 0$ ($> 0$) as borne out by our results. However, our experiments show (see Fig. 8) that with increasing excitation density, an initial blue shift of 0.5 meV for both ±1 states, without any splitting, is followed by a red shift of the −1 polaritons, while the +1 remain at the same energy. Therefore, the coupling strength of the +1 polaritons does not decrease, invalidating the previous argument. Note that the appearance of the splitting occurs concomitantly with the developing of negative values of the polarization and that both the splitting and $\wp$ saturate simultaneously. The fact that the splitting increases with excitation power density indicates that it could originate from exciton-exciton interactions. An existing theory for bare excitons would qualitatively explain the splitting and the ±1 level ordering, as a result of exchange and vertex corrections to the self-energies [34], but only for $\Delta < 0$.

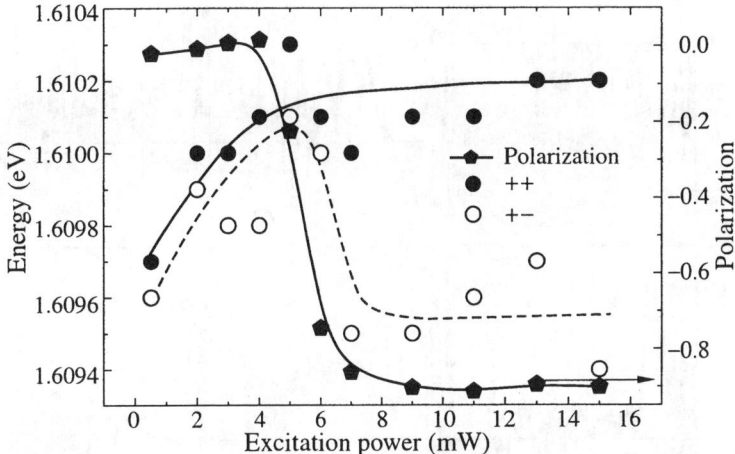

*Figure 8.* Energies of the light emitted from sample A at a negative detuning of −10 meV as a function of excitation power (left axis). Solid (open) points depict the $\sigma^+$ ($\sigma^-$) polarized component of the emission after excitation with a $\sigma^+$ pulse. The diamonds show the integrated polarization as a function of excitation power (right axis). The lines are a guide to the eye.

## 4. Angular Dependence

In order to obtain further insight on the origin of the negative degree of polarization we have studied the dynamics of the polarization as a function of the angle with respect to the growth axis of the microcavity. The data discussed so far, obtained with the up-conversion technique, were limited to **K** ∼ 0, due to the selectiveness of the frequency doubling in the non-linear crystal used in that technique, however the momentum of the polaritons is known to play a very important role in their relaxation and the stimulated scattering processes [11, 35–37], and could also be important in the determination of the spin dynamics. For these experiments, performed at the EPFL using a streak camera with a total time-resolution of 5 ps, sample B, with a Rabi splitting of 12 meV was used. The angular analysis of the emission was done with a small pinhole, which could be moved in a section of a parallel beam in the detection path, obtaining a resolution of 1°. We will focus on the non-linear regime (excitation density of 50 W cm$^{-2}$) and at a negative detuning of −13 meV.

Figure 9 depicts the dynamics of the polarization for different angles. The PL is also shown in Fig. 9(a) for co-polarized ($\sigma^+\sigma^+$, dashed line) and counter-polarized ($\sigma^+\sigma^-$, dot-dashed line) configurations, while $\wp$ is depicted with a solid line. Note that a 0° the polarization shows a dynamic very similar to that obtained from sample A at a comparable negative detuning (see Fig. 6), and the only significant difference is the value of the polarization with diminishes from −60% (sample A) to −30% (sample B). The temporal dependence of $\wp$ presents a striking behavior when the detection angle is increased to 6° (Fig. 9b): an initial negative peak at 35 ps

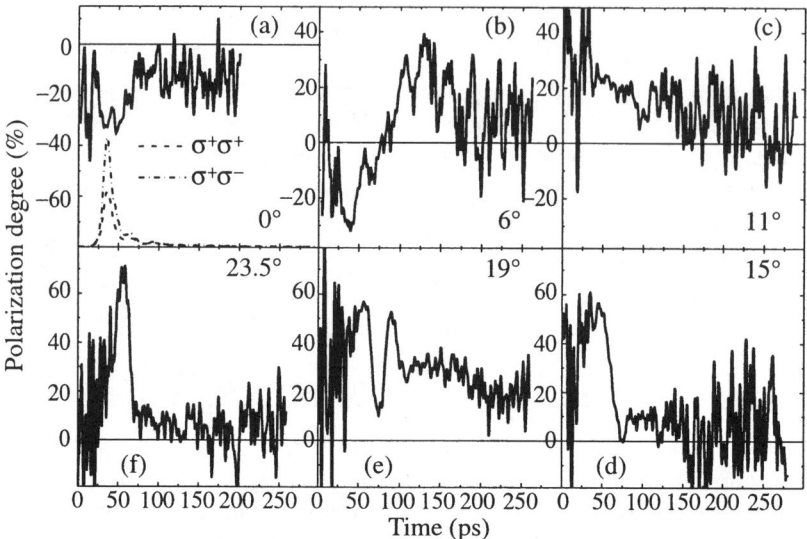

*Figure 9.* Time evolution of the polarization degree of the emission of the LPB of sample B for different angles of detection: (a) 0°, (b) 6°, (c) 11°, (d) 15°, (e) 19° and (f) 23.5°. The dashed (dot-dashed) lines in panel a) depict the time evolution of the $\sigma^+$-($\sigma^-$) polarized component of the PL after excitation with a $\sigma^+$ pulse. Thin solid lines indicate the zero of the degree of polarization for each trace.

is followed by positives values of the polarization, reaching a maximum of +35% at 150 ps. Such an oscillatory behavior of the polarization has never been observed in bare quantum wells, however a analogous effect has been reported in GaAs microcavities and attributed to the fast disappearance of the polariton population that undergoes stimulated emission and the concurrent slower dynamics of the opposite spin population [31]. The negative values of the polarization disappear for angles larger than 7°. Still marked differences are seen in the microcavities with respect to the polarization behavior in QWs. Figure 9(c) documents that, in contrast with the monotonically decreasing behavior of $\wp$ found in bare QWs [38, 39], a maximum is observed at a finite time after excitation. This fact is even more clearly seen at larger detection angles (Fig. 9d-f). A second type of oscillations are also observed in the dynamics of the polarization, as shown in Fig. 9(e) for a detection angle of 19°. These have a different origin as those shown in Fig. 9(b) and are related to strong oscillations in the time-resolved emission (for both polarizations); the period of the oscillations amounts to $\sim$ 30 ps independently of the angle of observation. These oscillations are also responsible for the structure seen in Fig. 3(b) at $t \sim$ 40 ps. Finally, at the largest emission angle used in our experiments (Fig. 9e), were the excitonic character of the polaritons has increased considerably, a very sharp peak in the dynamics of the polarization is obtained.

It is worthwhile to note that at the powers used in these experiments under non-resonant excitation conditions, the maximum intensity of the emission is obtained at $K \sim 2 \times 10^4$ cm$^{-1}$, close to the inflection point of the lower polariton branch, giving rise to a ring emission in a cone centered at $\sim 15°$. Therefore one could be tempted to attribute the negative values of the polarization observed at small angles to a population effect, as the one discussed above for the origin of the oscillations reported in GaAs microcavities. However, this cannot be the explanation because the cavity starts emitting with maximum intensity at $K \sim 0$ and subsequently the intensity is rapidly transferred to larger angles. In fact the emission at $K \sim 0$ becomes negligible at the times when the cavity begins to emit at $K \sim 2 \times 10^4$ cm$^{-1}$. This behavior is in contrast to that observed in resonant excitation experiments, were the ring is observed at the earlier times [36], and could be originated by the higher exciton densities at very short times in our experiments and the following decrease due to recombination and photon escaping of the cavity [40]. Therefore, the negative values of the polarization obtained at small angles originate from the stimulated scattering process to the final state that provides the seed for the stimulation and which is determined by its lower energy.

## 5. Summary

We have studied the polarization of the light emitted by a semiconductor microcavity as a function of the detuning between the cavity-mode and the exciton and as a function of the angle with respect to the cavity normal. Under high excitation conditions, when the cavity is in a non-linear regime, the emission originates from the cavity-like branch of the polaritons, i.e. the lower polariton branch (LPB) for negative detuning and the upper polaritons branch (UPB) for positive detuning. The time dependence of the polarization shows a very rich and novel behavior in this non-linear regime, as compared to that under low excitation conditions. In the latter case, the polarization decays exponentially to zero after a pulsed excitation, in a similar way to that known for bare excitons in quantum wells, while in the non-linear regime the polarization reaches its maximum at a finite time and furthermore, its sign is strongly dependent on the cavity-exciton detuning ($\delta = E_C - E_X$): it is positive for $\delta > 0$ and negative for $\delta < 0$. The negative polarization is directly related with an energy splitting between the $\sigma^+$- and $\sigma^-$-polarized components of the emission, which appears when the excitation density drives the cavity into the non-linear regime. The angular dependence of the emission, at negative detunings, shows that although the cavity starts emitting with maximum intensity at $K \sim 0$, the intensity is rapidly transferred to $K \sim 2 \times 10^4$ cm$^{-1}$, close to the inflection point of the lower polariton branch, giving rise to a ring emission in a cone centered at $\sim 15°$. Furthermore, at small angles the polarization of the emitted light reverses its sign with respect to that of the exciting pulses and steadily becomes positive for large angles.

## Acknowledgements

This work has been partially supported by the EU (TMR-Ultrafast Quantum Optoelectronics Network), the Spanish DGICYT (PB96-0085) and the CAM (07N/0064/2001).

## References

1. Weisbuch, C., Nishioka, M., Ishikawa, A., and Arakawa, Y. (1992) Observation of the coupled exciton-photon mode splitting in a semiconductor quantum microcavity, *Phys. Rev. Lett.* **69**, 3314–3317.
2. Haroche, S. and Kleppner, D. (1989) Cavity Quantum Electrodynamics, *Phys. Today* **42**, 24–30.
3. Houdre, R., Gibernon, J.L., Pellandini, P., Stanley, R.P., Osterle, U., Weisbuch, C., O'Gorman, J., Roycroft, B., and Ilegems, M. (1995) Saturation of the strong-coupling regime in a semiconductor microcavity: Free-carrier bleaching of cavity polaritons, *Phys. Rev. B* **52**, 7810–7813.
4. Senellart, P. and Bloch, J. (1999) Nonlinear emission of microcavity polaritons in the low density regime, *Phys. Rev. Lett.* **82**, 1233–1236.
5. Stevenson, R.M., Astratov, V.N., Skolnick, M.S., Whittaker, D.M., Emam-Ismail, M., Tartakovskii, A.I., Savvidis, P.G., Baumberg, J.J., and 5, J. S. R. (2000) Continuous Wave Observation of Massive Polariton Redistribution by Stimulated scattering in Semiconductor Microcavities, *Phys. Rev. Lett.* **85**, 3680–3683.
6. Dasbach, G., Baars, T., Bayer, M., Larionov, A., and Forchel, A. (2000) Coherent and incoherent polaritonic gain in a planar semiconductor microcavity, *Phys. Rev. B* **62**, 13076–13083.
7. Dang, L.S., Heger, D., Andre, R., Boeuf, F., and Romestain, R. (1998) Stimulation of Polariton Photoluminescence in Semiconductor Microcavity, *Phys. Rev. Lett.* **81**, 3920–3923.
8. Bleuse, J., Kany, F., Boer, A.P. d., Christianen, P.C.M., André, R., and Ulmer-Tuffigo, H. (1998) Laser emission on a cavity-polariton line in a II–VI microcavity, *J. Crystal Growth* **184-185**, 750–753.
9. Boeuf, F., André, R., Romestain, R., and Dang, L.S. (2000) Evidence of polariton stimulation in semiconductor microcavities, *Phys. Rev. B* **62**, 2279–2282.
10. Mueller, M., Bleuse, J., and Andre, R. (2000) Dynamics of the cavity polariton in CdTe-based semiconductor microcavities: Evidence for a relaxation edge. *Phys. Rev. B* **62**, 16886–16892.
11. Savvidis, P.G., Baumberg, J.J., Stevenson, R.M., Skolnick, M.S., Whittaker, D.M., and Roberts, J.S. (2000) Angle-resonant stimulated polariton amplifier, *Phys. Rev. Lett.* **84**, 1547–1550.
12. Pikus, G.E. and Ivchenko, E.L. (1982) Optical Orientation and Polarized Luminescence of Excitons in Semiconductors, in E.I. Rashba and M.D. Surge (eds.), *Excitons*, North-Holland Publishing Company, pp. 209–266.
13. D'yakonov, M.I. and Perel', V.I. (1984) Theory of optical spin orientation of electrons and nuclei in semiconductors, in F. Maier and B.P. Zakharchenya (eds.), *Optical Orientation*, Elsevier Science Publishers Amsterdam, pp. 11–72.
14. Pikus, G.E. and Titkov, A.N. (1984) Spin relaxation under optical orientation in semiconductors, in F. Maier and B.P. Zakharchenya (eds.), *Optical Orientation*, Elsevier Science Publishers Amsterdam, pp. 73–132.
15. Johnson, E.J., Seymour, R.J., and Alfano, R.R. (1984) Photoluminescence of spin-polarized electrons in semiconductors, in R.R. Alfano (eds.), *Semiconductors Probed by Ultrafast Laser Spectroscopy* Vol. II, Academic Press, pp. 200–241.
16. Elliot, R.J. (1954) Theory of the effect of spin-orbit coupling on magnetic resonance in some semiconductors, *Phys. Rev.* **96**, 266–279.

17. Yafet, Y., Ed. (1963) *Solid State Physics* New York, Academic Press.
18. D'yakonov, M.I. and Perel', V.I. (1971) On spin orientation of electrons in interband absorption of light in semiconductors, *Sov. Phys. JETP* **33**, 1053–1065.
19. D'yakonov, M.I. and Perel', V.I. (1972) Spin relaxation of conduction electrons in noncentrosymmetric semiconductors, *Sov. Phys.-Solid State* **13**, 3023–3027.
20. Dymnikov, V.D., D'yakonov, M.I. and Perel', V.I. (1976) Anisotropy of the momentum distribution of photo-excited electrons and the polarization of hot luminescence in semiconductors, *Sov. Phys.-JETP* **44**, 1252–1259.
21. Bir, G.L., Aronov, A.G., and Pikus, G.E. (1976) Spin relaxation of electrons due to scattering by holes, *Sov. Phys. JETP* **42**, 705–712.
22. Damen, T.C., Vina, L., Cunningham, J.E., and Shah, J. (1991) Subpicosecond spin relaxation dynamics of excitons and free carriers in GaAs quantum wells, *Phys. Rev. Lett.* **67**, 3432–3435.
23. Vinattieri, A., Shah, J., Damen, T.C., Goossen, K.W., Pfeiffer, L.N., Maialle, M.Z., and Sham, L.J. (1993) Electric field dependence of exciton spin relaxation in GaAs/AlGaAs quantum wells, *Appl. Phys. Lett.* **63**, 3164–3166.
24. Viña, L. (1999) Spin relaxation in low-dimensional systems, *J. Phys.: Condens. Matt.* **11**, 5929–5952.
25. Malinowski, A., Britton, R.S., Grevatt, T., Harley, T., Ritchie, D.A., and Simmons, M.Y. (2000) Spin relaxation in GaAs/GaAlAs quantum wells, *Phys. Rev. B* **62**, 13034–13039.
26. Sham, L.J. (1993) Theory of spin dynamics of excitons and free carriers in quantum wells, in D.J. Lockwood and A. Pinczuk (eds.), *Optical Phenomena in Semiconductor Structures of Reduced Dimension*, Kluwer Academic Publishers B.V. The Netherlands, pp. 201–211.
27. Ferreira, R. and Bastard, G. (1994) "Spin"-flip of holes in asymmetric quantum wells, *Solid State Electron.* **37**, 851–855.
28. de Andrade e Silva, E.A. (1997) Exciton-bound electron-spin relaxation, *Phys. Rev. B* **56**, 9259–9262.
29. Tartakovskii, A.I., Kulakovskii, V.D., Krizhanovskii, D.N., Skolnick, M.S., Astratov, V.N., Armitage, A., and Roberts, J.S. (1999) Nonlinearities in emission from the lower polariton branch of semiconductor microcavities, *Phys. Rev. B* **60**, 11293–11296.
30. Renucci, P., Marie, X., Amand, T., Paillard, M., Senellart, P., and Bloch, J. (2001) Non-linear spin polarization dynamics in semiconductor microcavities, in N. Miura and T. Ando (eds.), *Springer Proceedings in Physics 87*, Springer-Verlag New York, pp. 653–654.
31. Martin, M.D., Viña, L., Son, J.K., and Mendez, E.E. (2001) Spin dynamics of cavity polaritons, *Solid State Commun.* **117**, 267–271.
32. Lagoudakis, P.G., Savvidis, P.G., Baumberg, J.J., Whittaker, D.M., Eastham, P.R., Skolnick, M.S., and Robert, J.S. (2002) Stimulated spin dynamics of polaritons in semiconductor microcavities, *Phys. Rev. B* **65**, 161310–161313.
33. Shah, J. (1996) *Ultrafast Spectroscopy of Semiconductors and Semiconductor Heterostructures*. Berlin, Springer Verlag.
34. Fernández-Rossier, J., Tejedor, C., Muñoz, L. and Viña, L. (1996) Polarized interacting exciton gas in quantum wells and bulk semiconductors, *Phys. Rev. B* **54**, 11582–11591.
35. Houdre, R., Weisbuch, C., Stanley, R.P., Oesterle, U., Pellandini, P., and Hegems, M. (1994) Measurements of cavity-polariton dispersion curve from angle-resolved photoluminescence experiments, *Phys. Rev. Lett.* **73**, 2043–2046.
36. Freixanet, T., Sermage, B., Bloch, J., Marzin, J.Y., and Planel, R. (1999) Annular resonant Rayleigh scattering in the picosecond dynamics of cavity polaritons, *Phys. Rev. B* **60**, 8509–8512.
37. Tartakovskii, A.I., Kulakovskii, V.D., Krizhanovskii, D.N., Skolnick, M.S., Astratov, V.N., Armitage, A., and Roberts, J.S. (1999) Nonlinearities in emission from the lower polariton branch of semiconductor microcavities, *Phys. Rev. B* **60**, 11293–11296.

38. Damen, T.C., Viña, L., Cunningham, J.E., Shah, J., and Sham, L.J. (1991) Dynamics of formation and relaxation of intrinsic excitons and relaxation and thermalization of exciton-spin in GaAs quantum wells *Phys. Rev. Lett.* **67**, 3432-3435.
39. Muñoz, L., Pérez, E., Viña, L. and Ploog, K. (1995) Spin relaxation in intrinsic GaAs quantum wells, influence of exciton localization, *Phys. Rev. B* **51**, 4247–4257.
40. Savvidis, P.G., Baumberg, J.J., Porras, D., Whittaker, D.M., Skolnick, S., and Roberts, J.S. (2002) Ring emission and exciton-pair scattering in semiconductor microcavities, *Phys. Rev. B* **65**, 73309–73312.

# SPIN DYNAMICS OF NEUTRAL AND CHARGED EXCITONS IN InAs/GaAs QUANTUM DOTS

M. SÉNÈS, B.L. LIU, X. MARIE, T. AMAND
*Laboratoire de Physique de la Matière Condensée, INSA-CNRS*
*Complexe scientifique de Rangueil, 31077 Toulouse cedex, France*

J.M. GÉRARD‡
*Laboratoire de Photonique et Nanostructures,*
*Route de Nozay, 91460 Marcoussis, France*
‡ *New address: DRFMC/CENG/CEA, avenue des Martyrs, 38000 Grenoble, France*

Abstract. We have studied the spin dynamics in non-intentionally doped and n-modulation doped self-organized InAs/GaAs quantum dots (QD) by time-resolved photoluminescence performed under strictly resonant excitation. We demonstrate that the carrier spins in these nanostructures are totally frozen on the neutral or charged exciton lifetime scale.

Key words: quantum dots, spin properties, exchange interaction

## 1. Introduction

The discrete energy levels in artificial atoms like semiconductor quantum dots and the corresponding lack of energy dispersion lead to a predicted modification of the spin relaxation dynamics compared to bulk or two-dimensional structures [1–3]. The absolute lack of energy states between QD energy levels is expected to suppress not only the elastic processes of spin relaxation but also the inelastic ones such as phonon scattering.

Recent optical pumping experiments have indeed given good indications of a slowing down of the carrier spin relaxation processes in QD compared to bulk or quantum wells (QW) structures [2, 4–7]. However most of these studies have been performed in non-resonant excitation conditions. The observed spin polarization dynamics of the QD ground state is then the result of all the spin relaxation mechanisms which have occurred in the bulk barrier, in the QD excited state and finally in the QD ground state, including any spin flip scattering due to the energy relaxation process itself. In order to study the *intrinsic* spin dynamics, a time-resolved strictly resonant excitation of the QD ground state is highly desirable [8, 9]. We present here a time-resolved investigation of the carrier spin dynamics in *non-intentionally doped* and *n-modulation doped* self-organized InAs/GaAs quantum dots (QD) performed under strictly resonant excitation.

The investigated structures are grown by molecular beam epitaxy on (001) GaAs substrates. The *non-intentionally doped* structure (Sample I) consists of 5 InAs lens-shaped QD planes embedded in a GaAs λ planar cavity, inserted between two GaAs/AlAs Bragg mirrors [10]. The QD layers are localized in the vicinity of the electromagnetic field antinodes into the microcavity which operates in the weak coupling regime. The QDs are obtained after a nominal deposition of 2.2 InAs monolayers and a 21 s growth interruption. These growth conditions lead to a QD density $\sim 4 \times 10^{10}$ cm$^{-2}$ per array, and to a QD ground state emission centered around 1.15 eV with a Full Width at Half Maximum (FWHM) of $\sim$ 60 meV. A variation of the cavity thickness along the radial direction of the wafer allows us to tune the cavity resonance by moving the laser beam across the sample surface. The microcavity is designed so that the cavity mode (FWHM $\sim$ 3 meV) can be tuned in the energy range of the QD ground state emission. Inserting the QD in a microcavity brings an increase in the detected QD emission signal with respect to bare QD structures: the narrowing of the radiation pattern emitted by the microcavity allows us to collect the QD emission efficiently in spite of the very small acceptance solid angle ($\sim 10^{-3}$ steradians) of the up-conversion detection set-up we have used [8]. Very similar experimental results have been obtained in QD structures without microcavities but with weaker signal to noise ratio [11].

The *n-modulation doped* structure (Sample II) consists of 20 layers of lens-shaped self-assembled InAs/GaAs QD separated by 40 nm thick GaAs layers. It contains a Si-delta doping layer located 2 nm below each wetting layer with a nominal donor concentration of $8 \times 10^{10}$ cm$^{-2}$. The average QD density is about $4 \times 10^{10}$ cm$^{-2}$. Two experimental results suggest that most of the QD contain a single doping electron:
(i) Observation of far-infrared intraband absorption around 43 meV (corresponding to the s-p conduction transition energy) indicates that the doping is effectual [12];
(ii) The observation of luminescence under strictly resonant excitation as shown in section 3 proves that the s-shell is not fully occupied

The samples are excited by 1.5 ps pulses generated by an Optical Parametric Oscillator, synchronously pumped by a mode-locked Ti-doped sapphire laser. For sample I, the time-resolved PL is then recorded by up-converting the luminescence signal in a LiIO$_3$ non-linear crystal with the picosecond pulses generated by the Ti:Sa laser [8, 13]. The time-resolution is limited by the laser pulse-width ($\sim$ 1.5 ps) and the spectral resolution is about 3 meV. For sample II, the time-resolved PL is recorded at the excitation energy using a S1 photocathode Hamamatsu Streak Camera with an overall time-resolution of 8 ps.

The linear and the circular polarization degrees of the luminescence are defined as $P_{\text{Lin}} = (I^X - I^Y)/(I^X + I^Y)$ and $P_c = (I^+ - I^-)/(I^+ + I^-)$ respectively. Here

$I^X$ ($I^Y$) and $I^+$ ($I^-$) denote respectively the $X(Y)$ linearly polarized and the right (left) circularly polarized luminescence components ($X$ and $Y$ are chosen parallel to the [110] and [1$\bar{1}$0] sample directions). The detection energy is strictly the same as the excitation one.

Let us recall that in a (001)-grown type I quantum well, the relevant symmetry is $D_{2d}$. If the growth direction $0z$ is chosen as the quantization axis for the angular momentum, the conduction band is s-like, with two spin states $s_{e,z} = \uparrow, \downarrow$; the upper valence band is split into a heavy-hole band with the angular momentum projection $j_{h,z} = \pm 3/2$ and a light-hole band with $j_{h,z} = \pm 1/2$ at the center of the Brillouin zone. The heavy-hole exciton states can be described using the basis set $|J_z\rangle = |j_{h,z}, s_{e,z}\rangle$, i.e. $|J_z = 1\rangle \equiv |3/2, \downarrow\rangle$, $|J_z = -1\rangle \equiv |-3/2, \uparrow\rangle$, $|J_z = 2\rangle \equiv |3/2, \uparrow\rangle$, $|J_z = -2\rangle \equiv |-3/2, \downarrow\rangle$. This basis set is diagonal with respect to the exciton exchange interaction and the twofold degenerate optically-active $J_z = |\pm 1\rangle$ states are split from the non-optically-active $J_z = |\pm 2\rangle$ states by the electron-hole exchange interaction energy [14].

In *non-intentionally doped* QD structures, the symmetry of the system is lowered and the exchange interaction is thus no longer isotropic [15]. The anisotropic exchange interaction splits the $|\pm 1\rangle$ radiative doublets into two eigenstates labeled $|X\rangle = (|1\rangle + |-1\rangle)/\sqrt{2}$ and $|Y\rangle = (|1\rangle - |-1\rangle)/(i\sqrt{2})$, linearly polarized along the [110] and [1$\bar{1}$0] directions respectively. Cw single dot spectroscopy experiments have clearly evidenced these two linearly polarized lines in self-organized InGaAs QD with an exchange splitting of $\delta \sim 150$ $\mu$eV [6]. This anisotropic exchange splitting may originate from QD elongation and/or interface optical anisotropy [5, 15].

## 2. Exciton Spin Dynamics in Non-Intentionally-Doped Quantum Dots

Figure 1(a) displays the time-resolved PL intensity with polarization parallel ($I^X$) and perpendicular ($I^Y$) to the linearly polarized $\sigma^X$ excitation laser in sample I ($T = 10$ K) (the initial peak on the $I^X$ luminescence components intensity close to $t = 0$ corresponds to backscattered laser light from the sample surface). The corresponding linear polarization ($P_{\text{Lin}}$) kinetics is also plotted.

The experiment is performed here at low excitation power ($\sim 7$ W cm$^{-2}$) which corresponds to an estimated average density of photoexcited carriers less than one electron-hole pair per QD. The laser excitation energy is set to 1.137 eV; it coincides with both the cavity mode and a given QD family ground state energies.

The PL intensity decays with a characteristic time $\tau_{\text{rad}} \sim 800$ ps. After the pulsed excitation, the QD emission exhibits a strong linear polarization ($P_{\text{Lin}} \sim 0.75$) which remains strictly constant within the experimental accuracy during the exciton emission (i.e. over $\sim 2.5$ ns). This behavior differs strongly from the exciton linear polarization dynamics in bulk or type I QW structures, characterized by a linear polarization decay time of a few tens of picoseconds [1, 13]. The experimental

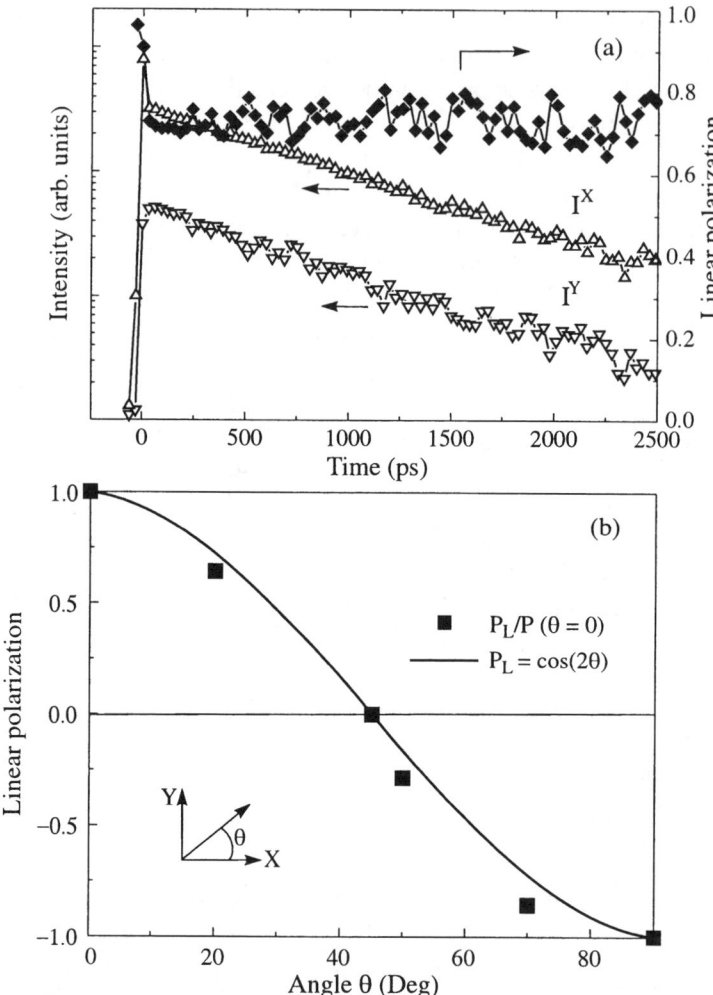

*Figure 1.* (Sample I) (a) Time resolved PL intensity with polarization parallel $I^X$ ($\Delta$) and perpendicular $I^Y$ ($\nabla$) to the linearly polarized ($\sigma^X$) excitation laser ($T = 10$ K); the time evolution of the corresponding linear polarization $P_{\text{Lin}}$ ($\diamond$) is also displayed. (b) Dependence of the luminescence linear polarization on the angle of the excitation laser field with respect to the [110] direction (see text).

observation of a QD exciton linear polarization which does not decay with time is the proof that neither the electron, nor the hole spin relax on the exciton lifetime scale. It shows also that the exciton spin coherence between $|1\rangle$ and $|-1\rangle$ states is maintained during the whole exciton lifetime. From our observation, we can infer that the exciton spin relaxation time is longer than 20 ns, i.e. at least 25 times larger than the radiative lifetime.

Figure 1(b) displays the dependence of the luminescence linear polarization on

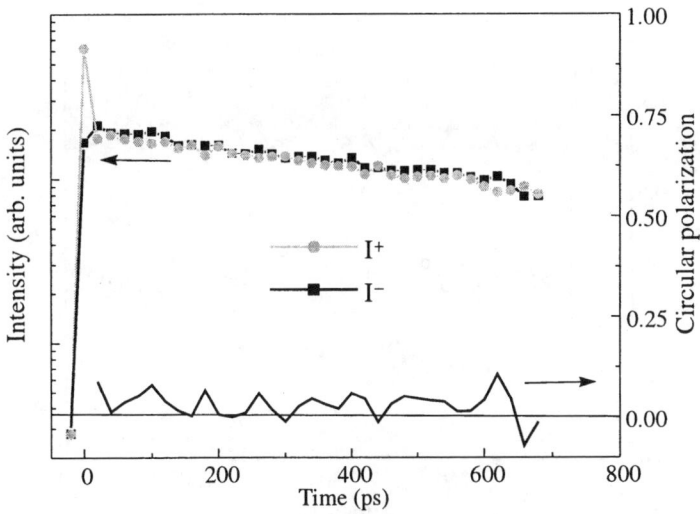

*Figure 2.* (Sample I). Time dependence of the photoluminescence components co-polarized ($I^+$) and counter-polarized ($I^-$) to the $\sigma^+$ polarized excitation laser and the corresponding circular polarization (full line).

the angle of the excitation laser field with respect to the [110] direction. The linearly polarized luminescence components are still detected along the [110] ($X$) and [1$\bar{1}$0] ($Y$) directions. As expected for exciton eigenstates polarized along the [110] and [1$\bar{1}$0] directions, the linear polarization follows a simple law with the angle $\theta$ given by $P_{\text{Lin}}(\theta) = \cos(2\theta)$ A linearly polarized laser excitation along the [100] or [010] directions ($45^{c}irc$ tilt with respect to the [110] and [1$\bar{1}$0] directions) should lead to the observation of $P_{\text{Lin}}$ beats at the pulsation corresponding to the exchange splitting $\delta$. We do not observe any linear polarization and as a consequence any beats in this configuration. Similarly a circularly polarized ($\sigma^+$) excitation should lead to the observation of circular polarization $P_c$ quantum beats at the pulsation $\delta/\hbar$. As we see in Fig. 2, after $\sim$ 15 ps, the time required for the QD PL signal to overcome the backscattered laser light, we do not observe any polarization. This absence of polarization in these two configurations is interpreted as a consequence of the exchange splitting energy statistical fluctuations among the detected QD [16].

If a magnetic field is applied along the growth direction ($B \parallel$ [001]) and the Zeeman splitting $\hbar\Omega_z = g\mu_B B$ is larger than the exchange energy $\delta$, the QD exciton eigenstates are no more the $|X\rangle$ and $|Y\rangle$ linearly polarized states but the $|+1\rangle$ and $|-1\rangle$ circular ones ($g$ is the exciton longitudinal Landé factor). As a consequence, the QD emission is circularly polarized after a ($\sigma^+$) polarized pulsed excitation (not shown here) and again the striking feature is the absence of any polarization decay on the exciton emission time scale [9].

All the results presented up to now have been obtained at low excitation intensity (i.e. when the number of photo-generated electron-hole pairs is small compared to

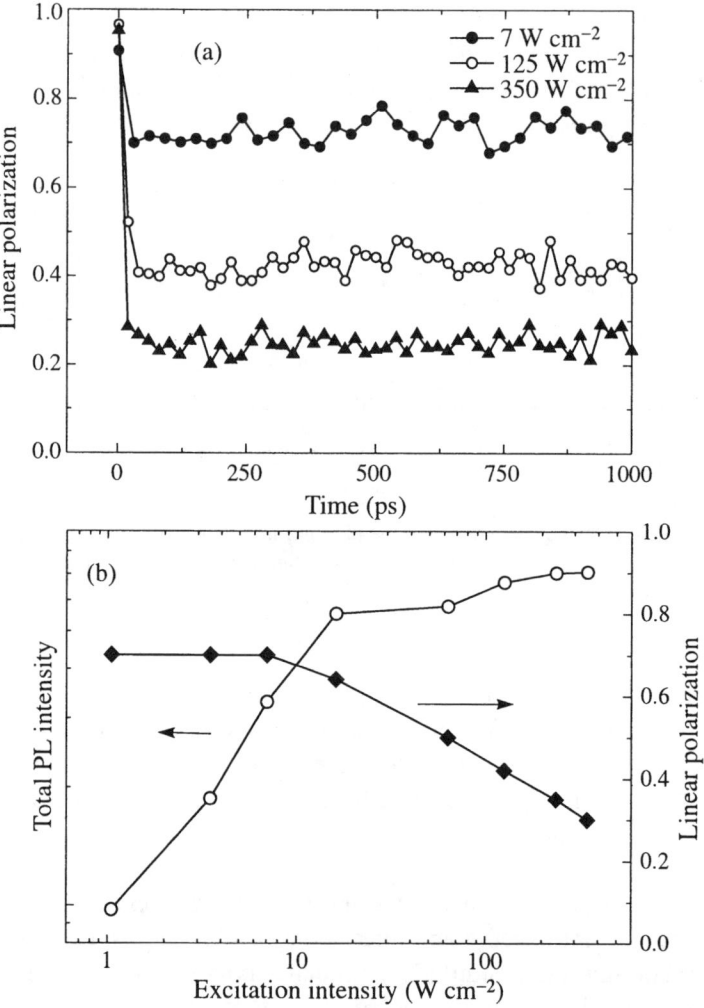

*Figure 3.* (Sample I): (a) Linear polarization dynamics for three excitation powers ($T = 10$ K) (b) Total luminescence intensity and linear polarization as a function of the excitation power.

the number of QDs). Figure 3(a) displays the linear polarization dynamics under resonant linear polarization excitation for three excitation intensities. We note that the linear polarization never decays with time but the amplitude of this polarization decreases when the excitation intensity increases. This effect is due to the photogeneration of biexciton states. Under linearly polarized excitation, a quantum dot in which the ground state is already occupied by an exciton $|X\rangle = (|1\rangle + |-1\rangle)/\sqrt{2}$ can absorb another photon, which yields the photogeneration of a biexciton state labeled $(|XX + YY\rangle)/\sqrt{2}$. This biexciton state will emit $\sigma^X$ or $\sigma^Y$ photons with the same probability (the QDs occupied only by an exciton will still emit fully polarized $\sigma^X$ luminescence). As a consequence, the amplitude of the measured linear polarization decays when the number of photogenerated biexciton states increases. We emphasize that the spectral resolution of our setup ($\sim 3$ meV) does not allow to resolve separately the exciton and the biexciton emission lines.

This interpretation in terms of biexciton photogeneration is confirmed by the observed dependence of the total luminescence intensity on the excitation intensity displayed in Fig. 3(b). We observe that the luminescence intensity increases almost linearly with the excitation intensity at low excitation ($P < 10$ W cm$^{-2}$). In this regime when the number of photogenerated electron-hole pairs is small compared to the number of QDs, the photogeneration of biexciton states is negligible and the linear polarization does not depend on the excitation intensity. At higher excitation intensities ($P > 10$ W cm$^{-2}$), we note in Fig. 3(b) a clear saturation of the luminescence intensity and a simultaneous drop of the measured linear polarization value. The non-linear dependence of the luminescence intensity comes from the saturation of the ground state absorption due to the photogeneration of biexciton states in most of the analysed QDs.

## 3. Charged Exciton Spin Dynamics in n-Modulation Doped Quantum Dots

Let us consider now the strictly resonant excitation of the QD ground state in the *n-modulation doped* structure (sample II) in which most of the QDs contain a single doping electron [17].

The strictly resonant excitation of the QD in this case with ($\sigma^+$) light photogenerates charged excitons X$^-$ in the state:

$$|X^-\rangle = \frac{1}{\sqrt{2}}\left(\left|\frac{3}{2}, \uparrow, \downarrow\right\rangle - \left|\frac{3}{2}, \downarrow, \uparrow\right\rangle\right)$$

(note the anti-symmetry with respect to the electron states). Figure 4 displays the time dependence of the photoluminescence components co-polarized ($I^+$) and counter-polarized ($I^-$) to the $\sigma^+$ polarized excitation laser and the corresponding circular polarization (full line). The temperature is $T = 10$ K. The initial peak corresponds to the intense backscattered laser light from the sample surface which hides the small QD signal at short time delay. The excitation and detection energy is $E = 1.148$ eV; it lies in the QD ground state region, as measured in cw PL spectroscopy.

In these conditions, the PL dynamics reflect the $X^-$ radiative lifetime. We find $\tau_{\text{rad}} \sim 800$ ps. We note in Fig. 4 that the QD emission exhibits a strong circular polarization degree $P_c = (I^+ - I^-)/(I^+ + I^-) \sim 80\%$ which remains *constant*, within the experimental uncertainty, during the charged exciton emission (i.e. over $\sim 2$ ns). The circular polarization simply reflects here the spin polarization of the hole within the charged exciton. This result demonstrates directly that the hole spin relaxation processes are quenched in QD, whereas they are very efficient in quantum wells or bulk structures [1, 18, 19]. Moreover, the observation of

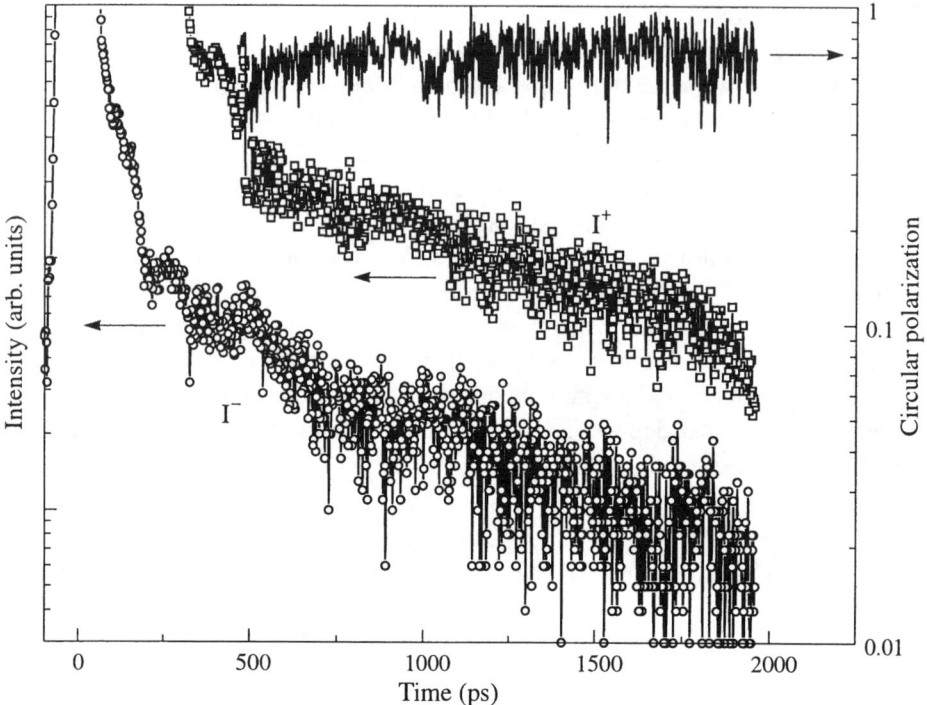

*Figure 4.* (Sample II) Time dependence of the photoluminescence components co-polarized ($I^+$) and counter-polarized ($I^-$) to the $\sigma^+$ polarized excitation laser and the corresponding circular polarization (full line). The initial peak corresponds to the intense backscattered laser light from the sample surface which hides the small QD signal at short time delay. The excitation and detection energy is $E = 1.148$ eV.

this strong circular polarization of the charged exciton emission contrasts with the measurement of a strong linear polarization of the neutral exciton emission in non-intentionally doped sample under the same resonant excitation conditions (see section 2). The effect of the anisotropic exchange interaction which is clearly observed in the QD neutral exciton fine structure is cancelled for the charged exciton as it is formed with two electrons of opposite spins.

## 4. Conclusion

We have studied the PL polarization dynamics in self-organized QD under strictly resonant excitation. We never observed at low temperature any measurable temporal decay of the luminescence polarization regardless of the investigated structure (*non-intentionally doped* or *n-modulation doped*). This evidences a spin relaxation quenching in these zero-dimensional structures at least on the carrier lifetime scale.

## Acknowledgements

We thank V. Thierry-Mieg for the sample growth. We are grateful to E.L. Ivchenko, K.V. Kavokin, O. Krebs and P. Voisin for fruitful discussions.

## References

1. Meier, F. and Zakharchenya, B. (1984) *Optical orientation*, edited by, Modern Problems in Condensed Matter Sciences Vol. 8 (North-Holland, Amsterdam).
2. Chamaro, M., Gourdon,C., Lavallard, P., Lublinskaya, O., and Eimov, A.I. (1996) Enhancement of electron-hole exchange interaction in CdSe nanocrystals: a quantum confinement effect, *Phys. Rev. B* **53**, 1336–1342.
3. Khaetskii, A.V. and Nazarov, Y. (2000) Spin relaxation in semiconductor quantum dots, *Phys. Rev. B* **61**, 12639–12642.
4. Gupta, J.A., Awschalom, D.D., Peng, X., and Alivisatos, A.P (1999) Spin coherence in semiconductor quantum dots, *Phys. Rev. B* **59**, R10421–R10424
5. Gammon, D., Snow, E.S., Shanabrook, B.V., Katzer, D.S., and Park, D. (1996) Fine structure splitting in the optical spectra of single GaAs quantum dots, *Phys. Rev. Lett.* **76**, 3005–3008.
6. Bayer, M., Kuther, A., Forchel, A., Gorbunov, A., Timofeev, V.B., Schäfer, F., Reithmaier, J.P., Reinecke, T.L., and Walck, S.N. (1999) Electron and Hole *g* factors and Exchange interaction from studies of the exciton fine structure in InGaAs quantum dots, *Phys. Rev. Lett.* **82**, 1748-1751.
7. Gotoh, H., Ando, H., Kamada, H., Chavez-Pirson, A., and Temnyo, J. (1998) Spin relaxation in zero-dimensional InGaAs quantum disks, *Appl. Phys. Lett.* **72**, 1341–1343.
8. Paillard, M., Marie, X., Vanelle, E., Amand, T., Kalevich, V.K., Kovsh, A.R., Zhukov, A.E., and Ustinov, V.M. (2000), Time-resolved photoluminescence in self-assembled InAs/GaAs quantum dots under strictly resonant excitation, *Appl. Phys. Lett.* **76**, 76–78.
9. Paillard, M., Marie, X., Renucci, P., Amand, T., Jbeli, A., and Gérard, J.M. (2001) Spin relaxation quenching in semiconductor quantum dots, *Phys. Rev. Lett.* **86**, 1634–1637.
10. Gérard, J.M., Sermage, B., Gayral, B., Legrand, B., Costard, E., and Thierry-Mieg, V. (1998) Enhanced spontaneous emission by quantum boxes in a monolithic optical microcavity, *Phys. Rev. Lett.* **81**, 1110–1113.
11. Paillard, M. (2001) Dynamique de relaxation de spin dans les boîtes quantiques auto-organisées InAs/GaAs, *PhD thesis*, Institut National des Sciences Appliquées, Toulouse, France.
12. Hameau, S., Guldner, Y., Verzelen, O., Ferreira, R., Bastard, G., Zeman, J., Lemaître, A., and Gérard, J.M., (1999) Strong electron-phonon coupling regime in quantum dots: evidence for everlasting resonant polarons, *Phys. Rev. Lett.* **83**, 4152–4155.
13. Marie, X., Le Jeune, P., Amand, T., Brousseau, M., Barrau, J., and Paillard, M. (1997) Coherent control of the optical orientation of excitons in quantum wells, *Phys. Rev. Lett.* **79**, 3222–3225.
14. Amand, T., Marie, X., Le Jeune, P., Brousseau, M., Robart, D., Barrau, J., and Planel, R. (1997) Spin quantum beats of 2D excitons, *Phys. Rev. Lett.* **78**, 1355–1358.
15. Dzhioev, R.I., Zakharchenya, B.P., Ivchenko, E.L., Korenev, V.L., Kusraev, Y.G., Ledentsov, N.N., Ustinov, V.M., Zhukov, A.E., and Tsatsul'nikov, A.F. (1998) Fine structure of excitonic levels in quantum dots, *Phys. Solid State* **40**, 790–793.
16. Flissikowski, T., Hundt, A., Lowisch, M., Rabe, M., and Henneberger, F. (2001) Photon beats from a single semiconductor quantum dot, *Phys. Rev. Lett.* **86**, 3172–3175.
17. Cortez, S., Jbeli, A., Marie, X., Krebs, O., Ferreira, R., Amand, T., Voisin, P., and Gérard, J.M. (2002) Spin polarization dynamics in n-doped InAs/GaAs quanttum dots, *Proceedings of the MSS10 conference*, Linz 23-27 July 2001 *Physica E* (in press).

18. Vina, L. (1999), Spin relaxation in low dimensional systems, *J. Phys., Cond. Matt.* **11**, 5929–5992.
19. Baylac, B., Marie, X., Amand, T., Brousseau, M., Barrau, J., Shekun, Y., (1995) Hole spin relaxation in intrinsic quantum wells, *Surface Science* **326**, 161–166.

# NOVEL MANY-BODY PROCEDURE FOR INTERACTING CLOSE-TO-BOSON EXCITONS

M. COMBESCOT, O. BETBEDER-MATIBET AND C. TANGUY
*GPS, Université Denis Diderot and*
*Université Pierre et Marie Curie, CNRS,*
*Tour 23, 2 place Jussieu, 75251 Paris Cedex 05, France*

Abstract. It is well known that excitons are not exact bosons. However up to now, their close-to-boson character, which results from Pauli exclusion between their electrons and holes, has been extremely tricky to handle properly.

We have recently developed a "commutation technique" which allows to include the effects of Pauli exclusion in the exciton-exciton interactions exactly. This "Pauli way" for excitons to feel each other turns out to be as important as Coulomb interaction.

By using our commutation technique, we have already shown that:
(i) The effective bosonic hamiltonian for excitons quoted by everyone up to now cannot be correct because it is not even hermitian. Purely "Pauli" terms are missing, which restore the hermiticity.
(ii) Worse, one has to give up the concept of effective bosonic hamiltonian itself, since whatever the interacting terms are, they cannot generate the X–X correlations correctly.
(iii) It is impossible to extract an exciton–exciton interacting potential from the semiconductor hamiltonian. As a major consequence, standard perturbative theories do not hold anymore: One has to invent novel many-body procedures from scratch. For instance, the Fermi golden rule is meaningless as well as the previous approaches to the X–X scattering rate.
(iiii) The Mott criterion for disappearance of excitons based on Coulomb scattering is by two orders of magnitude less restrictive than the disappearance of the bosonic character of excitons based on Pauli exclusion.

Key words: excitons, exciton–exciton interaction, many-body effects

## 1. Introduction

The interaction between excitons is a quite tricky concept because the object "exciton" is not physically defined. Indeed, the exciton being made of two (indistinguishable) fermions, if we have two electron-hole pairs $(e_1, e_2, h_1, h_2)$, one cannot say if a specific exciton is made of $(e_1, h_1)$ or of $(e_1, h_2)$. This immediately shows that, the Coulomb interaction $V_{e_1 h_2}$ between $e_1$ and $h_2$ is a part of the interaction *between* excitons if we consider the two excitons as being made of $(e_1, h_1)$ and $(e_2, h_2)$. On the opposite, if we think of these excitons as being made of $(e_1, h_2)$ and $(e_2, h_1)$, the same $V_{e_1 h_2}$ is the Coulomb interaction reponsible for the binding of the $(e_1, h_2)$ exciton, so that it is *inside* one of the two excitons.

Besides the contribution of the Coulomb potential to the exciton–exciton interaction, which is already quite subtle, there is another "interaction" between excitons, less obvious at first, which comes from Pauli exclusion and which *a priori* exists in the absence of any Coulomb process. Indeed, two excitons feel each other because they cannot be built on the same electron and hole states: This *has to* appear in some way in the interaction between excitons, without the necessity to have a $V_q = 4\pi e^2/\mathcal{V} q^2$ prefactor appearing along with it.

If we now turn to the mathematical aspect of the problem, the situation is rather dramatic: The semiconductor hamiltonian is known in terms of free electron and free hole creation operators, $a_{\mathbf{k}}^\dagger$ and $b_{\mathbf{k}}^\dagger$,

$$H = H_e + H_h + V_{ee} + V_{hh} - V_{eh} \,. \tag{1}$$

$H_e$ is the electron kinetic contribution,

$$H_e = \sum_{\mathbf{k}} \epsilon_{e\mathbf{k}} a_{\mathbf{k}}^\dagger a_{\mathbf{k}} \,, \tag{2}$$

with $a_{\mathbf{k}}$ replaced by $b_{\mathbf{k}}$ and $\epsilon_{e\mathbf{k}}$ replaced by $\epsilon_{h\mathbf{k}}$ for holes. $V_{eh}$ is the electron–hole Coulomb interaction,

$$V_{eh} = \sum_{\mathbf{k},\mathbf{k}',\mathbf{q}\neq 0} V_q\, a_{\mathbf{k}+\mathbf{q}}^\dagger b_{\mathbf{k}'-\mathbf{q}}^\dagger b_{\mathbf{k}'} a_{\mathbf{k}} \,, \tag{3}$$

with $b^\dagger$ replaced by $a^\dagger$ for $V_{ee}$ and $a^\dagger$ replaced by $b^\dagger$ for $V_{hh}$ (with an additional $1/2$ prefactor).

This semiconductor hamiltonian thus reads $H = H_0 + V$ with $H_0 = H_e + H_h$ being the free carrier hamoltonian, and $V = V_{ee} + V_{hh} - V_{eh}$ being the Coolomb interaction between carriers. All many-body procedures are then built on the iteration of the identity

$$\frac{1}{a-H} = \frac{1}{a-H_0} + \frac{1}{a-H} V \frac{1}{a-H_0} \,. \tag{4}$$

From it, we can in particular get the usual perturbative expansion of the energy of $N$ electron–hole pairs in the large density limit *i. e.* when the system is well represented by free electrons and holes.

In the low density limit, *i. e.* when the system looks like an exciton gas, it would be nice to be able to write $H$ as $H = H_x + V_{xx}$, with $H_x$ being a "free" exciton hamiltonian and $V_{xx}$ an exciton–exciton interaction, because the usual many-body procedures could then be used. However, $H_x$ is already not easy to think of because a "free" exciton intuitively means an exciton which does not feel the other excitons, neither by Coulomb, nor by Pauli. Consequently, the creation operator of such a "free" exciton cannot be the usual linear combination [1],

$$B_n^\dagger = \sum_{\mathbf{k}} \langle \mathbf{k}|x_{\nu_n}\rangle a_{\mathbf{k}+\alpha_e \mathbf{Q}_n}^\dagger b_{-\mathbf{k}+\alpha_h \mathbf{Q}_n}^\dagger , \qquad (5)$$

which verifies $H B_n^\dagger |v\rangle = E_n B_n^\dagger |v\rangle$, $n$ standing for $(\nu_n, \mathbf{Q}_n)$ with $\mathbf{Q}_n$ being the exciton center of mass momentum and $\nu_n$ the relative motion index; (the constants $\alpha_e$ and $\alpha_h$ are related to the electron and hole masses by $\alpha_{e,h} = m_{e,h}/(m_e + m_h)$). Indeed, these $B_n^\dagger$ operators are such that [2–4]

$$\left[B_m, B_i^\dagger\right] = \delta_{m,i} - D_{mi}, \qquad (6)$$

where the deviation-from-boson operator $D_{mi}$ originates from Pauli exclusion between the electrons and holes making the $m$ and $i$ excitons. As shown more precisely below, this $D_{mi}$ operator is at the origin of the Pauli way for two excitons to feel each other. Consequently, if we want to speak of "free" excitons, we should of course forget Coulomb interaction *between* them, but also drop this deviation-from-boson operator *i. e.* we should use boson-exciton creation operators $\overline{B}_n^\dagger$, with

$$\left[\overline{B}_m, \overline{B}_i^\dagger\right] = \delta_{m,i} . \qquad (7)$$

We thus expect the "free" exciton part of the hamiltonian to appear as

$$H_x = \sum_n E_n \overline{B}_n^\dagger \overline{B}_n . \qquad (8)$$

With respect to the exciton–exciton interaction, we could think of something like

$$V_{xx} = \sum_{mnij} V_{mnij} \overline{B}_m^\dagger \overline{B}_n^\dagger \overline{B}_i \overline{B}_j , \qquad (9)$$

with possibly higher order terms in $\overline{B}^\dagger \overline{B}^\dagger \overline{B}^\dagger \overline{B}\, \overline{B}\, \overline{B}$.

This is just what all "bosonization" procedures [5, 6] have tried to do for years. They essentially end with a scattering $V_{mnij}$ which contains direct and exchange Coulomb processes [6],

$$\begin{aligned}V_{mnij} = \int &d\mathbf{r}_{e_1}\, d\mathbf{r}_{e_2}\, d\mathbf{r}_{h_1}\, d\mathbf{r}_{h_2}\, \phi_m^*(e_1,h_1)\phi_n^*(e_2,h_2) \\ &\times \left(V_{e_1 e_2} + V_{h_1 h_2} - V_{e_1 h_2} - V_{e_2 h_1}\right) \\ &\times \left[\phi_i(e_1,h_1)\phi_j(e_2,h_2) - \phi_i(e_1,h_2)\phi_j(e_2,h_1)\right] .\end{aligned} \qquad (10)$$

The second term of the bracket, which comes from possible exchanges, is said to result from the "fermionic character of the excitons".

Although widely quoted, this $V_{mnij}$ cannot be correct. First, as $V_{mnij}^* \neq V_{ijmn}$, the corresponding interaction $V_{xx}$ is not even hermitian. Second, it contains Coulomb processes only, so that it misses the purely Pauli way for two excitons to feel each other.

With respect to the bosonization procedures, some of them appearing as extremely sophisticated, we have recently *proved* [7] that the close-to-boson character of the excitons has so subtle consequences that there is no way to hide it into a $V_{mnij}$ scattering. Whatever $V_{xx}$ is, there is no hope to produce the correlations between excitons correctly: The best thing which can be done is to adjust $V_{xx}$ in order to get the correct first order term in the exciton–exciton interaction. At higher orders, the way the exchange processes mix with Coulomb interactions is too subtle to be possibly reduced to one dressed Coulomb scattering only.

The fact that the semiconductor hamiltonian cannot be written as $H_x + V_{xx}$ in the low density limit is indeed dramatic as an identity like eq. (4) does not exist anymore, so that a new approach to many-body effects between excitons has to be invented from scratch!

Before outlining our new many-body approach to interacting excitons, we wish to say that the other existing procedure, besides bosonization, is based on the semiconductor Bloch equations [8] or possibly its more elaborate extension through a hierarchy of many-body density matrices [9–11]. By increasing the order of these density matrices, one can of course include more and more sophisticated effects. However, as these procedures use "clean" free electron and free hole fermion operators, they are *a priori* appropriate to systems well represented by free carriers, not by bound state excitons. So that they must suffer of more or less well controlled truncatures. This is why, besides the fact that they appear to us as far from transparent, we find them rather unsatisfactory in the low density limit, where the system is made of bound state excitons, at lowest order in their possible interactions.

## 2. Commutation technique for interacting close-to-boson excitons

As discussed in the introduction, the excitons feel each other (*i. e.* interact) through Coulomb interaction *and* Pauli exclusion, both being rather subtle. Let us start with Coulomb.

It is rather obvious that the Coulomb interaction must generate a scattering of two excitons from $(i, j)$ states to $(m, n)$ states. The difficulty is to formally extract from $V_{ee} + V_{hh} - V_{eh}$ an "inside" part which binds the electrons and holes into excitons and an "outside" part which makes these excitons to interact. For that, we have to identify an operator linked to *the interaction of a specific exciton* $(i)$ *with the rest of the system*. This can be done through the commutator

$$\left[H, B_i^\dagger\right] = E_i B_i^\dagger + V_i^\dagger . \tag{11}$$

Indeed, this equation also reads

$$H B_i^\dagger |\psi\rangle = E_i B_i^\dagger |\psi\rangle + B_i^\dagger H |\psi\rangle + V_i^\dagger |\psi\rangle , \tag{12}$$

with $|\psi\rangle$ being any semiconductor state. When $H$ acts on the state $B_i^\dagger|\psi\rangle$, the first term of the r. h. s. of eq. (12) corresponds to the energy of the $(i)$ exciton as if this exciton did not interact with $|\psi\rangle$; the second term corresponds to the energy of $|\psi\rangle$ as if this state did not interact with the exciton. The last term is there because, of course, $B_i^\dagger$ interacts with $|\psi\rangle$. $V_i^\dagger$ is thus directly linked to the interaction of the $(i)$ exciton with the rest of the system.

If we go one step further, and calculate $\left[V_i^\dagger, B_j^\dagger\right]$, we find

$$\left[V_i^\dagger, B_j^\dagger\right] = \sum_{mn} \xi_{mnij}^{\text{dir}} B_m^\dagger B_n^\dagger . \tag{13}$$

Among the various possible expressions of $\xi_{mnij}^{\text{dir}}$, the most transparent one for physical understanding is

$$\begin{aligned}\xi_{mnij}^{\text{dir}} &= \frac{1}{2} \int d\mathbf{r}_{e_1} d\mathbf{r}_{e_2} d\mathbf{r}_{h_1} d\mathbf{r}_{h_2} \, \phi_m^*(e_1, h_1) \phi_n^*(e_2, h_2) \\ &\quad \times \left(V_{e_1 e_2} + V_{h_1 h_2} - V_{e_1 h_2} - V_{e_2 h_1}\right) \\ &\quad \times \phi_i(e_1, h_1) \phi_j(e_2, h_2) + (m \leftrightarrow n).\end{aligned} \tag{14}$$

We do see that $\xi_{mnij}^{\text{dir}}$ contains all Coulomb interactions *between* the $(m, n)$ and $(i, j)$ excitons, these excitons being made on both sides with the same pairs $(e_1, h_1)$ and $(e_2, h_2)$. It is shown in fig. (1a).

If we turn to Pauli, an exciton–exciton scattering linked to Pauli has to be related to the close-to-boson operator $D_{mi}$. By calculating the commutator $\left[D_{mi}, B_j^\dagger\right]$, we find

$$\left[D_{mi}, B_j^\dagger\right] = 2 \sum_n \lambda_{mnij} B_n^\dagger , \tag{15}$$

where the coefficient $\lambda_{mnij}$ appears as

$$\begin{aligned}\lambda_{mnij} &= \frac{1}{2} \int d\mathbf{r}_{e_1} d\mathbf{r}_{e_2} d\mathbf{r}_{h_1} d\mathbf{r}_{h_2} \, \phi_m^*(e_1, h_1) \phi_n^*(e_2, h_2) \\ &\quad \times \phi_i(e_1, h_2) \phi_j(e_2 h_1) + (m \leftrightarrow n).\end{aligned} \tag{16}$$

Such a scattering exists because the excitons are composite particles which can be built in different ways. It just corresponds to cross the electrons and holes of these excitons (see fig. (1b)). This $\lambda_{mnij}$ is not a true scattering as it is dimensionless: it has to be "cooked" with quantities homogeneous to an energy, in order to possibly appear in an exciton–exciton scattering. Possible energy-like quantities we can think of, are $\xi_{mnij}^{\text{dir}}$ but also $(E_m + E_n) \pm (E_i + E_j)$.

The two parameters $\xi_{mnij}^{\text{dir}}$ and $\lambda_{mnij}$ are the two key ingredients of our commutation technique. Using them, we can calculate any quantity in which enter excitons exactly.

## 2.1. EXCITON STATES

Let us consider the two-exciton states $B_i^\dagger B_j^\dagger |v\rangle$. While it is always possible to normalize these states by adding a prefactor $\langle v|B_j B_i B_i^\dagger B_j^\dagger |v\rangle^{-1/2}$, these two-exciton states are quite unpleasant because they are not orthogonal. Indeed, using eqs. (6,15), we find

$$\langle v|B_m B_n B_i^\dagger B_j^\dagger |v\rangle = \langle v|B_m (B_i^\dagger B_n + \delta_{n,i} - D_{ni})B_j^\dagger |v\rangle$$
$$= \delta_{n,i}\delta_{m,j} + \delta_{m,i}\delta_{n,j} - 2\lambda_{mnij}, \qquad (17)$$

since, the one-exciton states being orthogonal, we do have $\langle v|B_m B_i^\dagger |v\rangle = \delta_{m,i}$, so that $D_{ni}|v\rangle = 0$. The above equation clearly shows that, due to $\lambda_{mnij}$, i. e. the close-to-boson character of the excitons, $\langle v|B_m B_n B_i^\dagger B_j^\dagger |v\rangle$ differs from zero even if $(m,n) \neq (i,j)$. It can be interesting to link this nonorthogonality to the fact that

$$B_i^\dagger B_j^\dagger = -\sum_{mn} \lambda_{mnij} B_m^\dagger B_n^\dagger, \qquad (18)$$

which results from the two possible ways to bind two electron–hole pairs into two excitons. Equation (18) physically means that $B_i^\dagger B_j^\dagger$ contains a "piece" of any $B_m^\dagger B_n^\dagger$ so that the $N$-exciton states cannot be orthogonal even if they are written with different $B^\dagger$.

Our commutation technique also allows [2, 12] to calculate the normalization factor of the $N$-ground-state-exciton state,

$$\langle v|B_0^N B_0^{\dagger N}|v\rangle = N! F_N, \qquad (19)$$

in a quite easy way. For that we just have to "push $B_0$ above each $B_0^\dagger$", using the $\left[B_0, B_0^\dagger\right]$ commutator, in order to end with $B_0|v\rangle$ which is zero. If the excitons were exact bosons, i. e. if $B_0$ was replaced by $\overline{B}_0$, we would find $F_N = 1$. Because of the close-to-boson operator $D_{00}$, $F_N$ differs dramatically from 1 as it is exponentially small with respect to $N^2 a_x^3/\mathcal{V}$, parameter which can be quite large in a macroscopic sample even at low density $N a_x^3/\mathcal{V}$. This superextensive behavior of $F_N$, which can be surprising at first, comes from the fact that Pauli exclusion generates a quite new type of many-body effects since it "at once" induces a link between the $(i)$ exciton and *all* the other excitons present in the system, while Coulomb interaction is a $2 \times 2$ interaction only.

While $F_N$ does not appear in physical effects, due to its superextensive character, ratios like $F_{N+1}/F_N$ enter in many of them. We have in particular shown [2] that, in order for $N$ ground-state excitons to behave as bosons, we should have their close-to-boson character small, i. e.

$$\langle D_{00}\rangle_N = \langle v|B_0^N D_{00} B_0^{\dagger N}|v\rangle / \langle v|B_0^N B_0^{\dagger N}|v\rangle \qquad (20)$$

small. Our commutation technique allows to show that $\langle D_{00} \rangle_N$ writes simply as $2(1 - F_{N+1}/F_N)$ and that $\langle D_{00} \rangle_N \ll 1$ if $100 N a_x^3 / \mathcal{V} \ll 1$. We thus find that the criterion for bosonic behavior of excitons is by two orders of magnitude more restrictive than the Mott criterion for the disappearance of bound states by (RPA) screening. This probably explains why the Bose condensation of excitons has not been seen within experimental conditions relying on the Mott criterion.

## 2.2. MATRIX ELEMENTS OF H BETWEEN EXCITON STATES

As the one-exciton state $B_i^\dagger |v\rangle$ is eigenstate of $H$, i. e. $(H - E_i) B_i^\dagger |v\rangle = 0$, eq. (11) immediately gives $V_i^\dagger |v\rangle = 0$. Using this eq. (11) we then find

$$\langle v | B_m B_n H B_i^\dagger B_j^\dagger |v\rangle = \langle v | B_m B_n (B_i^\dagger H + E_i B_i^\dagger + V_i^\dagger) B_j^\dagger |v\rangle$$
$$= (E_i + E_j) \langle v | B_m B_n B_i^\dagger B_j^\dagger |v\rangle$$
$$+ \sum_{i'j'} \xi^{\text{dir}}_{i'j'ij} \langle v | B_m B_n B_{i'}^\dagger B_{j'}^\dagger |v\rangle , \qquad (21)$$

with a similar equation if $H$ acts on the left. By using eq. (17) for these matrix elements, we see appearing two types of Coulomb exchange scattering by these two ways of calculating the $H$ matrix elements in the two-exciton subspace, namely $\xi^{\text{left}}_{mnij}$ and $\xi^{\text{right}}_{mnij}$ defined by

$$\xi^{\text{right}}_{mnij} = \sum_{i'j'} \xi^{\text{dir}}_{mni'j'} \lambda_{i'j'ij} . \qquad (22)$$

$$\xi^{\text{left}}_{mnij} = \sum_{i'j'} \lambda_{mni'j'} \xi^{\text{dir}}_{i'j'ij} , \qquad (23)$$

These two Coulomb exchange scatterings are linked by

$$\xi^{\text{left}}_{mnij} - \xi^{\text{right}}_{mnij} = (E_m + E_n - E_i - E_j) \lambda_{mnij} . \qquad (24)$$

It is physically interesting to note that $\xi^{\text{left}}_{mnij}$ reads

$$\xi^{\text{left}}_{mnij} = \frac{1}{2} \int d\mathbf{r}_{e_1} d\mathbf{r}_{e_2} d\mathbf{r}_{h_1} d\mathbf{r}_{h_2} \, \phi_m^*(e_1, h_2) \, \phi_n^*(e_2, h_1)$$
$$\times (V_{e_1 e_2} + V_{h_1 h_2} - V_{e_1 h_2} - V_{e_2 h_1})$$
$$\times \phi_i(e_1, h_1) \phi_j(e_2, h_2) + (m \leftrightarrow n) , \qquad (25)$$

with $(-V_{e_1 h_2} - V_{e_2 h_1})$ replaced by $(-V_{e_1 h_1} - V_{e_2 h_2})$ for $\xi^{\text{right}}_{mnij}$: We see that the electron–hole Coulomb terms are "between" the excitons on one side and "inside" the excitons on the other side. These exchange Coulomb scatterings $\xi^{\text{left}}_{mnij}$ and $\xi^{\text{right}}_{mnij}$ are shown in fig. (1c) and (1d).

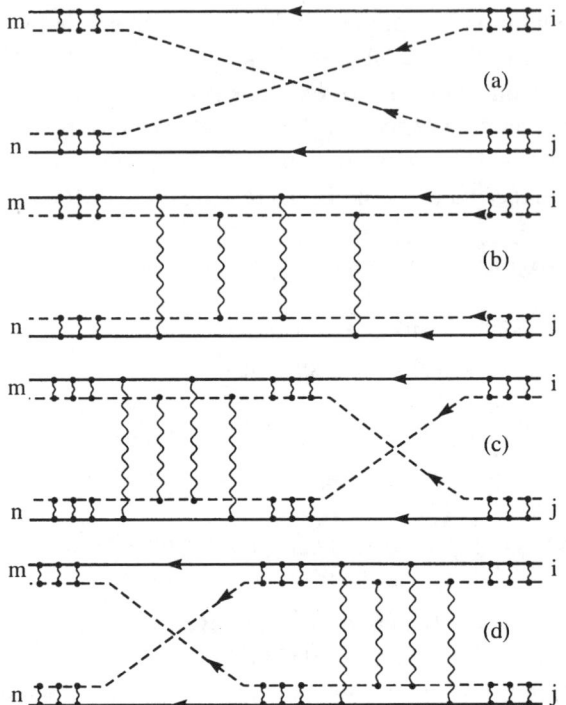

*Figure 1.* (a): The Pauli coefficient $\lambda_{mnij}$ as defined in eq. (16). (b), (c) and (d): The scatterings $\xi_{mnij}^{\text{dir}}$, $\xi_{mnij}^{\text{right}}$, $\xi_{mnij}^{\text{left}}$ as defined in eqs. (14,22,23).

Recently [13], we have also calculated the average energy in the $N$-ground-state-exciton state, namely

$$\langle H \rangle_N = \langle v | B_0^N H B_0^{\dagger N} | v \rangle \,/\, \langle v | B_0^N B_0^{\dagger N} | v \rangle \,. \tag{26}$$

Although Coulomb interaction appears at first order only in $\langle H \rangle_N$, by construction, we have shown that this quantity contains high order density dependent terms in $(Na_x^3/\mathcal{V})^n$ with $n \geq 2$: They result from subtle many-body effects induced by Pauli "interaction".

### 2.3. MATRIX ELEMENTS OF $1/(a - H)$ BETWEEN EXCITON STATES

If we now turn to the matrix elements of $1/(a - H)$ (with $a = \omega + i\eta$ in most physical problems), we cannot use the standard interaction expansion based on the identity (4) to get the correlations, as there is no way to write $H$ as $H_x + V_{xx}$ for excitons. We have succeeded in overcoming this difficulty by finding the equivalent of eq. (4) in the sense of commutator. Indeed, starting from eq. (11), it is easy to check that

$$\frac{1}{a-H} B_i^\dagger = B_i^\dagger \frac{1}{a-H-E_i} + \frac{1}{a-H} V_i^\dagger \frac{1}{a-H-E_i}. \quad (27)$$

Using this equation, we have calculated the matrix elements of $1/(a-H)$ in the two-exciton subspace at any order in the (Coulomb and Pauli) interactions. This calculation allowed us to prove [7] that there is no way to construct an effective boson-exciton interaction $V_{xx}$ which can reproduce these correlation terms, the interplay between Coulomb interaction and exchange processes being too subtle, even for two excitons.

## 3. Conclusion

Our commutation technique provides a clean and powerful approach to the problem of interacting close-to-boson excitons. It allows to take *exactly* into account Pauli exclusion between the components of the excitons and to derive the subtle exchange processes this Pauli exclusion induces in the bare Coulomb processes.

This commutation technique should provide a fresh view to all problems dealing with interacting excitons. Among them, we can think of the electron–hole plasma energy in the low density limit as calculated by Keldysh and Koslov, or the excitonic Fermi edge singularity for doped semiconductors, as calculated by Combescot and Nozières.

Very recently, we have generated a similar commutation technique for an exciton interacting with electrons. It allowed us to derive the electron–exciton ladder diagrams [14] giving rise to the trions. These diagrams are basically similar to the electron–hole ladder diagrams for excitons, although somewhat more complex due to the composite nature of the exciton. This allows to put the trion on the same footing as the exciton with respect to its possible interaction with doped semiconductors.

## References

1. Betbeder-Matibet, O., and Combescot, M. (2001) Electron–hole diagrams versus exciton diagrams: The simplest problem on interacting excitons, *Eur. Phys. J. B* **22**, 17-29.
2. Combescot, M., and Tanguy, C. (2001) New criteria for bosonic behavior of excitons, *Europhys. Lett.* **55**, 390–396.
3. Combescot, M., and Betbeder-Matibet, O. (2002) The effective bosonic hamiltonian for excitons reconsidered, *Europhys. Lett.* **58**, 87–92.
4. Betbeder-Matibet, O., and Combescot, M. (2002) Commutation technique for interacting close-to-boson excitons, *Eur. Phys. J. B* (in press).
5. For a review, see Klein, A., and Marshalek, E.R. (1991) Boson realization of Lie algebras with applications to nuclear physics, *Rev. Mod. Phys.* **63**, 375–558.
6. Haug, H., and Schmitt-Rink, S. (1984) Electron theory of the optical properties of laser excited semiconductors, *Prog. Quant. Elec.* **9**, 3–100.

7. Combescot, M., and Betbeder-Matibet, O. (2002) Effective bosonic hamiltonian for excitons: A too naïve concept, *Europhys. Lett.* (in press).
8. Haug, H., and Koch, S. (1990) *Quantum theory of the optical and electronic properties of semiconductors*, World Scientific, London.
9. Axt, W.M., and Mukamel, S. (1998) Nonlinear optics of semiconductor and molecular nanostructures: A common perspective, *Rev. Mod. Phys.* **70**, 145–174.
10. Schäfer, W., Kim, D.S., Shah, J., Damen, T.C., Cunningham, J.E., Goossen, K.W., Pfeiffer, L.N., and Köhler, K. (1996) Femtosecond coherent fields induced by many-particle correlations in transient four-wave mixing, *Phys. Rev. B* **53**, 16429–16443.
11. Meier, T., and Koch, S.W. (1999) Excitons versus unbound electron-hole pairs and their influence on exciton bleaching: A model study, *Phys. Rev. B* **59**, 13202–13208.
12. Combescot, M., Leyronas, X., and Tanguy, C. (2002) On the $N$-exciton normalization factor, *Eur. Phys. J. B* (submitted).
13. Betbeder-Matibet, O., and Combescot, M. (2002) Density expansion of the energy of $N$ close-to-boson excitons, *Eur. Phys. J. B* (submitted).
14. Combescot, M. and Betbeder-Matibet, O. (2002) Trion ladder diagrams, *Solid State Commun.* (submitted).

# DENSITY-MATRIX THEORY OF TRIONIC SPECTRA IN SEMICONDUCTOR NANOSTRUCTURES

AXEL ESSER
*Lyman Laboratory of Physics, Harvard University,
Cambridge, USA*

ROLAND ZIMMERMANN
*Institut für Physik der Humboldt-Universität zu Berlin,
Hausvogteiplatz 5-7, 10117 Berlin, Germany*

Abstract. A consistent density-matrix approach for the absorption spectra of semiconductor nanostructures in the presence of a background carrier density is presented. Trionic spectra contain bound and exciton-electron scattering states and exhibit low-energy tails due to energy recoil. The asymmetric trion lineshape is preserved even in the case of weak disorder and optical transitions occur between states of different localization degree.

Key words: three-particle correlations, oscillator strength transfer, recoil process, localization

## 1. Introduction

Since the first observation of trions in semiconductor quantum wells (QW) nearly a decade ago [1–3], the question of their localization has not been answered conclusively. At first glance, one expects in all semiconductor nanostructures a certain amount of disorder either due to rough interfaces or the presence of electrostatic disorder in doped samples. For excitons, it has been shown that disorder-induced localization can dominantly contribute to the secondary emission [4], the relaxation of spin-polarization [5], and the inhomogeneous broadening of exciton lines [6]. Thus it is not at all surprising that strong spatial fluctuations of the trion intensity are observed in near-field spectra [7, 8]. Four-wave mixing experiments on trions [9] and quantum-beats between trion and exciton [10] have also been discussed in terms of trion localization. In contrast, the observation of trion motion in electric field gradients [11] and a characteristic temperature-dependence of the radiative lifetime of trions [12] are most naturally explained in terms of delocalized trion states.

A particular property of trion spectra is that optical transitions are possible for all trion momenta $k$, in contrast to the exciton transition, where only the $k = 0$

component of the exciton states is relevant in optical spectra. Indeed, it has been shown in wide GaAs QWs, that trions transfer their momentum to the electron in the optical transition. This recoil process gives rise to asymmetric trion lineshapes [13, 14]. Moreover, trionic gain has been reported as a result of transitions at finite $k$ [15]. Thus signatures of less localized trion states may contribute to the trion lineshape as well. The present paper shows that these seemingly contradictory results can be traced back to the existence of trion and electron states of quite different localization lengths. In particular, it is shown that depending on the localization lengths of the involved states, transitions contribute to different parts of the observed trion line.

Besides the trion, strong contributions of excitons are present. Their relative ratio is density-dependent, and an oscillator strength transfer from exciton to trion is commonly observed in absorption and luminescence spectra in combination with a modification of the exciton lineshape [16, 17]. Continuum states of the trion spectrum play a decisive role in this process and have to be included into a theoretical calculation [18–20].

In this paper, we present a consistent density-matrix approach for the optical properties of semiconductor nanostructures in the presence of background carriers. In the first part we show how three-particle bound and scattering states change the exciton spectrum and discuss the origin of the density-dependent oscillator strength transfer between trion and exciton. In the second part, we include a weak disorder potential which gives rise to a localization of the center-of-mass (COM) trion wave function. Then, optical transition occur between localized trion and localized electron states. Depending on the localization degree, contributions from localized and extended states to the trion lineshape are distinguished. Negatively charged excitons are assumed throughout. Numerical results are given for a prototype quantum wire (QWR). Extension of the theory to positively charged trion and to other structures is straightforward.

## 2. Density Matrix and Three-Particle Correlations

We consider the time-dependent interband polarization $P_{k\sigma}(t) = \langle c^\dagger_{k\sigma} d^\dagger_{-k-\sigma} \rangle$ for a two-band semiconductor in the presence of a background density of electrons. Electrons (heavy holes) in the conduction (valence) band are created by $c^\dagger_{k\sigma}$ ($d^\dagger_{-k-\sigma}$), where $k$ and $\sigma$ denote the 2D (QW) or 1D (QWR) in-plane momentum and spin, respectively. The Heisenberg equation of motion for the interband polarization reads

$$\left(E^e_{k_1} + E^h_{k_1} + i\hbar\partial_t\right) P_{k_1\sigma_1}(t) - \sum_{\bar{q}} v^{eh}_{\bar{q}} P_{k_1-\bar{q}\,\sigma_1}(t)$$
$$- \sum_{k_2\bar{q}\sigma_2\nu} v^{ee}_{\bar{q}} T^\nu_{k_2}(k_1 - \bar{q}\sigma_1, k_2 + \bar{q}\sigma_2, t)$$

$$+ \sum_{k_2 \bar{q} \sigma_2 \nu} v_{\bar{q}}^{eh} T_{k_2}^{\nu}(k_1 \sigma_1, k_2 - \bar{q} \sigma_2, t)$$

$$= \frac{e\mathbf{A}(t)\mathbf{p}_{\sigma_1}^{cv}}{cm_0} \left(1 - f_{k_1 \sigma_1}^{e}\right). \quad (1)$$

Here, a parabolic band dispersion $E_k^c = \hbar^2 k^2 / 2m_c$ with effective masses $m_c$ ($c = \{e, h\}$) is assumed. The Coulomb form factors $v_{\bar{q}}^{cd}$ result from the convolution of the bare 3D Coulomb potential with the lowest sublevel states of the quantum confinement [14]. For realistic nanostructures we have $v_0^{ee} < v_0^{eh} < v_0^{hh}$ as a consequence of band offsets and different effective masses. Further, $\mathbf{p}_{\sigma_1}^{cv} = (\mathbf{e}_x + i\sigma_1 \mathbf{e}_y) p_{cv} / \sqrt{2}$ denotes the spin-dependent interband momentum matrix element, in which the weak $k$-dependence is neglected for simplicity. The interband polarization in Eq. (1) couples to symmetrized four-operator terms

$$T_q^{\nu}(k_1 \sigma_1, k_2 \sigma_2, t) = \frac{1}{2}\left[T_q(k_1 \sigma_1, k_2 \sigma_2, t) + s_{\nu} T_q(k_2 \sigma_1, k_1 \sigma_2, t)\right], \quad (2)$$

which are defined as

$$T_q(k_1 \sigma_1, k_2 \sigma_2, t) = \langle c_{k_1 \sigma_1}^{\dagger} c_{k_2 \sigma_2}^{\dagger} d_{q-k_1-k_2-\sigma_1}^{\dagger} c_{q \sigma_2} \rangle, \quad (3)$$

and give rise to trionic contribution in the absorption spectrum [13]. The index $\nu$ runs over singlet and triplet channels in the three-particle space, accounted for by the spin-dependent factor: $s_{\nu} = 1$ for $\nu =$'singlet', $s_{\nu} = -1$ for $\nu =$'triplet'. The Hartree–Fock approximation in Eq. (1) would decompose $T_q^{\nu}(k_1 \sigma_1, k_2 \sigma_2, t)$ into a product of electron density and interband polarization (semiconductor Bloch equation). Unfortunately, the trion is lost completely in this approximation. Consequently, the next level of the many-particle hierarchy will be considered here. We apply a generalized truncation scheme and keep contributions which are linear in the electromagnetic field ($\chi^{(1)}$) *and* linear in the electron density. Then, the trion amplitudes $T_q^{\nu}$ are found from

$$\left(E_{k_1}^e + E_{k_2}^e + E_{q-k_1-k_2}^h - E_q^e + i\hbar \partial_t\right) T_q^{\nu}(k_1 \sigma_1, k_2 \sigma_2, t)$$
$$+ \sum_{\bar{q}} v_{\bar{q}}^{ee} T_q^{\nu}(k_1 - \bar{q} \sigma_1, k_2 + \bar{q} \sigma_2, t)$$
$$- \sum_{\bar{q}} v_{\bar{q}}^{eh} \left[T_q^{\nu}(k_1 - \bar{q} \sigma_1, k_2 \sigma_2, t) + T_q^{\nu}(k_1 \sigma_1, k_2 - \bar{q} \sigma_2, t)\right]$$
$$= -\frac{e\mathbf{A}(t)\mathbf{p}_{\sigma_1}^{cv}}{2cm_0} f_{q\sigma_2}^e \left(\delta_{k_2,q} + s_{\nu} \delta_{k_1,q}\right)\left(1 - s_{\nu}\delta_{\sigma_1,\sigma_2}\right). \quad (4)$$

The coupling to quatronic ($3e + h$) and biexcitonic ($2e + 2h$) contributions are neglected in this approximation. Eq. (4) shows, that spin and momentum are decoupled and spin and triplet channels are distinguished by different source terms.

When taking $k_2 = q$ in Eq. (4) and summing over $k_2$ and $\sigma_2$, the hierarchy of equations is combined into a single equation for the sum $P + \sum T$, which is

$$\left(E^e_{k_1} + E^h_{k_1} + i\hbar\partial_t\right)\left(P_{k_1\sigma_1}(t) + \sum_{k_2\sigma_2\nu} T^\nu_{k_2}(k_1\,\sigma_1, k_2\,\sigma_2, t)\right)$$

$$-\sum_{\bar{q}} v^{eh}_{k_1-\bar{q}}\left(P_{\bar{q}\sigma_1}(t) + \sum_{k_2\sigma_2\nu} T^\nu_{k_2}(\bar{q}\,\sigma_1, k_2\,\sigma_2, t)\right)$$

$$= \frac{e\mathbf{A}(t)\mathbf{p}^{cv}_{\sigma_1}}{cm_0}\left(1 - \sum_{k_2\sigma_2} f^e_{k_2\sigma_2}\right). \tag{5}$$

This equation is similar to the original exciton polarization equation. Only the inhomogeneity is modified and contains the total electron number $N_e = \sum_{k_2\sigma_2} f^e_{k_2\sigma_2}$ within the normalization volume $\Omega$. Therefore, the standard exciton Green's function $G^X(k_1, k_2, \omega)$ can be used for the formal solution of Eq. (5). The trion Green's function $G^T_{q\nu}(k_1, k_2, t)$ may be defined as

$$T^\nu_q(k_1 + \xi q\,\sigma_1, k_2 + \xi q\,\sigma_2, t)$$
$$= -\frac{eA_0 p^{cv}_{\sigma_1}}{2cm_0\Omega} f^e_{q\sigma_2}\left(1 - s_\nu \delta_{\sigma_1,\sigma_2}\right) G^T_{q\nu}(k_1, k_2, t), \tag{6}$$

where a short (delta-like) pulse $A(t) = A_0\delta(t)$ will be assumed from now. Furthermore, the mass ratios $\xi = m_e/M_T$ and $\eta = M_X/M_T$ are found from the exciton mass $M_X = m_e + m_h$ and trion mass $M_T = 2m_e + m_h$. The symmetry $G^T_{q\nu}(k_1, k_2, t) = s_\nu G^T_{q\nu}(k_2, k_1, t)$ reflects the symmetry of the channel-dependent source terms in

$$\left[\frac{\hbar^2}{2\mu}(k_1^2 + k_2^2) + \frac{\hbar^2}{m_h}k_1 k_2 - W_q + i\hbar\partial_t\right] G^T_{q\nu}(k_1, k_2, t)$$
$$+ \sum_{\bar{q}} v^{ee}_{\bar{q}} G^T_{q\nu}(k_1 - \bar{q}, k_2 + \bar{q})$$
$$- \sum_{\bar{q}} v^{eh}_{\bar{q}} \left[G^T_{q\nu}(k_1 - \bar{q}, k_2) + G^T_{q\nu}(k_1, k_2 - \bar{q})\right]$$
$$= \left(\delta_{k_2,\eta q} + s_\nu\,\delta_{k_1,\eta q}\right)\Omega\,\delta(t). \tag{7}$$

The recoil energy $W_q = E^e_q - \hbar^2 q^2/2M_T = \hbar^2 q^2 M_X/2m_e M_T$ comprises trion COM energy and electron recoil and yields a momentum-dependent low-energy shift in the kinetic energy.

Fourier transformation of the time-dependent polarization yields the optical susceptibility

$$\chi(\omega) = \frac{4\pi ec}{m_0\omega^2\Omega}\sum_{k_1\sigma_1}\frac{p^{he}_{\sigma_1} P_{k_1\sigma_1}(\omega)}{A_0}, \tag{8}$$

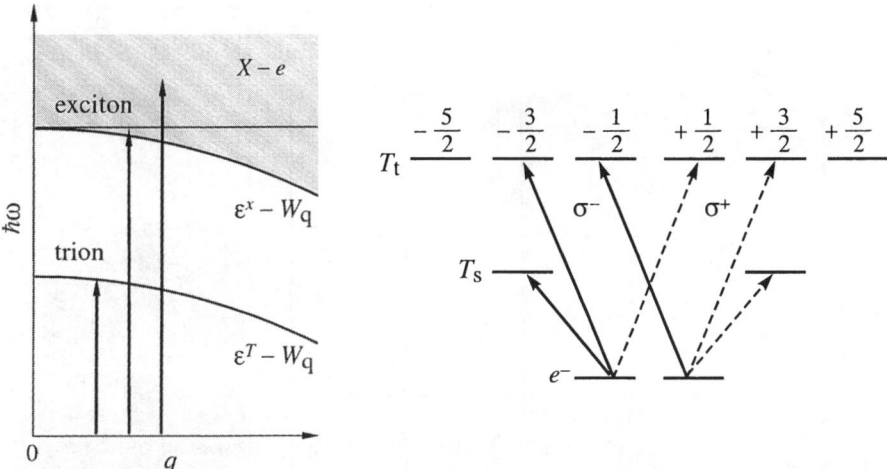

*Figure 1.* Optical transitions in (a) momentum and (b) spin space: (a) Optical interband transitions to trion, exciton, and exciton-electron scattering states. The initial electron momentum $q$ is transferred to the final states and gives rise to low-energy wings at the trion and exciton (energy recoil). Furthermore, high-energy wings appear at the exciton (even at $q = 0$) due to the excitation of exciton-electron states. (b) Term scheme of electron ($e^-$), singlet trion ($T_s$) and triplet trion ($T_t$, all unbound). No dark states appear in the singlet channel, but the $S_z = \pm 3/2$ states are excited with different light polarization. Dark states exist in the triplet channel ($S_z = \pm 5/2$). For $\sigma^+$ polarized light, triplet states are accessible independently on the polarization of the background electrons, whereas the appearance of the singlet state strongly depends on the polarization of the background electrons.

where the summation over $\sigma_1$ reflects the properties of the light excitation. We consider linear and circular polarized light in the following. (i) Linear polarized light: Optical transitions occur for both allowed exciton states ($\sigma_1 = \uparrow, \downarrow$) and the optical susceptibility reads

$$\chi(\omega) = \frac{\chi_0}{\Omega^2}\left[(1-N_e)\sum_{k_1 k_2} G^X(k_1, k_2, \omega) \right. $$
$$\left. + \sum_q \left(\frac{1}{2}\sum_{\sigma_2} f^e_{q\sigma_2}\right) \sum_{k\nu} S^\nu G^T_{q\nu}(k, \eta q, \omega)\right]. \quad (9)$$

The prefactors have been collected into $\chi_0 = 8\pi e^2 p_{cv}^2/m_0^2\omega^2$. The summation over spin $\sigma$ gives the symmetry-dependent factor $S^\nu = 1 - s_\nu/2$ ($S^\nu = 1/2$ for singlet, $S^\nu = 3/2$ for triplet). This holds independently of spin polarization of the background electrons. It has been shown in Ref. [20], that the trion Green's function contains a contribution which exactly compensates the $N_e$ part in the exciton susceptibility. In diagrammatic language, this term with the 'incorrect' volume dependence $N_e$ represents a subclass of the three-particle ladder diagrams which remain disconnected in the two-particle polarization. (ii) Circular polarized light: In the experimental situation of a circular polarized light ($\sigma^+$), only the spin

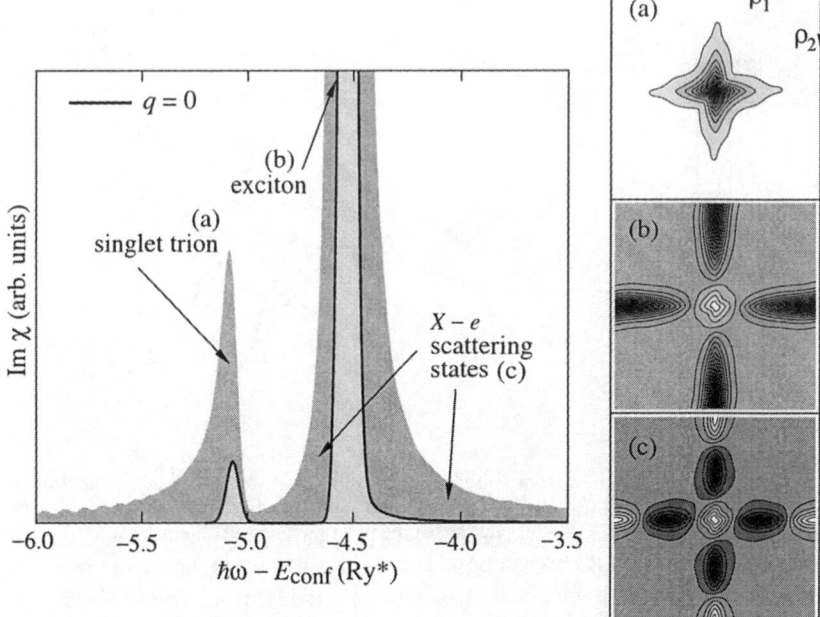

*Figure 2.* Linear optical absorption of a QWR with background electrons in the singlet channel. The mass ratio is $\sigma = 0.3$ (GaAs), the exciton binding energy is $E_{BX} = 4.5$ Ry*. Calculated spectra for different momenta $q$ of the initial electron are summed up with equal weight. A phenomenological Gauss broadening (FWHM = 0.07 Ry*) is assumed and smooths the $(\hbar\omega - E_T^{\rm rel})^{-1/2}$ trion-singularity in 1D. Examples for the respective trion wave functions are given (a-c). The exciton-electron scattering states are characterized by a nearly undisturbed $\exp(iq\rho)$ envelope, where $q$ is related to the quantization volume. The small oscillations result from the discrete set of energy levels therein and have no physical meaning.

component $\sigma_1 = \downarrow$ in Eq. (8) contributes to the optical susceptibility and gives

$$\chi(\omega) = \frac{\chi_0}{\Omega^2}\left[\frac{1}{2}(1 - N_e)\sum_{k_1 k_2} G^X(k_1, k_2, \omega) \right.$$
$$\left. + \sum_{qk} \frac{1}{4}\left(f_{q\uparrow}^e G_{qs}^T(k, \eta q, \omega) + (2f_{q\downarrow}^e + f_{q\uparrow}^e)G_{qt}^T(k, \eta q, \omega)\right)\right]. \quad (10)$$

As can be seen from this expression, the trion part of the susceptibility $\chi(\omega)$ and in particular the relative contributions of singlet and triplet channels now strongly depend on the polarization of the background electrons. In the case of completely polarized electrons $f_{q\uparrow}^e = 0$, only triplet states are available, whereas in the opposite case of $f_{q\downarrow}^e = 0$, both singlet and triplet channels contribute with equal statistical weight. Fig. 1 highlights this behaviour and shows the eight spin states, which can be constructed from the antisymmetric ($T_s$) or symmetric ($T_t$) combinations of two electron spins with an additional heavy hole spin. Two dark states with $S_z = \pm 5/2$

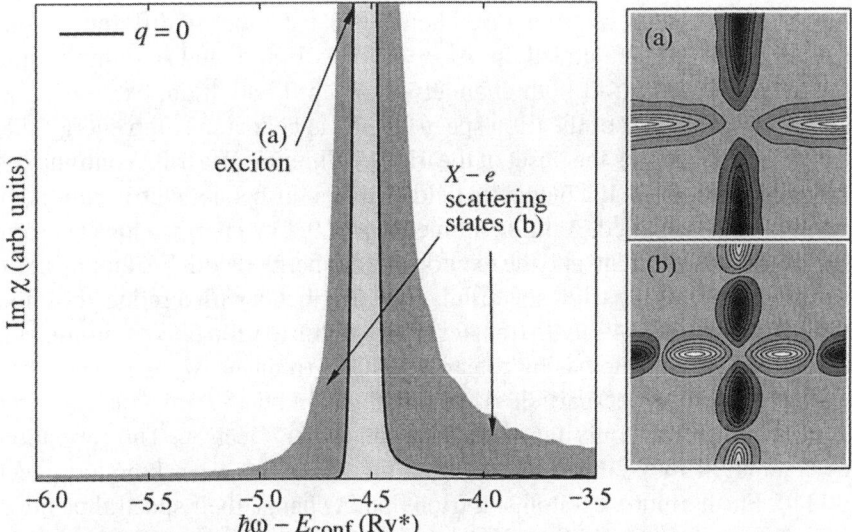

*Figure 3.* Linear optical absorption of a QWR with background electrons in the triplet channel. Same parameter as in Fig. 2. No bound state is found, but the onset of continous exciton-electron scattering states at the exciton. Examples of these wave functions are given in (a-b). Note their electron-exchange antisymmetry $\varphi(\rho_1, \rho_2) = -\varphi(\rho_2, \rho_1)$ and their inversion symmetry $\varphi(\rho_1, \rho_2) = \varphi(-\rho_1, -\rho_2)$. We remark that those triplet states with $\varphi(\rho_1, \rho_2) = -\varphi(-\rho_1, -\rho_2)$, which are not shown here, are dark at $q = 0$, see the discussion in [20].

appear in the triplet channel, which cannot be reached from initial electron states ($S_z = \pm 1/2$) by $\sigma^\pm$ polarized light of angular momentum $\pm 1$.

## 3. Trion Absorption Spectrum

The momentum dependence of trion spectra (Fig. 1a) is found from the time-dependent Schrödinger equation (7), which we solve in real space for a prototype QWR with circular cross section and fixed exciton binding energy, see Ref. [20] for details. We put $N_e = 1$ in the numerical calculation of Eq. (9). Thus one additional electron is present in the normalization volume $\Omega$, which limits our method to low background densities. More specifically, the box size has to be large compared to the size of the trion bound state. This procedure avoids to calculate the exciton problem in addition and the full optical susceptibility is found from the trion Green's function $G^T_{qv}(\rho_1, \rho_2, \omega)$ with

$$\chi(\omega) = \frac{\chi_0}{\Omega} \sum_{qv} S^v f^e_q \int d\rho_2 \, G^T_{qv}(0, \rho_2, \omega) \, e^{i\eta q \rho_2}. \quad (11)$$

Here, the relative coordinates $\rho_1 = \rho_{e1} - \rho_h$ and $\rho_2 = \rho_{e2} - \rho_h$ are used. The imaginary part of $\chi(\omega)$ (absorption spectrum) is shown for singlet and triplet channels in Fig. 2 and 3, respectively, where all contributions of different momenta ($q \leq q_{\max} = 1.5/a_B^*$) are added up. At $q = 0$, the trion bound state in the singlet channel and the exciton in both channels show up. In addition, exciton-electron scattering states are part of the trion spectrum and appear at the high energy side of the exciton, which marks the onset of the trion continuum. The trion continuum will split into discrete states, if a magnetic field [22] or a further geometric confinement to quantum dots is applied. At finite momenta $q \neq 0$, low-energy wings are formed below the trion bound state and the exciton due to energy recoil. We emphasize that the exciton is part of the trion spectrum, and contributes with a reduced oscillator strength. This oscillator strength transfer from exciton to trion plus exciton-electron scattering states with increasing electron density (reduced $\Omega$) is a result of the orthogonality of the three-particle wave functions. In other words, the exciton can be formed undisturbed only far away from the 'other' electron. The formation of trions compete if they come close to each other, see the wave function panel in Fig. 2(a,b). Furthermore, exciton and trion do not change their spectral position in the low-density regime. Similar results were observed in GaAs QWs [16].

## 4. Disorder

The presence of disorder is an unavoidably result of the growth process of the nanostructure. Indeed strong spatial fluctuations of the trion intensity were observed in near-field spectra of modulation-doped wide QWs and have been related to the localization of trions [7, 8]. Here, we will consider in a first step the influence of interface roughness, for which, on atomic scale, a spatially uncorrelated disorder potential $\langle V(r)V(r')\rangle = \sigma^2 a_0^D \delta(r-r')$ is assumed. $a_0$ is a typical island size at the QW ($D = 2$) or QWR ($D = 1$) interfaces. It is straightforward to take electrostatic disorder from the random position of ionized donors into account but this is beyond the scope of the present paper. The disorder strength is assumed to be weak, i.e. the inhomogeneous width of the exciton line is less than the trion binding energy $B_T$. This is the case in wide QWs of high quality. As an example, a change of one monolayer in a 20 nm-wide GaAs yields a fluctuation of the confinement energy of $\Delta E_{\text{ML}} = 0.25$ meV, which is only a fraction of $B_T = 1$ meV. In this limit, the in-plane trion wave function $\Psi_T(R, \rho_1, \rho_2)$ may be separated into relative and COM part, i.e.

$$\Psi_T(R, \rho_1, \rho_2) = \Phi_\alpha^T(R)\varphi_T(\rho_1, \rho_2). \tag{12}$$

The relative Schrödinger equation has already been solved, see the trion wave function in Fig. 2a, and gives the internal trion energy $E_T^{\text{rel}}$. The resulting COM equation in the presence of the disorder potential reads

$$\left[-\frac{\hbar^2}{2M_T}\Delta_R + V_T^{\text{eff}}(R) + E_T^{\text{rel}}\right]\Phi_\alpha^T(R) = E_\alpha^T \Phi_\alpha^T(R) \tag{13}$$

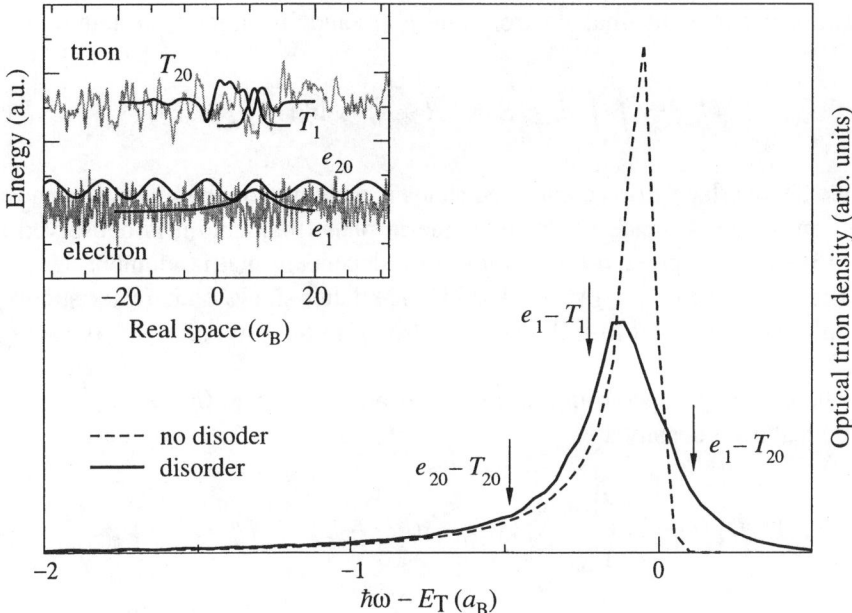

*Figure 4.* 1D optical trion density $D_T(\hbar\omega)$ in the presence of disorder. The profile is asymmetric and has a low-energy tail, which coincides with the disorder-free limit and is shown for comparison. Three different regimes corresponding to (with increasing energy) transitions (i) between weakly localized states, (ii) between strongly localized states, and (iii) between states of different localization character can be distinguished. The inset shows the ground and 20th excited state of both the electron and the trion COM in 'their' disorder potentials.

$$V_T^{\text{eff}}(R) = \int\int d\rho_1 d\rho_2 \, |\varphi_T(\rho_1, \rho_2)|^2 \\ \times \left[ 2V\left(R + \eta\rho_1 - \xi\rho_2\right) + V\left(R - \xi(\rho_1 + \rho_2)\right) \right].$$

In photoluminescence (absorption) experiments, the optical transition from (to) the disordered trion state $\Phi_\alpha^T(R)$ occurs into (from) a disordered electron state $\phi_\beta^e(r)$, for which an one-particle Anderson-like equation with the original disorder potential $V(r)$

$$\left[ -\frac{\hbar^2}{2m_e}\Delta_r + V(r) \right] \phi_\beta^e(r) = E_\beta^e \phi_\beta^e(r) \tag{14}$$

has to be solved in addition. The effective disorder potential $V_T^{\text{eff}}$ is smoothed by the relative trion wave function and exhibits a correlation length close to the trion Bohr radius. This behaviour is shown in the inset of Fig. 4. However, trion states experience a stronger influence of the disorder potential since $M_T > m_e$, and the corresponding localization length is smaller. The matrix element for the optical transition between the final electron state $\phi_\beta^e(r)$ and the initial trion state $\Phi_\alpha^T(R)$ is

$$M_{\alpha\beta}^T = \int\int d\rho_2 dR\, \phi_\beta^e(R + \eta\rho_2)\, \Phi_\alpha^T(R)\varphi_T(0,\rho_2). \tag{15}$$

Hence the overlap of two disordered states is weighted by the relative trion wave function. In the absence of disorder, the in-plane momentum is conserved and Eq. (15) can be expressed by a momentum-dependent matrix element $M_{\alpha\alpha}^T \longrightarrow M^T(q) = \int d\rho_2 \varphi_T(0,\rho_2) e^{iq\rho_2\eta}$ [14, 21]. The latter shows a characteristic exponential low-energy tail in QWs [12, 14], see also the QWR low-energy wing in Fig. 2.

The part of the spectrum close to the trion bound state ($\hbar\omega = E_T^{\text{rel}}$) is called the optical trion density

$$D_T(\hbar\omega) = \left\langle \sum_{\alpha\beta} N_\beta |M_{\alpha\beta}^T|^2\, \delta(\hbar\omega - E_T^{\text{rel}} - E_\alpha^T + E_\beta^e) \right\rangle, \tag{16}$$

where $\langle\ldots\rangle$ means averaging over many realizations of the disorder potential. Here, $N_\beta$ is the electron occupation function, which we may approximate in the weak disorder regime with a quasi-equilibrium Boltzmann distribution.

The optical trion density (16) has been calculated for the prototype QWR and is shown in Fig. 4. For comparison, also the disorder-free lineshape is given. As $\int d\omega D_T(\hbar\omega) = $ const., we observe a redistribution of the oscillator strength in the presence of disorder. The optical trion density results from transitions between two localized states. It depends on the respective localization length which part of the spectrum is relevant. In particular, the transitions between strongly localized states shift the main peak of the optical trion density to lower energies ($\hbar\omega < E_T^{\text{rel}}$). In contrast, optical transitions between more extended states are the source for the low-energy tail, which is found to be very close to the disorder-free limit. In strong contrast to the well-known exciton case we find the contributions of extended states *below* the main trion peak. The third contributions are transition between strongly localized electron states *and* extended trion states, which give rise to contributions to the optical trion density at energies above $\hbar\omega > E_T^{\text{rel}}$, which are excluded in the disorder-free case by conservation of momentum.

In conclusion, trion lineshapes keep their asymmetry even in the presence of disorder. In particular, the low-energy tail has two components and we predict a gradual change between two slopes in experimental trion spectra. The main part is related to optical transitions between localized states, which turns into a second slope determined by the transition between extended states and $|M^T(q)|^2$. The transition between the two regime occurs at an energy $\hbar\omega < E_T^{\text{rel}} - \sigma$. In contrast to the observation of motional narrowing for excitons, we find a trion profile which is broader than the distributions of the disorder $V_T^{\text{eff}}$.

## Acknowledgements

A.E. is supported by the Postdoc-Programme of the German Academic Exchange Service (DAAD). We grateful acknowledge Israel Bar-Joseph, Yossi Yayon, Marc-André Dupertuis, and Erich Runge for helpful discussions.

## References

1. Kheng, K., Cox, R.T., Merle d'Aubigné, Y., Bassani, F., Saminadayar, K., and S. Tatarenko (1993) Observation of negatively charged excitons $X^-$ in semiconductor quantum wells, *Phys. Rev. Lett.* **71**, 1752–1755.
2. Finkelstein, G., Shtrikman, H., and Bar-Joseph, I. (1995) Optical spectroscopy of a two-dimensional electron gas near the metal-insulator transition, *Phys. Rev. Lett.* **74**, 976–979.
3. Shields, A.J., Osborne, J.L., Simmons, M.Y., Pepper, M., and Ritchie, D.A. (1995) Magneto-optical spectroscopy of positively charged excitons in GaAs quantum wells, *Phys. Rev. B* **52**, R5523–R5526.
4. Hegarty, J., M.D. Sturge, C. Weisbuch, A.C. Gossard, and W. Wiegmann (1982) Resonant Rayleigh scattering from an inhomogeneously broadened transition: A new probe of the homogeneous linewidth, *Phys. Rev. Lett.* **49**, 930–932.
5. Nickolaus, H., Wünsche, H.-J., and Henneberger, F. (1998) Exciton spin relaxation in semiconductor quantum wells: The role of disorder, *Phys. Rev. Lett.* **81**, 2586–2589.
6. Hess, H.F., Betzig, E., Harris, T.D., Pfeiffer, L.N., and West, K.W. (1994) Near-field spectroscopy of the quantum constituents of a luminescent system, *Science* **264**, 1740–1744.
7. Eytan, G., Yayon, Y., Rappaport, M., Shtrikman, H., and Bar-Joseph, I. (1998) Near-field spectroscopy of a gated electron gas: A direct evidence for electron localization, *Phys. Rev. Lett.* **81**, 1666–1669.
8. Yayon, Y., Rappaport, M., Umansky, V., and Bar-Joseph, I. (2001) Excitonic emission in the presence of a two-dimensional electron gas: A microscopic view, *Phys. Rev. B* **64**, 081308.
9. Brinkmann, D., Kudrna, J., Gilliot, P., Hönerlage, B., Arnoult, A., Cibert, J., and Tatarenko, S. (1999) Trion and exciton dephasing measurements in modulation-doped quantum wells: A probe for trion and carrier localization, *Phys. Rev. B* **60**, 4474–4477.
10. Gilliot, P., Brinkmann, D., Kudrna, J., Crégut, O., Lévy, R., Arnoult, A., Cibert, J., and Tatarenko, S. (1999) Quantum beats between trion and exciton transitions in modulation-doped CdTe quantum wells, *Phys. Rev. B* **60**, 5797–5801.
11. Sanvitto, D., Pulizzi, F., Shields, A.J., Christianen, P.C.M., Holmes, S.N., Simmons, M.Y., Ritchie, D.A., Maan, J.C., and Pepper, M. (2001) Observation of charge transport by negatively charged excitons, *Science* **294**, 837–839.
12. Ciulin, V., Kossacki, P., Haacke, S., Garnière, J., Deveaud, B., Esser, A., Kutrowski, M., and Wojtowicz, T. (2000) Radiative behavior of negatively charged excitons in CdTe-based quantum wells: A spectral and temporal analysis, *Phys. Rev. B* **62**, R16310–R16313.
13. Esser, A., Runge, E., Zimmermann, R., and Langbein, W. (2000) Electron and hole trions in wide GaAs quantum wells, *phys. stat. sol. (b)* **221**, 281–286.
14. Esser, A., Runge, E., Zimmermann, R., and Langbein, W. (2000) Photoluminescence and radiative lifetime of trions in GaAs quantum wells, *Phys. Rev. B* **62**, 8232–8239.
15. Puls, J., Mikhailov, G.V., Henneberger, F., Yakovlev, D.R., and Waag, A. (2002) Trionic gain in $\delta$-doped ZnSe quantum wells, *phys. stat. sol. (b)* **229**, 637–641.
16. Yusa, G., Shtrikman, H., and Bar-Joseph, I. (2000) Onset of exciton absorption in modulation-doped GaAs quantum wells, *Phys. Rev. B* **62**, 15390–15393.

17. Huard, V., Cox, R.T., Saminadayar, K., Arnoult, A., and Tatarenko, S. (2000) Bound states in optical absorption of semiconductor quantum wells containing a two-dimensional electron gas, *Phys. Rev. Lett.* **84**, 187–190.
18. Bronold, F.X. (2000) Absorption spectrum of a weakly n-doped semiconductor quantum well, *Phys. Rev. B* **61**, 12620–12623.
19. Suris, R.A., Kochereshko, V.P., Astakhov, G.V., Yakovlev, D.R., Ossau, W., Nuernberger, J., Faschinger, W., Landwehr, G., Wojtowicz, T., Karczewski, G., and Kossut, J. (2001) Excitons and trions modified by interaction with a two-dimensional electron gas, *phys. stat. sol (b)* **227**, 343–352.
20. Esser, A., Zimmermann, R., and Runge, E. (2001) Theory of trion spectra in semiconductor nanostructures, *phys. stat. sol. (b)* **227**, 317–330.
21. Stébé, B., Feddi, E., Ainane, A., and Dujardin, F. (1998) Optical and magneto-optical absorption of negatively charged excitons in three- and two-dimensional semiconductors, *Phys. Rev. B* **58**, 9926–9932.
22. Yakovlev, D.R., Kochereshko, V.P., Suris, R.A., Schenk, H., Ossau, W., Waag, A., Landwehr, G., Christianen, P.C.M., and Maan, J.C. (1997) Combined exciton-cyclotron resonance in quantum well structures, *Phys. Rev. Lett.* **79**, 3974–3977.

# CORRELATION BETWEEN TRION AND HOLE IN FERMI DISTRIBUTION IN PROCESS OF TRION PHOTO-EXCITATION IN DOPED QWs

R.A. SURIS
*Ioffe Physico-Technical Institute RAS,*
*26 Polytekhnicheskaya, St Petersburg 194021, Russia*

Abstract. The problem of correlation between a trion and a hole in the electron Fermi distribution created in the process of trion photo-excitation in doped quantum wells is under consideration. The hole in the Fermi distribution appears in the trion creation process consisting of picking of the Fermi Sea electron up by the exciton created in virtual state due to photon absorption. It is demonstrated that the interaction results in formation of a correlated state of the trion and the hole in the Fermi Sea. The state has excitation energy which is less then trion energy minus Fermi energy that can be obtained as a lower edge of trion excitation band using the simple energy conservation low in the picture of independent trion and electrons. The wave function of the correlated state is real and decreases with increase of distance between the trion and the Fermi Sea hole, $r$, as $1/r^{3/2}$. The wave function can be normalized to unity and it corresponds to correlated state of the trion and Fermi Sea hole. In contrast to this state, the states with excitation energies in the absorption band between the trion energy and the trion energy minus Fermi energy have complex wave functions that decrease as $1/r^{1/2}$. These states correspond to the trion and the Fermi Sea hole that is running away from the trion.
The correlated state described above is supposed to be responsible for the narrow trion absorption line that was observed experimentally.

Key words: quantum well, exciton, trion, electron, hole, photon, optical spectrum, singularity, wave function, susceptibility, Fermi distribution, Fermi energy, Green function, scattering amplitude, 2DEG

## 1. Introduction

The problem of optics of doped quantum wells (QW) attracts attention in last decade. The main reason of that is the possibility to observe the manifestation of many-body effects under controllable concentration of carriers in rather wide region of temperature and magnetic field values. Especially, it is of great interest to observe so called trions that are analogous of well-known negatively charged hydrogen atoms, $H^-$, or molecule $H_2^+$. For bulk semiconductors, $H^-$-like complexes were predicted in 1958 [1]. However, their binding energy in bulk semiconductors is too small to make them observable at reasonable temperatures. The reduction of the dimensionality down to 2D was of crucial importance for their observation due to strong increase of the binding energy for interacting confined particles. The first experimental observation of trions in CdTe/CdZnTe quantum wells contained two-dimensional electron gas was done in 1993 [2].

Nowadays, the trions are well studied in QW structures with a two-dimensional electron gas (2DEG) of low density in QWs based on $A^{II}B^{VI}$ as well as $A^{III}B^{V}$ semiconductor compounds. Negatively charged- and positively charged exciton complexes with heavy-holes and light-holes were found experimentally [3–6]. In addition to singlet trion states, triplet states were observed in high magnetic fields [5].

However, up to now the properties of the trion states in 2DEG are far from complete understanding. In [7] we presented a simple theory of optical spectra near trion and exciton resonances. The purpose of the paper is to develop the theory and to give a picture of the photo-excited trion states in weakly doped QWs. Special attention will be paid to analysis of the nature of the state corresponding to a sharp low-energy peak of the trion absorption. We will show that the peak arises due to strong correlation between the trion and the hole in the electron Fermi Sea. The hole is created due to picking Fermi Sea electron up by exciton in virtual state in process of trion formation.

## 2. QW Susceptibility

### 2.1. MAIN PROCESSES AND APPROXIMATIONS

Our task is to calculate optical spectrum of QWs modified by a dilute 2D electron gas. For this consideration, we will be assumed that the sheet electron concentration $n_e$ is so small that $n_e a_{tr}^2 \ll 1$ (here $a_{tr}$ is the trion radius). It means that i.e. the Fermi energy of 2DEG, $F$, is much smaller than the trion binding energy $E_{tr}$, $F \ll E_{tr}$. Under this condition, the electron gas does not modify the wave functions of trion and, of course, exciton states and does not affect their binding energies. Here we will limit ourselves with consideration of zero temperature, $T = 0$ K. A trion contribution to optical spectrum is essential in the energy region near the exciton resonance because the trion binding energy is much smaller than the binding energy of the exciton.

A naive picture of trion photo-excitation is as follows. A photon creates an exciton in a virtual state, which picks a background electron up forming a trion in final state (see diagram (1) below). It is clear that the photon energy measured relative to the exciton energy should be $\Omega = -E_{tr} - \varepsilon$ (we use units $\hbar \equiv 1$). Here $\varepsilon$ is energy of the electron and $\Omega$ is the photon energy minus the excitation energy of unperturbed exciton. Therefore, due to the Fermi statistics for electrons one can assume that the photon energy required for trion creation belongs to the energy band ranging from $-E_{tr}$ to $-E_{tr} - F$ and absorption spectrum is:

$$abs \propto \int_0^F d\varepsilon\, \delta(\Omega + \varepsilon + E_{tr}).$$

However, we will see further that this picture is not complete and we will find that a sharp absorption peak will appear slightly below this band.

Two types of processes caused by exciton-electron interaction are important for our considerations:

(i) Trion formation involving an intermediate state "exciton + electron":

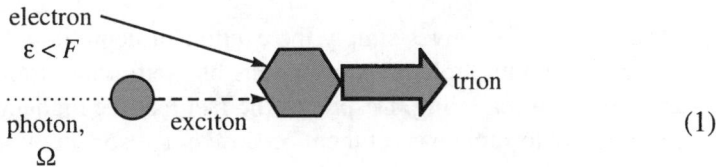
(1)

(ii) Scattering of a photo-excited exciton on a background electron:

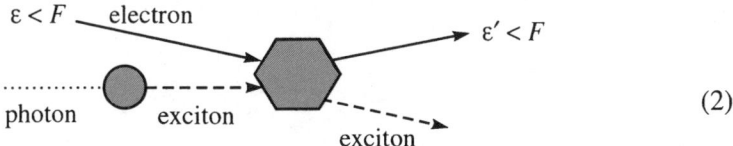
(2)

Because of $\varepsilon' > \varepsilon$, processes (2) result in high-energy tail of the exciton resonance.

Since the trion binding energy $E_{tr}$ is much less than the exciton binding energy $E_{exc}$, and as we are interested in the energy range close to the exciton and trion resonances, these processes should be the most effective. The reason is that the processes have an exciton as intermediate state. As a result a resonant denominator from the exciton Green's function appears in their amplitudes. So, these amplitudes are by factor $(E_{exc}/E_{tr})$ larger than the amplitudes of all other processes.

Besides, we should consider of course the process of exciton creation:

$$\cdots\!\bullet\text{---}\!\bullet\cdots \quad (3)$$

## 2.2. DIAGRAMS FOR QW SUSCEPTIBILITY

The complete information on optics of the QW gives its susceptibility.

The susceptibility of the QW in the vicinity of the exciton resonance is given by the sum of the following diagrams:

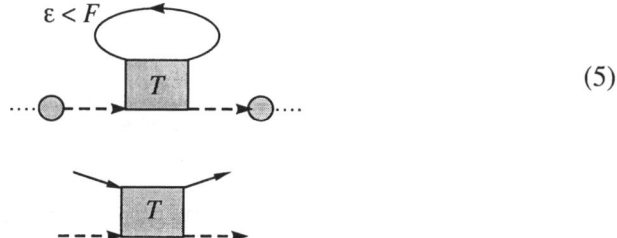

(4)

The first diagram gives simply the exciton susceptibility. Other diagrams account for mutual multiple transformations of exciton into trion and back and for exciton-electron scattering. Despite of the fact that the higher order diagrams are small, we have to sum up all of them because of a resonance contribution from the denominator in the exciton Green function.

The two diagrams in the second line could be represented in a common form as:

(5)

The square

is the amplitude of exciton-electron scattering.

In the terms of the scattering amplitude, the diagrams for the QW susceptibility can be presented in the following form:

$$\chi(\Omega) = \quad (\propto \tfrac{1}{\Omega})$$

$$+ \quad (\propto \tfrac{1}{\Omega^2})$$

$$+ \quad (\propto \tfrac{1}{\Omega^3})$$

$$+ \ldots$$

$$= \frac{1}{\Omega + i\gamma - \Pi(\Omega)}$$

(6)

Here we introduce self-energy operator

$$\Pi(\Omega) = \int_0^F \frac{d\varepsilon}{2\pi} \; T(\Omega+\varepsilon) \qquad (7)$$

Let us emphasize again that, despite of the fact that the higher order diagrams are small as corresponding power of concentration, we have to sum up all of them because of a resonance contribution from the exciton Green function that is proportional to $1/\Omega$.

We supposed the scattering amplitude independent on electron and exciton wave numbers for small relative electron-exciton momentum. That is true as far as Fermi momentum of the electron 2D gas is much smaller then the trion radius as we assumed at very beginning. However, we have to take into account the difference in amplitudes of electron-exciton scattering in singlet and triplet states. Therefore from (7) we obtain:

$$\chi_\uparrow(\Omega) = \chi_\downarrow(\Omega)$$
$$= \frac{|D_{CV}|^2 \, \Psi(0)^2}{\Omega + i\gamma - \int_0^F \left\{ \frac{1}{2} T_{\text{sngl}}\left(\Omega+\varepsilon\left[1 - \frac{m_e}{2m_e+m_h}\right]\right) + \frac{3}{2} T_{\text{trpl}}\left(\Omega+\varepsilon\left[1 - \frac{m_e}{2m_e+m_h}\right]\right) \right\} \left(1 + \frac{m_e}{m_e+m_h}\right) d\varepsilon}.$$
$$(8)$$

Here $D_{CV}$, is a conducting band–valence band dipole matrix element and $\Psi(0)$ is electron-hole relative motion wave function of the exciton at zero electron-hole distance.

Arrows in Eq. (8) indicate the electron spin orientation in the exciton defined by the circular polarization of incident photon. $T_{\text{sngl}}$ and $T_{\text{trpl}}$ are amplitudes of an electron-exciton scattering resulting in a singlet- and triplet state, respectively. Factors 1/2 and 3/2 account for the statistical weight of these states. We assume in Eq. (6) that the 2D electrons have no spin polarization. In case of such a polarization (e.g. induced by external magnetic fields) the polarizabilities $\chi_\uparrow(\Omega)$ and $\chi_\downarrow(\Omega)$ are different, which means that the QW optical properties are different in two circular polarizations of light. The second term in the brackets $\left(1 - \frac{m_e}{2m_e+m_h}\right)$ and $\left(1 + \frac{m_e}{m_e+m_h}\right)$ account the exciton and trion kinetic energies.

## 2.3. SCATTERING AMPLITUDES AND THE TRION SINGULARITY OF THE QW SUSCEPTIBILITY

Let us analyze now the scattering amplitudes in Eq. (5). It is known that at zero or weak magnetic fields the trion singlet state is bound whereas its triplet state is unbound. Consequently, the amplitude $T_{\text{sngl}}(\Omega)$ has a pole at $\Omega = -E_{\text{tr}}$, whereas the amplitude $T_{\text{trpl}}(\Omega)$ has no pole in the energy range $|\Omega| \propto E_{\text{tr}}$. Due to the short-range character of the exciton-electron interaction only the amplitude of the

$S$-scattering will be calculated at energies less than the trion binding energy. In this case one can show that for $\Omega \ll E_{\text{exc}}$ (see Appendix)

$$T_{\text{sngl}}(\Omega) \approx \frac{2\pi}{\overline{m}} \bigg/ \ln\left(\frac{E_{\text{tr}}}{-\Omega}\right) \quad \text{and} \quad T_{\text{trpl}}(\Omega) \approx \frac{2\pi}{\overline{m}} \bigg/ \ln\left(\frac{E_{\text{exc}}}{-\Omega} \frac{E_{\text{exc}}}{E_1}\right). \tag{9}$$

Here, $E_1$ is an energy of the order of $E_{\text{tr}}$, and $\overline{m}$ is the reduced effective mass of the system "exciton + electron", $1/\overline{m} = 1/m_e + 1/(m_e + m_h)$. $E_B$ is a 2D exciton Rydberg.

The imaginary parts of the amplitudes $T_{\text{sngl}}(\Omega)$ and $T_{\text{trpl}}(\Omega)$ are nonzero at $\Omega > 0$ only. This means that the imaginary part of the integral in Eq. (5) is equal to zero if $\Omega < -E_{\text{tr}} - F\left(1 - \frac{m_e}{2m_e + m_h}\right)$. One can be convinced that the integral is positive and diverges logarithmically when $\Omega$ approaches to $-E_{\text{tr}} - F\left(1 - \frac{m_e}{2m_e + m_h}\right)$ from the low energy side. As a result, the polarizabilities $\chi_\uparrow(\Omega)$, $\chi_\downarrow(\Omega)$ have a pole at a real value of the energy, $\Omega < -E_{\text{tr}} - F\left(1 - \frac{m_e}{2m_e + m_h}\right)$. As we will see in the next section, this pole corresponds to a correlated state of the trion and a hole in the Fermi Sea, which appears due to the capture of an electron by an exciton in the process of the trion formation.

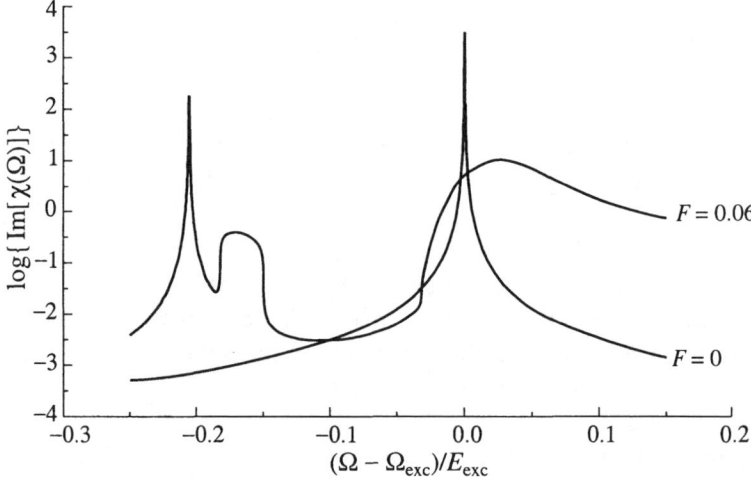

*Figure 1.* Dependence of $\log(\text{Im}(\chi))$ on the photon energy, $\Omega$, at Fermi energies $F = 0$ and $F = 0.06$. ($E_{\text{tr}} = 0.15$, $m_h/m_e = 3$). All energies are measured in $E_{\text{exc}}$ units.

The pole is clearly seen as a sharp line in Fig. 1 which gives dependence of $\log(\text{Im}(\chi))$ on the photon energy, $\Omega$, at Fermi energies $F = 0$ and $F = 0.06$. Fig. 2 illustrates evolution of the spectrum with increase of $F$ from 0 up to 0.06. Fermi energies are measured in the units of the 2D exciton binding energy. In these calculations we assumed the following values of parameters: $E_{\text{tr}}/E_{\text{exc}} = 0.15$, $m_h/m_e = 3$.

One can see that the gap between the pole and the left edge of the trion band, $-E_{\text{tr}} - F\left(1 - \frac{m_e}{2m_e + m_h}\right) < \Omega < -E_{\text{tr}}$, increases with increase of Fermi energy.

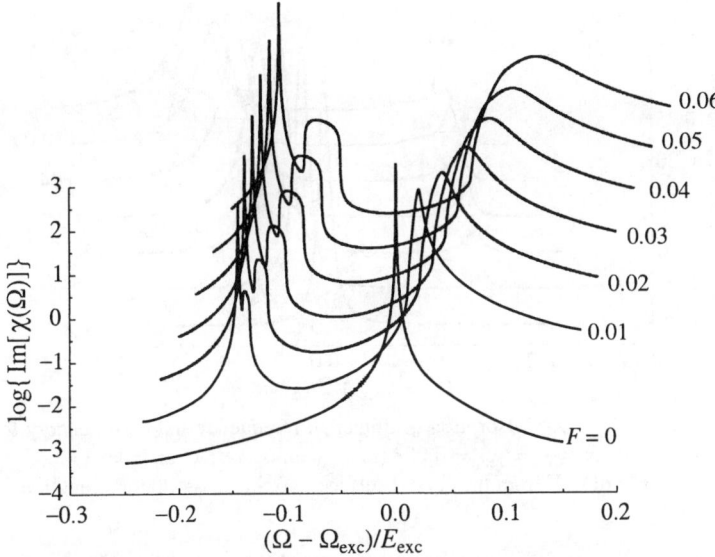

*Figure 2.* Evolution of the spectrum with increase of Fermi energy $F$ from 0 up to 0.06 (in $E_{exc}$ units) ($E_{tr} = 0.15$ (in $E_{exc}$ units), $m_h/m_e = 3$).

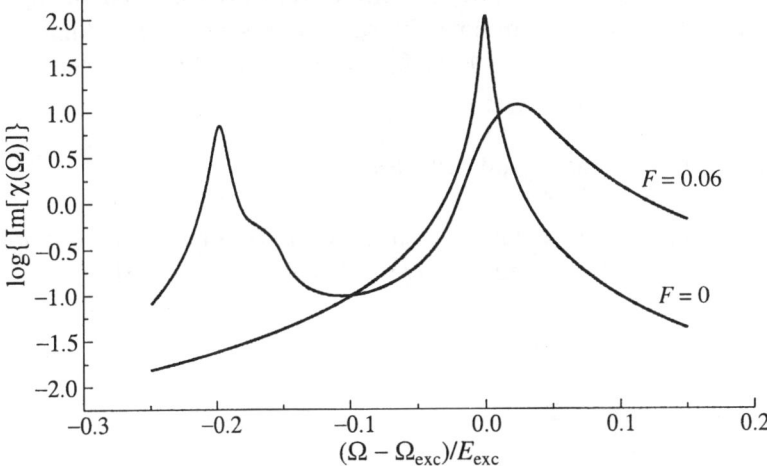

*Figure 3.* Evolution of the spectrum with increase of Fermi energy $F$ from 0 up to 0.06 (in $E_{exc}$ units) regarding the exciton and electron damping: $\gamma_{exc} = \gamma_{el} = 0.003$ (in $E_{exc}$ units) ($E_{tr} = 0.15$ (in $E_{exc}$ units), $m_h/m_e = 3$).

One can see also the high-energy tail to the right of the exciton peak. The tail is due to the combined processes presented in diagram (2) and it increases with increase of the electron concentration.

If we take into account the electron and exciton scattering in simplest way, introducing damping parameters for both of them, the peak on the left of the trion

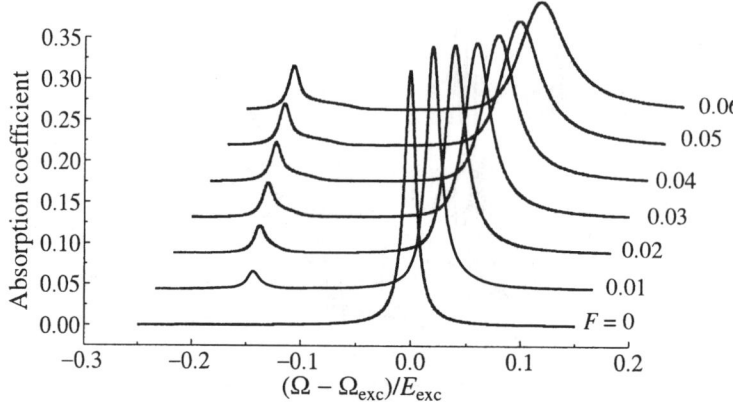

*Figure 4.* Dependence of QW absorption coefficient on frequency and Fermi energy for normally incident light ($E_{tr} = 0.15$, $\gamma_{exc} = \gamma_{el} = 0.003$ (in $E_{exc}$ units), refraction index = 2.65, wave vector = 1.4 $\mu m^{-1}$, distance of QW from the crystal surface = 0.5 $\mu m$, oscillator strength = 0.005).

band becomes wider and the gap between it and the band disappears (Fig. 3). For this example, we took the damping parameters as $\gamma_{exc} = \gamma_{el} = 0.003$ (in $E_{exc}$ units). The results presented in Fig. 1 and 2 were calculated at $\gamma_{exs} = \gamma_{el} = 10^{-4}$.

For normally incident light, dependence of QW absorption coefficient on frequency and Fermi energy is presented in Fig. 4 at the same values of mass, $E_{tr}$ and damping. We used in this example the following values of parameters:
Refraction index = 2.65,
Wave vector = 1.4 $\mu m^{-1}$,
Distance of QW from the crystal surface = 0.5 $\mu m$,
Oscillator strength = 0.005.

The absorption coefficient is obtained be means of solution of the simple electrodynamics problem supposing the QW width much less then the light wavelength.

## 3. Nature of the singularity

In this section, we will give the simple picture of the state responsible for the left-hand delta-function peak in the spectrum described above.

Due to very small value of the trion binding energy, we have to consider a mixture of the trion and exciton wave functions. Mixing of the states is because of interaction with Fermi Sea electrons. The following diagram describes the process:

$$\begin{array}{c} \text{electron} \\ \varepsilon < F \\ \text{------} \\ \text{exciton} \end{array} \Rightarrow \text{trion} \qquad (10)$$

Actually, we deal with the exciton on the background of Fermi Sea and with the trion on the background of Fermi Sea with a Fermi Sea hole. This can be illustrated

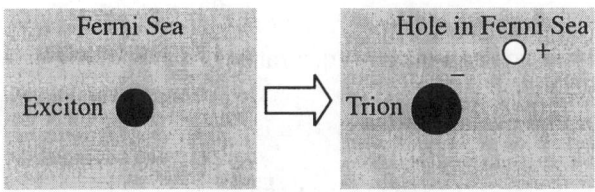

*Figure 5.* Schematic illustration of the trion creation on the Fermi Sea background.

as it is shown in Fig. 5. The hole arises due to picking of an electron of Fermi Sea up by the exciton in the virtual state to create the trion (see the diagram above). Therefore, we have to consider the four-particle state:

Trion (two electrons + a valence band hole) + Hole in Fermi Sea.

The wave function (WF) of the mixture of the state with the exciton state is

$$|\Psi\rangle = Exc\, |\mathbf{Exc}\rangle + \int_{\mathbf{k}} Tr(\mathbf{k})\, |\mathbf{Tr_k}\rangle. \qquad (11)$$

The WF of the exciton on the background of Fermi Sea is

$$|\mathbf{Exc}\rangle = \int_{\mathbf{R},\mathbf{r}} \varphi_{\text{exc}}(\mathbf{r} - \mathbf{R})\, a_s^+(\mathbf{r})\, b_\sigma^+(\mathbf{R})\, |\Phi\rangle. \qquad (12)$$

Here $|\Phi\rangle$ is Fermi Sea WF, $a_s^+(\mathbf{r})$ is the creation operator of the electron with spin S at the point with the coordinate $\mathbf{r}$, $b_\sigma^+(\mathbf{R})$ is the creation operator of the hole with spin $\sigma$ at the point $\mathbf{R}$. The function $\varphi_{\text{exc}}(\mathbf{r} - \mathbf{R})$ is the WF of the electron-hole relative motion in the free exciton with zero momentum.

WF of the trion + the hole with wave number k in Fermi Sea is

$$|\mathbf{Tr_k}\rangle = \int_{\mathbf{R},\mathbf{r}} \varphi_{\text{tr}}(\mathbf{r}_1 - \mathbf{R}, \mathbf{r}_2 - \mathbf{R})\, a_{-s}^+(\mathbf{r}_1)\, a_s^+(\mathbf{r}_2)\, b_\sigma^+(\mathbf{R})\, a_{\mathbf{k},-s}\, |\Phi\rangle. \qquad (13)$$

Here $\varphi_{\text{tr}}(\mathbf{r}_1 - \mathbf{R}, \mathbf{r}_2 - \mathbf{R})$ is the free trion WF and $a_{\mathbf{k},-s}$ is the annihilation operator of Fermi Sea electron with momentum $\mathbf{k}$ and spin–$S^1$. Due to annihilation operator $a_{\mathbf{k},-s}$ in Eq. (13), the trion WF is zero at $|\mathbf{k}| > k_F$, where $k_F$ is the Fermi momentum.

Generally, the approximate states (12) and (13) are not orthogonal. However, in the low-density limit, $n_e a_{\text{tr}}^2 \ll 1$, we are eligible to disregard the point.

The equation for amplitudes Exc and $Tr(\mathbf{k})$ are

$$\begin{aligned}\Omega\, Exc - \int_{|\mathbf{k}|<k_F} g(\mathbf{k})\, Tr(\mathbf{k}) &= 0 \\ (\Omega + E_{\text{tr}} - (-\varepsilon(\mathbf{k})))\, Tr(\mathbf{k}) - g(\mathbf{k})\, Exc &= 0\end{aligned}. \qquad (14)$$

---
[1] Let us remind that operators $a(\mathbf{r})$ and $a_{\mathbf{k}}$ are connected by Fourier-transformation.

Here $(-\varepsilon(\mathbf{k})) = -\mathbf{k}^2/2m_{el}$ is the energy of the Fermi Sea hole with momentum $-\mathbf{k}$. Function $g(\mathbf{k})$ is the vortex part in the diagram (13). The function can be expressed in terms of the trion WF.

Equation for eigenvalues of the system is

$$\Omega - \int_{|\mathbf{k}|<k_F} \frac{g(\mathbf{k})^2}{\Omega + E_{tr} + \varepsilon(\mathbf{k})} = 0. \tag{15}$$

Actually, this equation is the same as the equation for pole of the $\chi$ (Eq. (8)) obtained by expanding the first term in self-energy operator $\Pi$ near the trion pole in the limit $m_e/m_h \to 0$.

Due to our initial assumption, $k_F \ll a_{tr}$, we can neglect the $\mathbf{k}$-dependence of $g(\mathbf{k})$ and find the value of $g(0)$ from Eqs. (8) and (9). Therefore, we have the dispersion equation in the simple form

$$\Omega - \int_0^F \frac{E_{tr}}{\Omega + E_{tr} + \varepsilon} d\varepsilon = 0 \tag{16}$$

or

$$\frac{\Omega}{E_{tr}} - \ln\left(\frac{1 + \frac{\Omega}{E_{tr}} + \frac{F}{E_{tr}}}{1 + \frac{\Omega}{E_{tr}}}\right) = 0. \tag{16a}$$

The dispersion equation should be considered only at $\Omega < -E_{tr} - F$. For example, the solution of the equation for $F = 0.06$ and $E_{tr} = 0.15$ (in the units of the 2D exciton energy, $E_{exc}$) $\Omega = -E_{tr} - F - 0.017$. This pole is clearly seen as a sharp line in Fig. 1. In the energy region $-E_{tr} - F < \Omega < -E_{tr}$ we deal with continuous spectrum.

Using Eq. (14), the trion amplitude $Tr(\mathbf{k})$ can be presented as

$$Tr(\mathbf{k}) = \frac{g(\mathbf{k})}{\Omega + E_{tr} + \varepsilon(\mathbf{k})} Exc. \tag{17}$$

The spatial correlation of the photo-created trion and the hole in Fermi Sea is given by Fourier-transformation of the function $Tr(\mathbf{k})$:

$$\Phi(\mathbf{r}) \propto \int_{|\mathbf{k}|<k_F} Tr(\mathbf{k}) e^{i\mathbf{k}\cdot\mathbf{r}} \frac{d^2\mathbf{k}}{(2\pi)^2}. \tag{18}$$

Actually, this function WF describes the relative motion of the Fermi Sea hole and the trion.

The behavior of the function radically different in the two energy regions:
1. $\Omega < -E_{tr} - F$
2. $-E_{tr} - F < \Omega < -E_{tr}$.

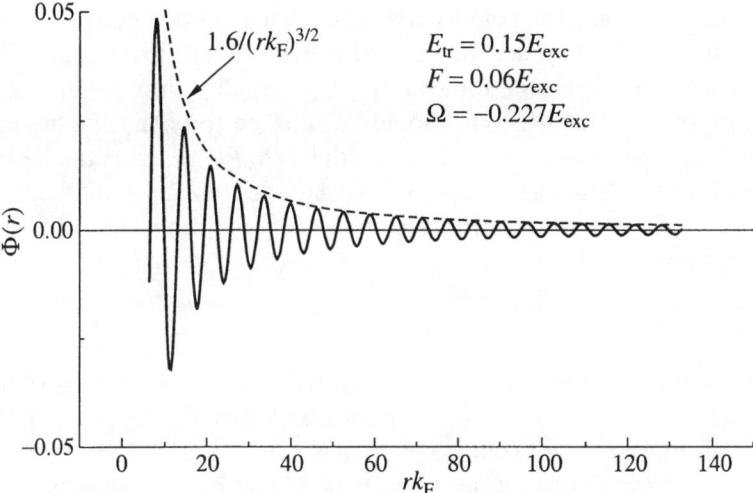

*Figure 6.* Coordinate dependence of wave function of the trion + Fermi Sea hole correlated state in the case of the Fermi Sea hole localization.

In the first region of energies the function (18) is real and, oscillating, it approaches to zero in the limit $|\mathbf{r}| \to \infty$ as $|\mathbf{r}|^{-3/2}$ where $|\mathbf{r}|$ is a distance between the Fermi Sea hole and the trion. Let us stress that, due to the strong decrease with distance, the function can be normalized to unity. *It corresponds to the Fermi Sea hole localized around the trion.* The behavior described above is illustrated in Fig. 6.

By contrast to this behavior, the function (18) in the second region is a complex

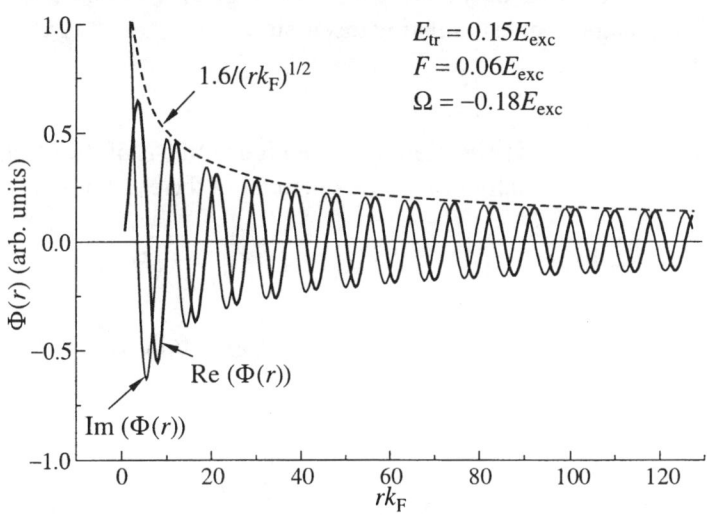

*Figure 7.* Coordinate dependence of wave function of the trion + Fermi Sea hole correlated state for the delocolized Fermi Sea hole.

function and its real and imaginary parts decrease with distance as $|\mathbf{r}|^{-1/2}$. In order to find the function in this energy region we need to define the integral (18) adding to denominator of expression for $Tr(\mathbf{k})$ (18) an negative infinitesimal imaginary term, $-i\gamma$. The sign of the term should be selected to obtain the hole running out from the photo-excited trion[2]. *The function (18) for the second energy region corresponds to the delocalized hole.* These states describe the photo-excited trion and the Fermi Sea hole that is running away from the trion. The behavior of the function is shown in Fig. 7.

## 4. Conclusions

In vicinity of the exciton and trion resonances the process of exciton – trion mutual transformation with creation of the hole in the electron Fermi Sea plays crucial role in optics of QWs filled with free electrons.

The trion absorption spectrum consists of a band Fermi energy wide and the narrow peak at low energy side. The peak corresponds to correlated state of the trion and the hole in Fermi Sea.

The sharp trion line was observed experimentally [7]. It can be supposed to be due to the trion–Fermi Sea hole correlated state discussed above.

## Acknowledgements

This work was supported in part by the Russian Foundation for Basic Research (Grant No. 02-02-17610), by Program "Low-Dimensional Quantum Structures" of the Presidium of the Russian Academy of Sciences and by Programs of the Ministry of the Science, Technology and Industry of Russian Federation.

It is my pleasure to thank my colleagues, Dr. V.P. Kochereshko and Dr. D.R. Yakovlev, for very helpful and stimulating discussions.

## Appendix

Let us consider Schrödinger equation for Green function, $G$, of 2D electron interacting with short-range axially symmetric potential of the exciton, $V(r)$:

$$\left(\Omega - \frac{\mathbf{p}^2}{2\bar{m}}\right) G\left(\mathbf{r}, \mathbf{r}' | E\right) - V(r) G\left(\mathbf{r}, \mathbf{r}' | E\right) = \delta\left(\mathbf{r} - \mathbf{r}'\right). \tag{A.1}$$

Scattering amplitude, $T\left(\mathbf{r}, \mathbf{r}' | E\right)$, is defined by the following symbolic equation:

$$G = g + gTg,$$

---

[2] Generally, we can consider our system inserted in a sphere of a large radius R. In this case we have to construct our WF combining the running away and incident cylindrical waves. Normalizing of the function to unity, due to its $|\mathbf{r}|^{-1/2}$ dependence on $\mathbf{r}$ we will get the WF amplitude proportional to $1/\sqrt{R}$.

where the free-electron Green function is

$$g(\mathbf{r}-\mathbf{r}'|\Omega) \equiv \int_0^\infty \frac{d^2k}{(2\pi)^2} \frac{\exp(i\mathbf{k}\cdot(\mathbf{r}-\mathbf{r}'))}{\Omega - \frac{k^2}{2\bar{m}}}. \qquad (A.2)$$

Then a symbolic expansion for the scattering amplitude, $T(\mathbf{r},\mathbf{r}')$, is

$$T = V + VgV + VgVgV + \ldots = V + VgT. \qquad (A.3)$$

For $\delta$-function-like potential

$$V(r) = \lambda \cdot \delta(\mathbf{r})$$

with a "potential strength" $\lambda \equiv 2\pi \int_0^\infty V(r)\, r dr$ the solution of Eq. (A.3) for $T$ is

$$(\mathbf{k}|T(E)|\mathbf{k}') = \frac{1}{\frac{1}{\lambda} - g(0|\Omega)}. \qquad (A.4)$$

However, the integral in Eq. (4) diverges on the upper limit at $|\mathbf{r}-\mathbf{r}'| \to 0$. It is due to neglect of real behavior of the potential at small distances. In order to take into account the issue, we have to replace $g(0|\Omega)$ with

$$\tilde{g}(0|\Omega) = \int_{k<\tilde{k}} \frac{d^2k}{(2\pi)^2} \frac{1}{\Omega - \frac{k^2}{2\bar{m}}} \qquad (A.5)$$

where $\tilde{k}$ is a cut off momentum that is about the inverse radius of the potential.

Performing integration in Eq. 5, we have

$$\tilde{g}(0|\Omega) = \frac{\bar{m}}{2\pi} \ln\left(\frac{-\Omega}{\tilde{E}}\right).$$

Here the cut off energy is $\tilde{E} = \frac{\tilde{k}^2}{2\bar{m}}$. Obtaining the expression for $\tilde{g}(0|\Omega)$, we supposed the actual energy values, $\Omega$, being much less than $\tilde{E}$.

Therefore we have for the scattering amplitude

$$(\mathbf{k}|T(\Omega)|\mathbf{k}') = T(\Omega) = \frac{2\pi}{\bar{m}} \frac{1}{\ln\left(\frac{\tilde{E}\, e^{2\pi/\bar{m}\lambda}}{-\Omega}\right)}. \qquad (A.6)$$

For attracting potential (singlet state of trion in our case) we have $\lambda < 0$ and the scattering amplitude has a pole at small negative energy:

$$\Omega_0 \equiv -E_{\text{binding}} = -\tilde{E} e^{-2\pi/\bar{m}|\lambda|}. \qquad (A.7)$$

The pole gives binding energy of the particle bound state in potential $V$. The result is valid for $E_{\text{binding}} \ll \tilde{E}$ e.g. for $\frac{|\lambda|\bar{m}}{2\pi} < 1$. In the case of trion in singlet state, the characteristic length, $\tilde{k}^{-1}$, is of order of the exciton radius and the characteristic energy, $\tilde{E}$, is of order of the exciton binding energy, $E_{\text{exc}}$ and $E_{\text{binding}} = E_{\text{tr}}$. Therefor, the first of equations (9) for the scattering amplitude is valid as soon as the trion binding energy is much less than the exciton binding energy. Comparing Eq. (A.7) with Eq (7), we have that

$$e^{2\pi/\bar{m}|\lambda|} \approx E_{\text{exc}}/E_{\text{tr}}. \qquad (A.8)$$

For repulsive potential (triplet state of trion), we have $\lambda > 0$. Using Eq. (A.8) and Eq. (A.6), we obtain the second expression of (9).

Generally, the value of the parameter $\lambda$ for triplet state differs from the absolute value of $\lambda$ for singlet state. However, having in mind the weak logarithmic dependence of the scattering amplitude on parameters and energy, we can consider the second equation of (9) for the scattering amplitude as a reasonable approximation. It should be stressed that, independently on the value of parameters, the logarithmic behavior of the both of the scattering amplitudes (9) are true for small energy values when $\Omega \ll \tilde{E}$.

## References

1. Lampert, M.A. (1958) *Phys. Rev. Lett.* **1**, 450.
2. Kheng, K., Cox, R.T., Merle d'Aubigne, Y., Bassani, F., Seminadayar, K., and Tatarenko, S. (1993) *Phys. Rev. Lett.* **71**, 1752.
3. Finkelstein, G., Shtrikman H., and Bar-Joseph, I. (1995) *Phys. Rev. Lett.* **74**, 976.
4. Shields, A.J., Pepper, M., Ritchie, D.A., Simmons M.Y., and Jones, G.A. (1995) *Phys. Rev.* **B51**, 18049.
5. Finkelstein, G., Shtrikman, H., and Bar-Joseph, I. (1996) *Phys. Rev.* **B53**, R1709.
6. Astakhov, G.V., Yakovlev, D.R., Kochereshko, V.P., Ossau, W., Nürnberger, J., Faschinger, W., and Landwehr, G. (1999) *Phys. Rev.* **B60**, R8485.
7. Suris, R.A., Kochereshko, V.P., Astakhov, G.V., Yakovlev, D.R., Ossau, W., Nürnberger, J., Faschinger, W., Landwehr, G., Wojtowicz, T., Karczewski, G., and Kossut, J. (2001) *phys. stat. sol. (b)* **227**, 343–352.

# COMBINED EXCITON-ELECTRON OPTICAL PROCESSES IN OPTICAL SPECTRA OF MODULATION DOPED QWs

V.P. KOCHERESHKO, D.R. YAKOVLEV†, R.A. SURIS
*Ioffe Physico-Technical Institute, RAS,*
*St Petersburg, Russia*
† *Physikalisches Institut der Universität Würzburg,*
*97074 Würzburg, Germany*

G.V. ASTAKHOV, W. FASCHINGER, W. OSSAU, G. LANDWEHR
*Physikalisches Institut der Universität Würzburg,*
*97074 Würzburg, Germany*

T. WOJTOWICZ, G. KARCZEWSKI, J. KOSSUT
*Institute of Physics, Polish Academy of Sciences,*
*PL-02608 Warsaw, Poland*

Abstract. We present a detailed study of combined exciton–electron processes in modulation doped quantum well structures contain a 2DEG of low density. We observed resonance exciton–electron processes in magnetic fields reveal in a separate narrow spectral lines, as well as nonresonant processes reveal in the exciton line broadening. Combine processes with participate of the excitons have been observed as well as with participate of trions (negatively charged exciton–electron complexes).

Key words: semiconductor nanostructures, excitons, two dimensional electron gas, trions

## 1. Introduction

Elementary processes such as photoexcitation of electrons between discrete levels of impurity centers, photocreation and radiative annihilation of an exciton, exciton scattering by impurity and est. are usually under consideration in semiconductor optics. In addition to such simplest processes one can consider combined processes, which involve several elemental ones. One of the known examples of such combined process is an exciton absorption in an indirect gap semiconductor in which the exciton photocreation accompanied by phonon absorption.

We should subdivide resonant and nonresonant combined optical processes. The nonresonant processes are characterized by continuum energy spectrum, it is, for example, an exciton absorption (reflection) line broadening due to the exciton–

phonon scattering. Resonant combined processes, which characterized by a discrete energy spectrum, are under our special interest. An example of such resonant process could be Raman scattering.

In a doped semiconductors and semiconductor heterostructures which contain 2DEG new elemental and combined processes with participation of additional electrons could be observed. For example, optical transitions to bound exciton–electron states (trions) were found recently in quantum well structures [1]. Since that time many papers devoted to trions have been published. The trions were observed in a number of QW structures based on different semiconductor compounds, such as, e.g., CdTe/CdZnTe, CdTe/CdMgTe, CdTe/CdMnTe, GaAs/AlGaAs, ZnSe/ZnMgSSe [2–4]. Negatively charged trions as well as positively charged trions, related to the heavy holes as well as to the light holes were found and studied experimentally [5, 6]. Trion triplet state, in addition to normally observed singlet state, was observed in high magnetic fields [5]. Pico- and femto-second trion dynamics in the presence of magnetic fields was studied [7]. In support of the experimental studies theoretical papers devoted to trions appear [8, 9].

In addition to bound exciton–electron states, resonance combined exciton–electron process involving three-particles have been observed in the presence of magnetic fields [10]. Such combined processes were called *Combined Exciton Cyclotron Resonances* (ExCR), meaning that in an external magnetic field an incident photon creates an exciton in its ground state and simultaneously it excites one of the resident electrons from the lowest to one of the higher Landau levels. The energy of these transitions is equal to the sum of the exciton energy and a multiple of cyclotron energy.

Present paper is concern with systematic study of combined processes with participating of excitons and electrons in modulation doped quantum well structures.

## 2. Experiment and Used Structures

Optical spectra were measured from two types of structures grown by molecular-beam epitaxy on (100)-oriented GaAs substrates. We used modulation-doped CdTe/(Cd,Mg)Te and ZnSe/(Zn,Mg)(S,Se) QW structures with a 2DEG of low and moderate density (the electron density varied from $n_e = 10^{10}$ to $10^{12}$ cm$^{-2}$). Typical structure contains a single QW of 80–120 Å width and was modulation doped in the barrier layer at a 100 Å distance from the QW (see Fig. 1). A special design of the structures made possible to control the electron concentration keeping all other QW parameters (QW width, barrier height, background impurity concentration, etc.) constant [11]. We studied reflectivity, photoluminescence and photoluminescence excitation spectra in magnetic fields applied in the Faraday geometry in $\sigma^+$ and $\sigma^-$ circular polarizations.

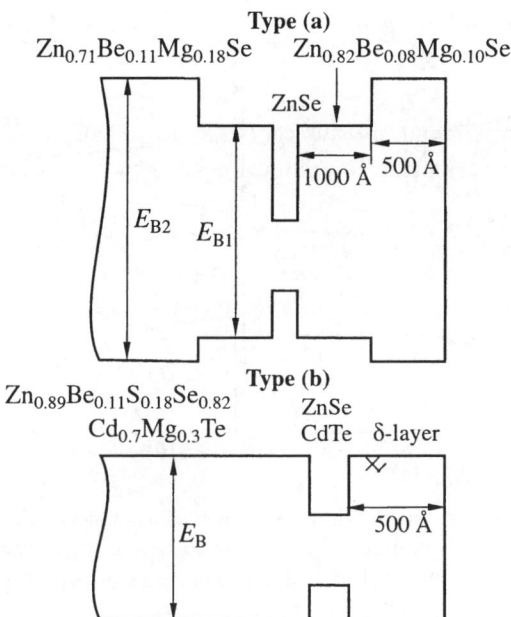

*Figure 1.* Schema of the structures studied. Type (a) is ZnSe/Zn$_{0.82}$Be$_{0.08}$Mg$_{0.10}$Se single QW surrounded by additional Zn$_{0.71}$Be$_{0.11}$Mg$_{0.18}$Se barrier. Type (b) is ZnSe/Zn$_{0.89}$Mg$_{0.11}$S$_{0.18}$Se$_{0.82}$ or CtTe/Cd$_{0.7}$Mg$_{0.3}$Te modulation doped single QW.

## 3. Combined Processes Reveal in Reflectivity

Figure 2(a) shows reflectivity spectra taken from 80 Å wide CdTe/Cd$_{0.7}$Mg$_{0.3}$Te single QWs with different electron concentration taken at 1.6 K. The upper spectrum corresponds to nominally undoped structure with the electron concentration in the QW less than $10^{10}$ cm$^{-2}$. The lowest spectrum corresponds to the electron concentration of $1.8 \times 10^{11}$ cm$^{-2}$. Only the sharp exciton reflectivity line X reveals in the spectrum taken from the undoped structure. With the electron concentration increasing, a negatively charged exciton line (T) appears in the spectra. Such states being bound exciton–electron complexes, can be observed in modulation-doped QWs with relatively low electron concentration. The exciton line becomes very broad at high enough electron concentrations and is difficult to distinguish.

We also observed similar spectra in ZnSe — based QW structures. Figure 2(b) shows a set of reflectivity spectra taken from modulation doped QW structures ZnSe/Zn$_{0.89}$Mg$_{0.11}$S$_{0.18}$Se$_{0.82}$ with different electron concentrations. Due to larger Coulomb energy the effects of the exciton–electron interactions are even more pronounced in these structures as compared to CdTe — based QW structures.

One can observe a very large width of the X line in the PLE spectrum, which is broadened at the high-energy site. This high-energy tail of the line appears due to a combined exciton–electron process in which the photo-creation of an exciton

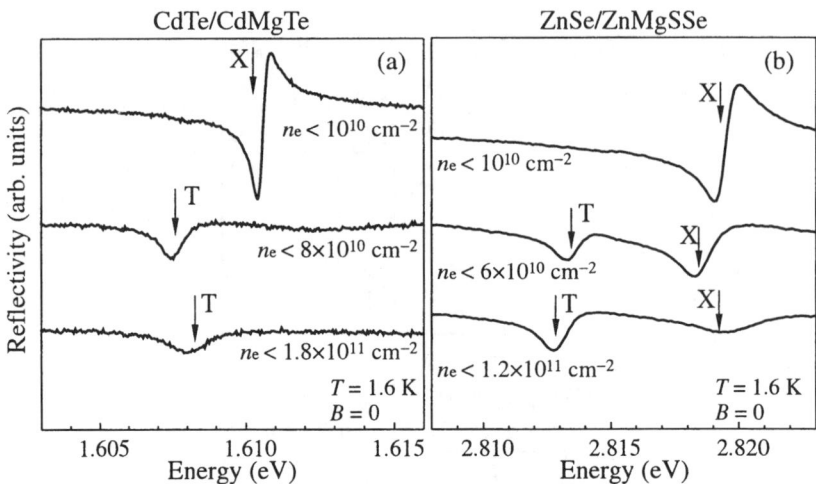

*Figure 2.* (a) Reflectivity spectra taken at 1.6 K from modulation-doped CdTe/Cd$_{0.7}$Mg$_{0.3}$Te 80 Å wide single QW structures with different electron concentrations $n_e$ in the QW. (b) Reflectivity spectra taken from modulation doped ZnSe/Zn$_{0.89}$Mg$_{0.11}$S$_{0.18}$Se$_{0.82}$ 80 Å single QW structures with different electron concentration.

is accompanied by the excitation of an additional electron from states below to states above the Fermi level of the 2DEG. A similar combined process in which the photo-generation of trions is accompanied by the excitation of an additional electron also results in an additional broadening of the trion line.

Strong modifications of the reflectivity and PLE spectra are observed in magnetic fields. Figure 3(a) shows these spectra taken from the QW structure with the electron concentration $8 \times 10^{10}$ cm$^{-2}$ in the magnetic field of 3 T. The trion reflectivity/PLE line becomes completely circularly polarized in the presence of the magnetic fields. It can be observed in $\sigma^+$ polarization only. This strong polarization of the line reflects the fact that the trion ground state is a singlet. Consequently, in the magnetic field, in which all the background electrons are spin-polarized, the trion can be created for one circular polarization of the incident photon only, namely, for those in which the background electron and the electron in the exciton have opposite spins.

Similar behavior is observed for ZnSe based QW structures. Figure 3(b) shows the reflectivity spectra taken from 80 Å ZnSe/ZnMgSSe QW structure with the electron concentration of $1.5 \times 10^{11}$ cm$^{-2}$ in the magnetic field $B = 4.5$ T and $B = 8$ T. The exciton (X), the trion (T) and the (TrCR) line were revealed in these spectra.

We used the effect of the strong circular polarization of the trion reflection line in the magnetic field to determine the electron concentration in the QW. The idea is based on the fact that the circular polarization of the trion reflectivity is controlled by the spin polarization of 2DEG. On the other hand, the polarization

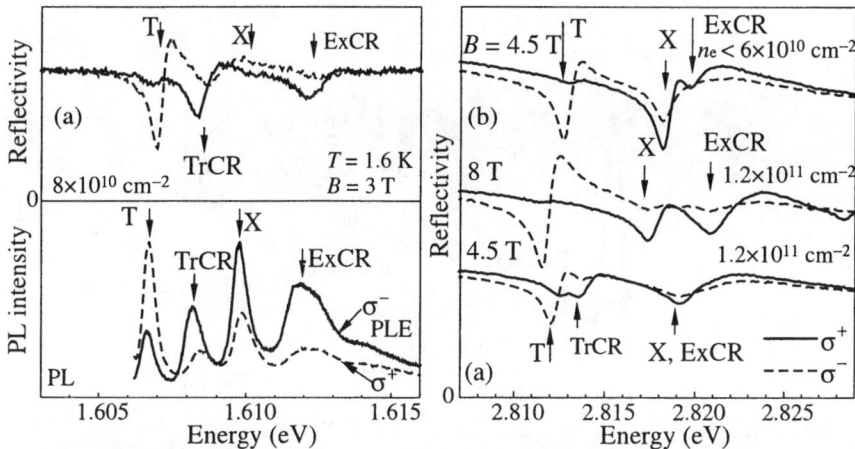

*Figure 3.* (a) Reflectivity (upper panel) and PLE (low panel) spectra taken from CdTe/Cd$_{0.7}$Mg$_{0.3}$Te SQW with the electron concentration of $8 \times 10^{10}$ cm$^{-2}$ in a magnetic field of 3 T in $\sigma^+$ (dotted) and $\sigma^-$ (solid) circular polarizations. (b) Reflectivity spectra taken from ZnSe/Zn$_{0.89}$Mg$_{0.11}$S$_{0.18}$Se$_{0.82}$ single QW with the electron concentration of $1.2 \times 10^{11}$ cm$^{-2}$ and $6 \times 10^{10}$ cm$^{-2}$ in the magnetic field of 4.5 T and 8 T in $\sigma^+$ (dotted) and $\sigma^-$ (solid) circular polarizations.

of 2DEG appears when the lowest Landau level crosses the Fermi energy, i.e. at filling factors $\nu < 1$. (The filling factor defined as $\nu = n_e c\hbar/eB$, with $e$ is the electron charge, $B$ is the magnetic field, and $n_e$ is the electron concentration). Thus, the polarization of the trion feature appears under the same conditions. We performed a fit of the calculated polarization of 2DEG (assuming the Fermi–Dirac distribution of the electrons between spin-split sublevels) to this experimentally observed circular polarization of the trion reflectivity line. Such fits allowed us to determine the electron concentration residing in the QW [12].

In addition to the exciton (X) and the trion (T) reflectivity lines two new lines appear in external magnetic fields, i.e., the *Combined Exciton Cyclotron Resonance line* (ExCR) and the *Combined Trion Cyclotron Resonance line* (TrCR). The ExCR line appears in the spectra as a transformation of the high-energy tail of the exciton line with increasing magnetic field. In magnetic fields the exciton line becomes narrow and the high-energy tail of the line splits into several ExCR lines (see Fig. 4). We attribute these lines with combined exciton–electron processes in which the photo-creation of the exciton is accompanied by the excitation of an additional electron from the lowest filled to an empty Landau levels.

The TrCR line is observed between the exciton and the trion reflectivity features. This line can be observed in the spectra of QWs with rather high electron concentrations and at filling factors between 2 and 1 only [13]. The TrCR line appears as a "splitting" of a broad band at $B = 0$ into two lines (trion line and TrCR line) when a magnetic field is applied. The magnetic field dependence of the observed line positions is presented in Fig. 5(b), (c).

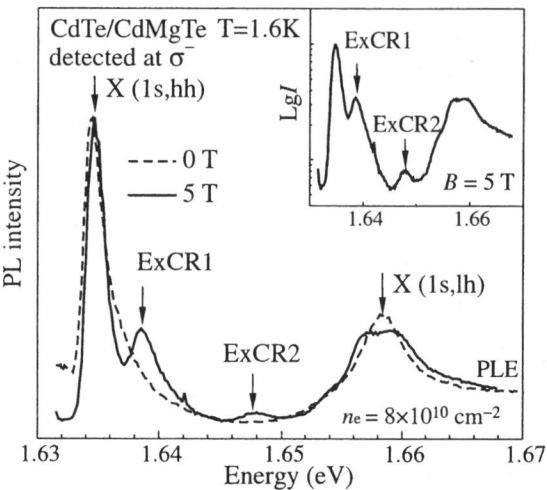

*Figure 4.* Modification of the high-energy tail of the exciton absorption line in magnetic field.

The exciton line and the trion line display a normal diamagnetic behavior at high magnetic fields. The ExCR line shifts linearly with a slope of 1.1 meV/T in CdTe and of 0.89 meV/T in ZnSe based QWs. The expected slope of the ExCR line for the studied QW is equal to $\hbar\omega_c[1 + m_e/(m_e + m_h)] = 1.35$ meV/T in CdTe and 0.88 meV/T in ZnSe, where $\hbar\omega_c = 1.16$ meV/T is the electron cyclotron energy, $m_e = 0.1 m_0$ and $m_h = 0.5 m_0$ in CdTe and $\hbar\omega_c = 0.77$ meV/T is the electron cyclotron energy, $m_e = 0.15 m_0$ and $m_h = 0.8 m_0$ in ZnSe correspondingly. We see an excellent agreement between the observed and theoretically expected values of the ExCR behavior.

The TrCR line is situated between the trion and exciton resonances (see Fig. 3a,b). It becomes resolvable from the trion resonance at a certain magnetic field corresponding to the 2DEG filling factor $\nu = 2$. The TrCR looses its intensity and disappears in higher magnetic fields when the filling factor becomes smaller than 1. The slope of the TrCR line of 0.6 meV/T in CdTe and 0.4 meV/T in ZnSe is equal to $\hbar\omega_c/2$ (0.58 meV/T in CdTe and 0.39 meV/T in ZnSe). In analogy with the ExCR line we attribute the TrCR line to a combined process in which the trion participates. Namely an incident photon creates a trion and promotes another electron to be excited between the Landau levels.

## 4. Discussion of the Reflectivity

The trion absorption (reflection) line is formed by the binding of one photo-created exciton and one electron from the 2DEG. In the initial state of this process we have an electron $e$ and in the final state we have a trion $Tr$. This means that an electron and the absorbed photon ($ph$) forms a trion:

$$e + ph \rightarrow Tr. \tag{1}$$

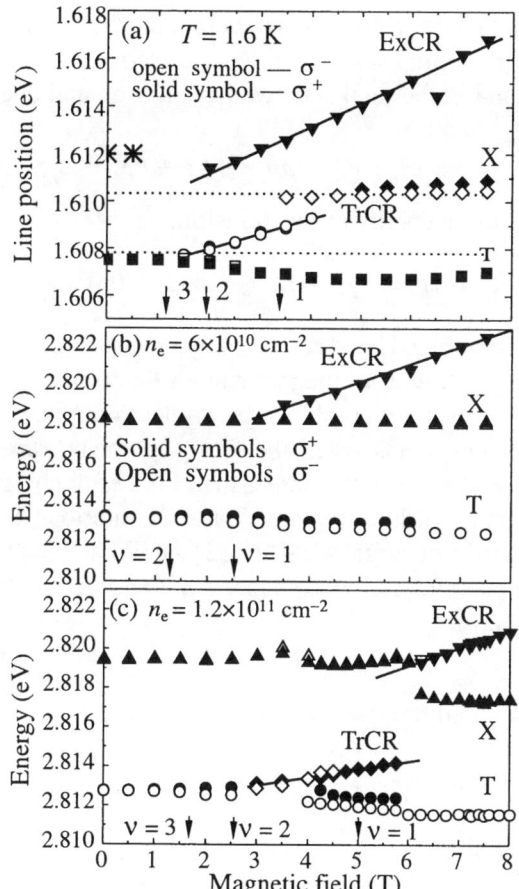

*Figure 5.* (a) Magnetic field dependencies of the energy position of the exciton X, the trion T, ExCR and TrCR lines for CdTe/Cd$_{0.7}$Mg$_{0.3}$Te single QW with the electron concentration $n_e = 8 \times 10^{10}$ cm$^{-2}$. Arrows indicate the values of filing factor. Open symbols correspond to $\sigma^-$ and closed to $\sigma^+$ polarization. (b), (c) Magnetic field dependencies of the energy position of the exciton X, trion X$^-$, ExCR and TrCR lines for ZnSe/Zn$_{0.89}$Mg$_{0.11}$S$_{0.18}$Se$_{0.82}$ SQW with the electron concentrations: $1.2 \times 10^{11}$ cm$^{-2}$ and $0.6 \times 10^{11}$ cm$^{-2}$. Arrows indicate the values of the filing factor. Open symbols correspond to $\sigma^-$ and closed to $\sigma^+$ polarization.

The energy of this transition is therefore:

$$E_{ph} = E_{Tr} - E_e. \qquad (2)$$

As electrons in the 2DEG have energies (measured from the bottom of conduction band) in the range from zero to the Fermi energy: $0 \leq E_e \leq E_F$, the trion absorption (reflection) band should be spread within the energies from $(E_{Tr} - E_F)$ to $E_{Tr}$.

One can easily see that the same energy range has to correspond to the trion combined process in which an incident photon creates a trion and simultaneously excites an additional electron from states below the Fermi level to states above that. This means, in the initial state we have two electrons $e1$ and $e2$ with energies $0 \leq E_{e1(e2)} \leq E_F$ and in the final state we have a trion and one electron $e3$ with an energy $E_{e3} \geq E_F$:

$$e1 + e2 + ph \rightarrow Tr + e3. \quad (3)$$

Therefore, the photon energy of this transition is

$$E_{ph} = E_{Tr} + E_{e3} - (E_{e1} + E_{e2}). \quad (4)$$

Hence, the trion combined band can be spread in the energy range from $(E_{Tr} - E_F)$ to $+\infty$. So, in the absence of magnetic fields the trion absorption (reflection) band overlaps with the band of combined trion–electron process.

In external magnetic fields when the filling factor becomes less than 2 (or when the cyclotron energy becomes larger than the Fermi energy) these the trion and TrCR bands cannot overlap anymore. At these conditions the trion band shifts with magnetic fields to low energies as $-\hbar\omega_c/2$ [see Eq. (2)] and the trion combined band shifts to high energies as $+\hbar\omega_c/2$ [see Eq. (4)]. For the trion transitions in magnetic fields we have:

$$E_{ph} = E_{Tr} - \hbar\omega_c/2. \quad (5)$$

And for the combined transitions in the presence of a magnetic field in which the photo-creation of the trion is accompanied by the excitation of an additional electron between Landau levels we have:

$$E_{ph} = E_{Tr} + (3/2)\hbar\omega_c - [(1/2)\hbar\omega_c + (1/2)\hbar\omega_c] == E_{Tr} + \hbar\omega_c/2. \quad (6)$$

Therefore, the energy distance between these two lines increases as $\hbar\omega_c$. In high magnetic fields the combined process line disappears from the spectra at the filling factor $\nu \approx 1$, because at $\nu \leq 1$ the probability of finding two electrons in one place becomes small. The trion line at these fields starts to shift to high energies due to diamagnetic effects.

So, resonant lines corresponding to combined optical transitions in which the photo-generation of excitons as well as of trions is accompanied by the excitation of an additional electron from states below- to states above the Fermi level were observed and analyzed as a function of the magnetic field strength.

## 5. Combined Processes Reveal in Photoluminescence

In the PL spectrum (Fig. 6) of the studied structures a wide band with two maxima was observed at zero magnetic field. The lowest in energy maximum of the line is located close to the trion resonance energy in weakly doped structure ($\hbar\Omega_{tr}$). The main, high energy, maximum is located at the energy ($\hbar\Omega_{tr} - E_F$), (where $E_F$ is the Fermi energy of the 2D electron gas).

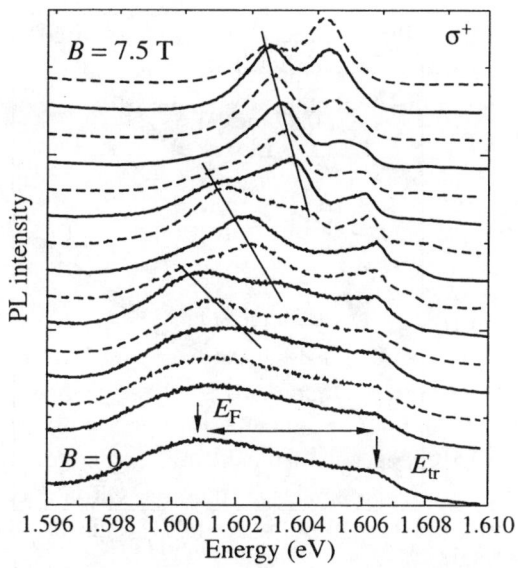

*Figure 6.* Photoluminescence spectra taken from a 120 Å CdTe/Cd$_{0.7}$Mg$_{0.3}$Te SQW with electron concentration of $2.5 \times 10^{11}$ cm$^{-2}$ in magnetic fields from zero to 7.5 T.

In the presence of external magnetic fields the PL spectra modified strongly. First, the wide PL band starts narrowing and its low energy maximum moves to high energy one (i.e. to $\hbar\Omega_{tr}$). A weakly pronounced structure of maxima and minima appears on the line contour. This structure depends strongly on magnetic fields: the maxima shift to the low energies, the intensity of the maxima increases initially and then goes down at higher fields. At high enough magnetic fields instead of the wide PL band an arbitrary narrow PL line was observed. Such behavior was observed in CdTe-based structures as well as in ZnSe-based structures. Figure 6 illustrates the described above on an example of CdTe/CdMgTe SQW with electron density $2.5 \times 10^{11}$ cm$^{-2}$.

The PL spectra were studied also from ZnSe/Zn$_{0.71}$Be$_{0.11}$Mg$_{0.18}$Se structures consists a 67-Å-thick ZnSe SQW embedded between 1000-Å-thick Zn$_{0.82}$Be$_{0.08}$Mg$_{0.10}$Se barriers. The structures contain a $\delta$-layer at a distance of 100 Å from the QW doped by Chlorine.

In Fig. 7 we show an example of the magnetic field dependence of the main (i.e. most intensive) PL maximum for ZnSe/ZnBeMgSe SQW structure. Here we can see that the main PL maximum moves to the high energies and has leaps at integer filling-factors. The similar behavior of the main maximum was observed also in QW structures based on GaAs [14, 17], CdTe [15] and ZnSe [16, 18].

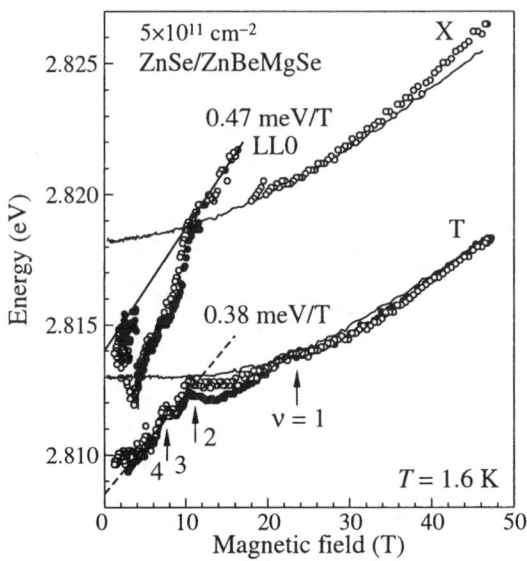

*Figure 7.* Energies of PL maximums vs magnetic field for an 67-Å-thick ZnSe/ $Zn_{0.71}Be_{0.11}Mg_{0.18}Se$ SQW $5 \times 10^{11}$ cm$^{-2}$ detected in $\sigma^+$ (open circles) and $\sigma^-$ (solid circles) polarizations. (X) and (T) show exciton and trion lines in high magnetic fields [18]. Their diamagnetic shifts shown by solid lines are moved to lower energies by 7.7 meV.

## 6. Discussion of the Photoluminescence

We explain these modifications of the PL spectra by the following way: In the absence of magnetic fields we have a trion in an initial state. The binding energy of this trion, of cause, depends on the electron density because of screening, but in 2D case a weakly bound state is still remains. After the trion annihilation we have a photon plus one electron in the final state: $Tr \to \hbar\omega_{tr} + e^*$. This residual electron $e^*$ can be placed the empty states above the Fermi level only. Consequently, the energy of the emitted photon is: $\hbar\omega_{tr} = E_{Tr} - E^*$, here: $E^* \in (E_F, \infty)$ is the energy of the residual electron, $E_{Tr}$ is the energy of the trion in the initial state. Obviously, in weakly doped samples the energy of the photon emitted in the trion annihilation ($\hbar\omega_{tr}$) is equal to the trion energy ($E_{tr}$). Although the residual electron after the trion annihilation can has any energy from the Fermi energy to infinity, the probability of such process decreases fast with increasing the electron energy. So, the maximum of the PL intensity is expected to be located at the energies close to ($E_{Tr} - E_F$).

Therefore, the PL maximum at the trion annihilation in heavily doped samples is shifted to lower energies from the trion PL line position in weakly doped samples. The value of this shift is of the order of the Fermi energy.

In the presence of magnetic fields the energy of the emitted photon is $\hbar\omega_c = E_{Tr}(H) - E^*$, here: $E^* = (N + 1/2)\hbar\omega_c$, $N$ is integer. In the range of magnetic

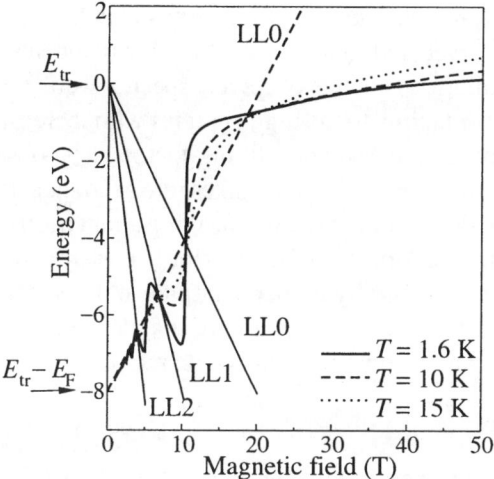

*Figure 8.* Scheme of magnetic field dependence of the main PL peak position, which corresponds to the case of the trion annihilation when the residual electron is left on the Fermy level. LL0, LL1, ... shows the Landau level fan.

fields when the magnetic length is smaller the trion radius but higher the exciton radius the trion energy goes up as: $E_{Tr}(H) \approx E_{Tr}(0) + (1/2)\hbar\omega_c$.

So that, at low temperature a fan of Landau PL lines should be observed shifting linearly to *low energies* from the trion resonance with increasing magnetic fields (these lines are so-called "shake-up" lines).

As the upper Landau levels become empty the maximum of the PL jumps closer to the trion energy. It shifts jumping from the higher Landau levels to the lower Landau levels as higher Landau levels become empty until it reaches the trion energy. It happens just when the filling-factor is equal to 2. These leaps are washed out when the temperature of the 2DEG increases. The similar behavior will be observed due to a broadening of the Landau levels caused by a scattering processes or inhomogenity.

At sufficiently high temperature the jumps will disappear completely and the PL line will shifts to the high energies as $(1/2)\hbar\omega_c$. This situation is illustrated in Fig. 8 for three different temperatures and for the ZnSe/ZnBeMgSe sample with the same parameters as for the presented in the Fig. 7 (see [18]). Qualitative coincidence of Fig. 8 and Fig. 7 is clearly seen.

So, the observed experimental behavior can be interpreted in a model, which takes into account a dependence of Fermi level on magnetic fields.

In conclusion, photoluminescence, photoluminescence excitation and reflectivity spectra of modulation-doped QW structures based on II–VI semiconductors (CdTe/CdMgTe, ZnSe/ZnMgSSe and ZnSe/ZnBeMgSe) were studied at high magnetic fields. Resonant reflectivity lines corresponding to combined optical transitions in which the photo-generation of excitons as well as of trions is accompanied by the excitation of an additional electron from states below- to states above the Fermi level were observed and analyzed as a function of the magnetic field strength. The following peculiarities of the observed spectra were found in PL spectra: (i) shift of the PL line maxima to the high energies with increasing magnetic fields, (ii) leaps in this dependence at integer filling-factors. The observed experimental behavior is interpreted in a model, which takes into account a dependence of Fermi level on magnetic fields. Although we are speaking about trion processes we should have in mine that at high 2D electron densities the trions are screened and we have to consider them as an exponentially weak bound state of three particles. Nevertheless the processes considered above will conserve in a dense 2DEG until the excitons are unscreened and can be considered as whole particles.

**Acknowledgements**

Financially it was supported in part by grants: RFBR 01-02-16990 and "Nanostructures" of Russian Ministry of Sciences and Russian Academy of Sciences.

**References**

1. Kheng, K., Cox, R.T., Merle d'Aubigne, Y., Bassani, F., Saminadayar, K., and Tatarenko, S. (1993) *Phys. Rev. Lett.* **71**, 1752.
2. Kheng, K., Cox, R.T., Kochereshko, V.P., Saminadayar, K., Tatarenko, S., Bassani, F., and Franciosi, A. (1994) *Superlatt Microstruct.* **15**, 253.
3. Finkelstein, G., Shtrikman, H., and Bar-Joseph, I. (1995) *Phys. Rev. Lett.* **74**, 976.
4. Shilds, A.J., Pepper, M., Ritchie, D.A., Simmons, M.Y., and Jones, G.A.C. (1995) *Phys. Rev. B* **51**, 18049.
5. Finkelstein, G., Shtrikman, H., and Bar-Joseph, I. (1996) *Phys. Rev.* **B53**, R1709.
6. Haury, A., Arnoult, A., Chitta, V.A., Cibert, J., Merle d'Aubigne, Y., Tatarenko, S., and Wasiela, A., (1998) *Superlatt. Microstrict.* **23**, 1097.
7. Finkelshtein, G., Umansky, V., Bar-Joseph, I., Culin, I., Haacke, S., Ganiere, J.-D., and Deveaud, B. (1998) *Phys. Rev. B* **58**, 12637.
8. Stebe, B., Feddi, E., Anane, A., and Dujardin, F. (1998) *Phys. Rev. B* **58**, 9926.
9. Dzyubenko, A.B. and Sivachenko, A.Yu. (1993) *Phys. Rev. B* **48**, 14690.
10. Yakovlev, D.R., Kochereshko, V.P., Suris, R.A., Schenk, H., Ossau, W., Waag, A., Landwehr, G., Christianen, P.C.M., and Maan, J.C. (1997) *Phys. Rev. Lett.* **79**, 3974.
11. Wojtowicz, T., Kutrowski, M., Karchewski, G., and Kossut, J. (1998) *Acta Physica Polonica A* **94**, 199.
12. Astakhov, G.V., Kochereshko, V.P., Yakovlev, D.R., Ossau, W., Nürnberger, J., Faschinger, W., Landwehr, G., Wojtowicz, T., Karczewski, G., Kossut, J. (2002) *Phys. Rev. B* **65**, 115310.
13. Ossau, W., Kochereshko, V.P., Astakhov, G.V., Yakovlev, D.R., Landwehr, G., Wojtowicz, T., Karczewski, G., and Kossut, J. (2001) *Physica B* **298**, 315–319.
14. Gekhtman, D., Cohen, E., Arza Ron, and Pfeiffer, L.N. (1996) *Phys. Rev.* **B54**, 10320.
15. Teran, F.J., Sadowski, M.L., Potemski, M., Karczewski, G., Mackowski, S. and Jaroszinski, J. (1998) *Physica* **B256-258**, 577.
16. Crooker, S.A., Rickel, D.G., Lyo, S.K., Samarth, N., and Awschalom, D.D. (1999) *Phys. Rev.* **B60**, R2173.
17. Rashba, E.I. and Sturge, M. (2001) *Phys. Rev.* **B63**, 045305.
18. Ossau, W., Yakovlev, D.R., Astakhov, G.V. et al. (2002) *Physica E*

# II–VI QUANTUM WELLS WITH HIGH CARRIER DENSITIES AND IN HIGH MAGNETIC FIELDS

D.R. YAKOVLEV[1,2,3], G.V. ASTAKHOV[2,3], W. OSSAU[2], S.A. CROOKER[4] AND A. WAAG[5]
[1] *Experimentelle Physik 2, Universität Dortmund,*
*44227 Dortmund, Germany*
[2] *Physikalisches Institut der Universität Würzburg,*
*97074 Würzburg, Germany*
[3] *Ioffe Physico-Technical Institute, Russian Academy of Sciences,*
*194021 St.Petersburg, Russia*
[4] *National High Magnetic Field Laboratory,*
*Los Alamos, New Mexico 87545, USA*
[5] *Abteilung Halbleiterphysik, Universität Ulm,*
*89081 Ulm, Germany*

Abstract. Optical properties of a two-dimensional electron gas in ZnSe/(Zn,Be,Mg)Se quantum well structures have been examined by means of photoluminescence and reflectivity techniques in external magnetic fields up to 50 T. For the studied structures, the Fermi energy of the two-dimensional electron gas falls in the range between the trion binding energy and the exciton binding energy, which keeps the dominating role of Coulombic interactions between electrons and photoexcited holes. Characteristic peculiarities of the optical spectra are discussed.

Key words:

## 1. Introduction

It is a textbook knowledge that the energy spectrum of a quantum well potential (e.g. in a semiconductor quantum well structure) is modified by the presence of free carriers. In the case of a single electron-hole pair in an undoped quantum well (QW), the ground state is an exciton (X), formed via the Coulomb interaction. In QWs with a dilute two-dimensional electron gas (2DEG), where the Fermi energy of the 2DEG taken from the bottom of conduction band $\varepsilon_F$ is much smaller than the exciton binding energy ($\varepsilon_F < 0.1\, E_B^X$), a negatively charged exciton complex can exist [1, 2]. Charged excitons, or trions (T), are constructed of two electrons and a hole (or two holes and an electron), and appear in optical spectra a few meV below the exciton resonance. The energy separation between T and X in the limit of zero electron concentration gives the trion binding energy $E_B^T$, which usually varies from 5 to 20% of the exciton binding energy depending on the thickness of the quantum well (see e.g. [3] and references therein). In QWs with a very high electron concentration, where the Fermi energy exceeds the exciton binding energy ($\varepsilon_F > E_B^X$), excitons are screened and the collective behavior of the correlated electrons becomes evident in the optical spectra [4, 5]. Emission lines broaden, exhibiting a linewidth determined by the Fermi energy. Also, a Fermi-edge singularity (FES) appears in optical spectra [6], resulting from the interaction of photocreated holes with the Fermi-sea of electrons, especially with electrons in vicinity of the Fermi energy.

The evolution of optical spectra from excitonic to FES, i.e. at electron densities where the Fermi energy falls in the range between the trion binding energy and the exciton binding energy ($E_B^X > \varepsilon_F > E_B^T$), is an interesting field for investigation, where a number of open questions remain. In this regime trion states are strongly modified due to screening, but excitons are still robust and their interaction with a 2DEG reveals new peculiarities in optical properties.

We will show in this paper that the commonly used scenario of optical spectra modification due to increasing electron density, which is developed to explain experimental appearances in III–V semiconductor heterostructures (GaAs, InGaAs, etc.), can not satisfactorily describe the evolution of optical spectra in ZnSe-based QWs with a stronger Coulomb interaction. The exciton binding energy in bulk ZnSe is 20 meV compared with 4.2 meV in GaAs and 10 meV in CdTe. We give here an experimental insight into modification of photoluminescence (PL) and reflectivity spectra of a modulation-doped ZnSe-based quantum well in high magnetic fields (to 50 T), and discuss the inconsistencies with existing models.

## 2. Experimentals

A modulation-doped QW structure (cb1040) and an undoped reference structure (cb1041) [7] were grown by molecular-beam epitaxy on (100)-oriented GaAs substrates. Each structure consists of a 67-Å-thick ZnSe single quantum well embedded between 1000-Å-thick $Zn_{0.82}Be_{0.08}Mg_{0.10}Se$ barriers. To prevent the loss of carriers escaping into the substrate and recombining at the surface, the structures were confined by additional 500-Å-thick $Zn_{0.71}Be_{0.11}Mg_{0.18}Se$ barriers. Both barrier materials are lattice matched to the GaAs substrate. The modulation-doped layer with Iodine donors is separated from the QW by a 100-Å-thick spacer layer. Two optical methods have been used to evaluate the 2DEG density: one is based on the properties of trion absorption (reflection) [8] and another one exploits the nonmonotonic behavior of optical spectra in vicinity of integer filling factors [9–12]. Both methods give the 2DEG density $n_e = 5 \times 10^{11}\,cm^{-2}$ for cb1040. This corresponds to a Fermi energy of $\varepsilon_F = 7.7$ meV, with electron effective mass $m_e = 0.15 m_0$. For comparison, the trion binding energy in this QW is $E_B^T = 5.3$ meV and the exciton binding energy is $E_B^X = 30$ meV [3, 12]. Therefore the condition $E_B^X > \varepsilon_F > E_B^T$ pertains to our experimental situation. The electron concentration in the reference sample cb1041 evaluated by the first method is $n_e = 3 \times 10^{10}\,\mathrm{cm}^{-2}$.

Photoluminescence and reflectivity spectra have been measured in external magnetic fields applied perpendicular to the QW plane (Faraday geometry). Circular polarized components of these spectra have been analyzed in order to get information on states with different spin configurations and spin orientations. Experiments in magnetic fields to 8 T have been performed in Würzburg in a split-coil superconducting solenoid and for the temperatures varied from 1.6 to 35 K. A long-pulse ($\sim$ 400-ms decay) magnet at the National High Magnetic Field Laboratory (Los Alamos, USA) was used in the experiments to 47 T and 1.6 K. Details of the pulsed magnet setup are given in Ref. [13]. Photoluminescence was excited with the UV lines of an Ar-ion laser (Würzburg) or by a He-Cd laser with a photon energy of 3.8 eV. In both cases the excitation energy exceeds the energy gaps of the barrier materials. Excitation density was kept below 10 W/cm$^2$, to exclude considerable heating of the 2DEG. In reflectivity experiments a halogen lamp was used as a light source.

## 3. Optical Spectra In The Absence Of Magnetic Fields

Let us first discuss properties of photoluminescence and reflectivity spectra in the absence of external magnetic fields. Typical spectra are given in Fig. 1.

The PL spectrum of the undoped sample consists of two narrow lines with a full-width at half-maximum (FWHM) of 1.2 meV. The linewidth at $T = 1.6$ K is due to inhomogeneous broadening caused by fluctuations of the QW width and alloy fluctuations of the barrier materials. An exciton line (X) has a peak at 2.826 eV and the trion line peak is shifted by 5.5 meV to lower energy from the exciton. Details of the line identification are given in Ref. [3]. It is based on observation of both resonances in reflectivity spectra at $T = 1.6$ K and fast disappearance of the trion state at higher temperatures of 30 K (see Fig. 1a).

The PL spectrum of a modulation-doped structure consists of a broad band with a maximum at 2.810 eV and with a full width at a half maximum (FWHM) of 7 meV (see Fig. 1b). The linewidth is very close to the value of the 2DEG Fermi energy 7.7 meV. This is typical for the emission spectra of modulation-doped QWs and is explained by the fact that all electrons from the Fermi sea can recombine with photoexcited holes. It is commonly suggested that the momentum conservation selection rules for optical transitions are partially relaxed in this case due to hole localization.

The reflectivity spectrum of the modulation-doped QW has a strong and narrow resonance at $T = 1.6$ K. We label it as C-line and the origin of this transition is

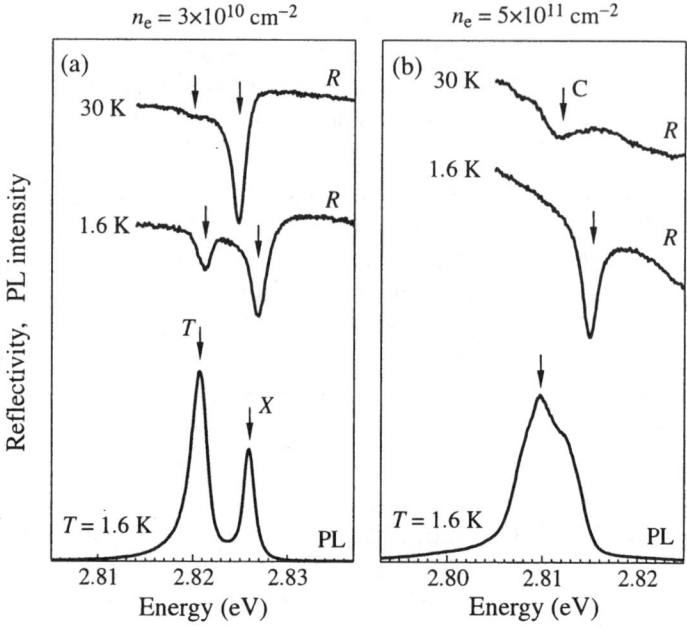

*Figure 1.* Photoluminescence (PL) and reflectivity (R) spectra of 67-Å-thick ZnSe/Zn$_{0.82}$Be$_{0.08}$Mg$_{0.10}$Se QWs, with different electron densities: (a) unintentionally doped reference sample with $n_e = 3 \times 10^{10}$ cm$^{-2}$, and (b) doped sample with $n_e = 5 \times 10^{11}$ cm$^{-2}$. $B = 0$ T.

the subject of investigation and discussion in this paper. The resonance frequency of the C-line at 2.815 eV is blue shifted with respect to the PL maximum by about 5 meV, a value comparable with the Fermi energy of 2DEG. Its position coincides with the high energy tail of the PL band. In modulation-doped QWs this spectral position is often ascribed to a "Fermi edge singularity (FES)". The FES is commonly associated with optical excitation of an electron from the valence band directly to the Fermi level in the conduction band and therefore is classified as a band-to-band absorption line. We will show below that this interpretation should be revised, at least for II–VI heterostructures with strong Coulombic interactions, i.e. in this regime where the exciton binding energy exceeds the Fermi energy. Thus, we do not apply a FES identification to the C-line.

Let us comment in more detail the properties of the C-line. It is rather surprising that the 2 meV width of this resonance is noticeably smaller than FWHM of the PL band. We would like to stress here that at $T = 1.6$ K the C-resonance looks very similar to the exciton resonance in undoped QWs (see e.g. [3]). Namely, it has the same linewidth and a very comparable oscillator strength (i.e. amplitude of the resonance). The different nature of both resonances is highlighted by their temperature dependences shown in Fig. 2. The exciton resonance in the undoped QW shows no temperature broadening for $T < 35$ K (exciton – LO-phonon

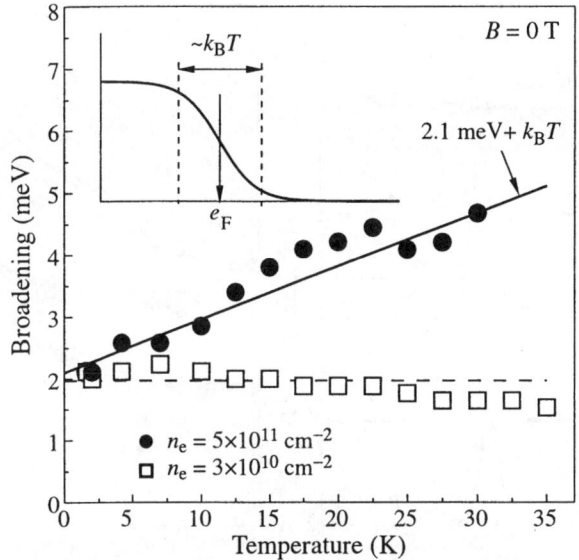

*Figure 2.* Temperature broadening of resonances in reflectivity spectra for an undoped and modulation-doped ZnSe-based QWs. Lines are guide to the eyes. Energy distribution of electrons in vicinity of the Fermi energy is shown schematically in the inset.

scattering becomes significant at temperatures exceeding 100 K, and exciton – LA-phonon scattering is very weak). In contrast, the linewidth of the C-line grows linearly with temperature with slope equal to $k_B T$. Such a behavior let us suggest that the C-line and its linewidth are determined by the properties of the 2DEG electrons with energies in the vicinity of the Fermi energy, shown schematically in the inset of Fig. 2.

The important issue related to the energy position of the C-line and respective interpretation of the origin of this resonance in the optical spectra will be discussed in Sec. 5 after the experimental results in high magnetic fields will be presented.

## 4. Modification Of Optical Spectra In High Magnetic Fields

It is instructive to analyze the behavior of the modulation-doped sample in high magnetic fields, and compare it with the properties of the reference undoped sample. Modification of the magneto-optical spectra in ZnSe-based QWs with low-density 2DEGs has been considered in great detail in Refs. [3, 14]. Among the interesting appearances are: pronounced diamagnetic shift of exciton and trion states, their Zeeman splitting, stabilization of a triplet trion state in high fields and its interaction with the singlet state, and specific polarization properties caused by the formation dynamics of the trion states. In this paper we illustrate only the properties which are essential for comparison with the doped sample. Figure 3(a) displays

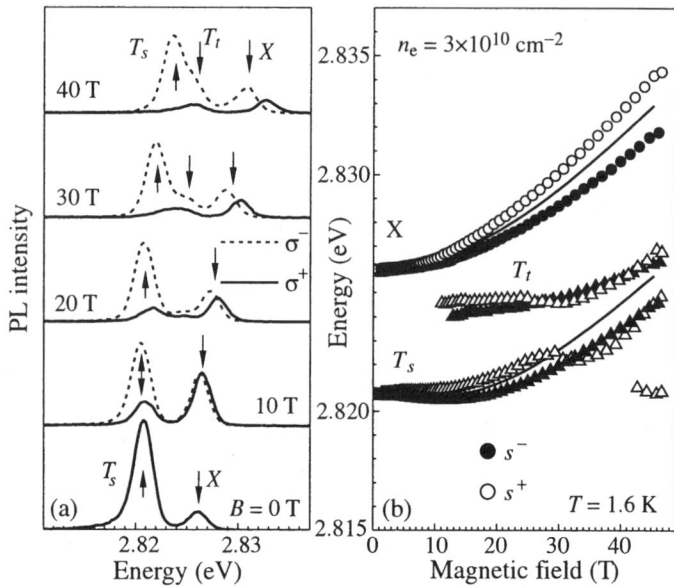

*Figure 3.* Photoluminescence spectra (a) and a fan-chart of magneto-excitons (b) in the reference structure a 67-Å-thick ZnSe/Zn$_{0.82}$Be$_{0.08}$Mg$_{0.10}$Se QW with $n_e = 3 \times 10^{10}$ cm$^{-2}$. Lines in panel (b) show the fit of the center-of-gravity for the exciton (X) and singlet trion (T$_s$) resonances.

the evolution of PL spectra taken in two circular polarizations with magnetic fields growing to 40 T. Singlet-(T$_s$) and triplet (T$_t$) states of trions and exciton (X) resonances are marked by arrows. Energy shifts of these lines are plotted in Fig. 3(b), where solid lines trace the shift of the centers of gravity for the exciton and trion resonances, i.e. they correspond to the shifts of these resonances in the absence of spin splitting of these states. The diamagnetic shift of these lines (i.e. quadratic with increasing magnetic fields) is characteristic for excitons and is well fitted with an electron effective mass $m_e = 0.15 m_0$ and the in-plane heavy-hole effective mass $m_{hh} = 0.44 m_0$. We note here that the energy spectrum of free particles (electrons or holes) exhibit linear shifts in magnetic fields. It is worthwhile to mention that the properties of the reference structure (energy shift, PL intensity, polarization degree) change very smoothly with magnetic field, showing none of the cusps or jumps that are characteristic for modulation-doped samples (as will be shown).

Now we turn to the modulation-doped structure with a dense 2DEG. The evolution of the photoluminescence and reflectivity spectra in magnetic fields are given in Fig. 4. The broad PL band that is characteristic at low magnetic fields (see also Fig. 1b) is transformed for $B > 20$ T into two narrow lines which are very similar to the exciton and trion spectra of undoped QWs. Details of the transformation are easier to follow on the fan-chart diagram given in Fig. 5(a). With increasing magnetic field the PL band splits into a set of lines, which shift linearly to higher energies. The lines show a nonmonotonic behavior with changing slopes.

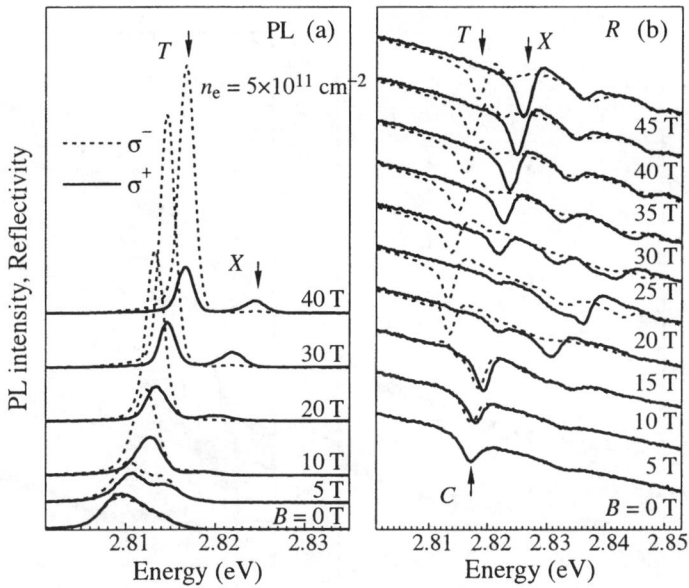

*Figure 4.* Photoluminescence (PL) and reflectivity (R) spectra of a 67-Å-thick ZnSe/Zn$_{0.82}$Be$_{0.08}$Mg$_{0.10}$Se QW, with a dense 2DEG of $n_e = 5 \times 10^{11}$ cm$^{-2}$ in magnetic fields to 45 T. $T = 1.6$ K.

The lowest energy peak reveals a linear shift with a slope of 0.38 meV/T in fields below 10.3 T, which converts at higher fields into a diamagnetic shift typical of excitonic states. Such a behavior has been reported recently for GaAs-based QWs [15, 16]. These authors demonstrated that the transition occurs at a filling factor $\nu = 2$, and at high magnetic fields the emission is indistinguishable from trions. For our structure the Landau level fan shown by dashed lines in Fig. 5(a) corresponds to the pure electron effective mass $m_e = 0.15 m_0$. It describes reasonably well the behavior of PL lines in low magnetic fields for filling factors $\nu > 2$. This property is rather general and has been noticed in GaAs QWs as well [16].

In magnetic fields above 10.3 T ($\nu < 2$), the PL spectrum consists of two lines having diamagnetic shifts similar to that of excitons and trions in undoped structures. Solid curves in Fig. 5(a,b) represent the behavior of X and T, measured in the reference sample. Small energetic shifts of these dependencies are included so that they coincide with experimental points for the doped QW in high magnetic fields (above 30 T) [17]. From these dependencies the transition energy of "bare" charged excitons at zero magnetic field is estimated to be 2.813 eV (shown by an arrow). It is only 3 meV higher than the PL maximum in doped QW; and is not equal to Fermi energy. We should note here that the energy of "bare" charged excitons in doped QWs might not coincide with the energy of charged excitons in undoped QW with the same design, due to effect of band-gap renormalization [12]. The nonmonotonic behavior of the energy shift in vicinity of the integer filling

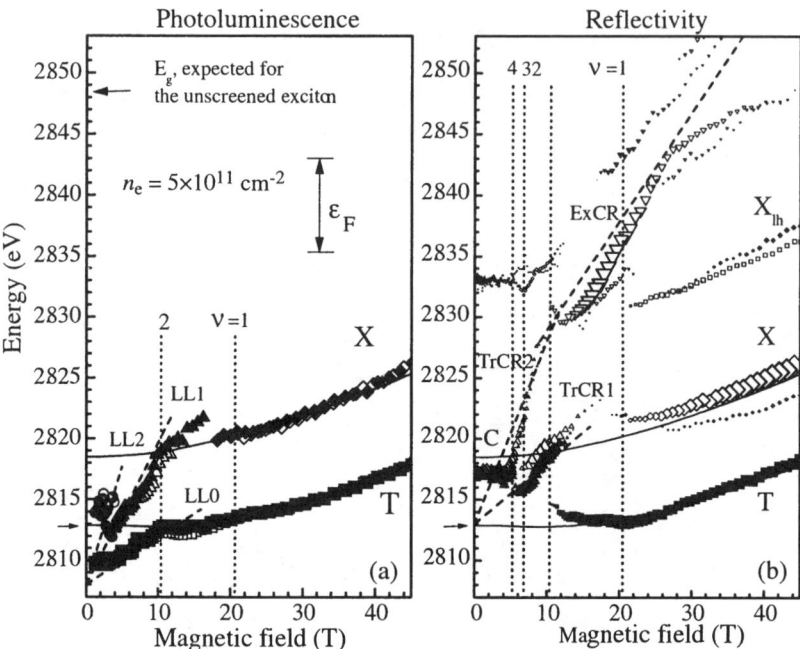

*Figure 5.* Experimental data of magnetic field behavior of optical spectra in a 67-Å-thick ZnSe/Zn$_{0.82}$Be$_{0.08}$Mg$_{0.10}$Se QW with $n_e = 5 \times 10^{11}$ cm$^{-2}$. (a) Energies of PL maxima *vs* magnetic field strength of detected in $\sigma^+$ (open symbols) and $\sigma^-$ (closed symbols) polarizations. Exciton (X) and trion (T) diamagnetic shifts for a reference structure are shown by solid lines in both panels. They are slightly shifted in energy positions for comparison with the relevant QW. Dashed lines show Landau level fan (LL0, LL1 and LL2) of electron with effective mass $m_e = 0.15m_0$. Dotted lines indicate magnetic fields of integer filling factors. (b) Energies of reflectivity lines *vs* magnetic field strength detected in $\sigma^+$ (open symbols) and $\sigma^-$ (closed symbols) polarizations. Intensities of resonances are displayed by a size of symbols. Dashed curves are linear fittings for TrCR1, TrCR2 and ExCR resonances.

factors will be discussed below in Sec. 6.

Let us now turn to the discussion of the reflectivity spectra in magnetic fields given in Fig. 4(b). In analogy with the emission spectra, the single C-resonance dominating in low magnetic fields is converted for $B > 20$ T into two resonances analogous to the exciton and trion states in the reference structure. Additionally, two strong resonances appeared on the high energy side of the exciton. They are identified as a light-hole exciton ($X_{lh}$) and a combined exciton-cyclotron resonance (ExCR) [18]. Figure 5(b) represents resonance energies of reflectivity lines *vs* magnetic fields (symbols). Solid lines in this figure are the data for the reference sample, the same as in the Fig. 5(a). It is clearly seen that the C-resonance at zero magnetic field is blue shifted by about 5 meV with respect to the energy of "bare" charged exciton. With growing magnetic fields the C-line transforms into a set of lines (TrCR1 and TrCR2). They shift linearly with slopes 0.5 meV/T and 1.5 meV/T (dashed lines in Fig. 5b). These values are close to one-half and three-halves of the electron cyclotron energy $\hbar\omega_c$, respectively. The energies of the TrCR1 and TrCR2 lines extrapolate (at zero field) to the energy of the "bare" charged exciton, and therefore have a trion origin. We ascribe these lines to a four-particle process, referred as a combined trion-cyclotron resonance (TrCR), in which the photogeneration of the trion is accompanied by excitation of an additional electron between Landau levels [19, 20].

At a filling factor $\nu = 2$, a low energy shoulder of the C-line appears in reflectivity spectra, it transforms to the trion resonance in high magnetic fields. The lines of neutral excitons X and $X_{lh}$ (related to heavy-hole and light-hole subbands, respectively) become detectable at a filling factor $\nu = 1$ and increase in intensity with growing magnetic fields.

At a magnetic field of 10.3 T ($\nu = 2$), the exciton-cyclotron resonance (ExCR) line appears in reflectivity spectra. This resonance is due to a process in which photocreation of a neutral exciton occurs simultaneously with a transition of a background electron between Landau levels [18]. The ExCR line shifts linearly with magnetic fields as 0.9 meV/T (dashed line). This value is very close to 0.96 meV/T taken from the theoretical approach of Ref. [18], as $[1 + m_e/(m_e + m_{hh})]\hbar\omega_c$ with heavy-hole mass $m_{hh} = 0.44 m_0$ [3]. Size of the symbols in Fig. 5(b) reflects the relative oscillator strength of the resonances. One can see that the ExCR process gains its maximum oscillator strength close to the filling factor $\nu = 1$, where the probability to find one (only one!) electron in the orbit of photogenerated exciton is maximal. We should note here that for the discussed experiments the exciton Bohr radius is still 2–2.5 times smaller than the magnetic length of electrons, which is 57 Å at 20 T.

## 5. Where Is The Fermi Edge (Singularity) ?

We have shown in Figs. 1 and 2 that the strong resonance in the reflectivity spectra of the modulation-doped structure (C-line) is related to the presence of the 2DEG in the quantum well. The temperature broadening of the C-line is characteristic for electron statistics in the vicinity of the Fermi energy. The problem which we are going to discuss in this section can be formulated as follows: "*In what respect does the energy of the optical transition of the C-line correspond to the Fermi energy of the 2DEG?*". We are not ready now to give a consistent description of the modification of optical spectra with increasing electron density, especially for "intermediate" regimes of exciton screening (Fermi energy is smaller or comparable with the exciton binding energy). This will require more experimental data for the transition range of carrier concentrations. However we would like to stress the fact that the experimental appearances in the studied structures can not be described satisfactorily in the frame of the approach commonly used so far.

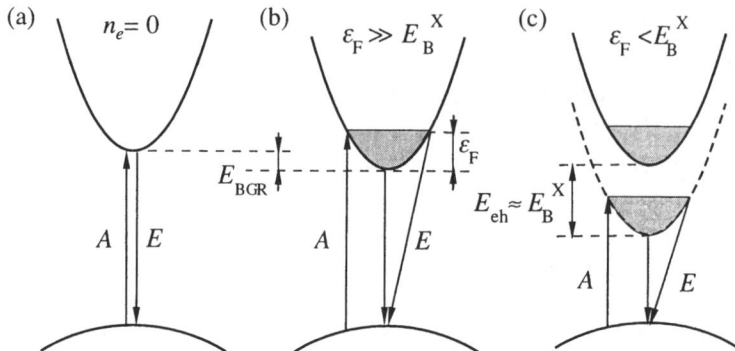

*Figure 6.* Schematic presentation of absorption (A) and emission (E) optical transitions in quantum wells with a 2DEG: (a) The case of empty QW ($n_e = 0$) and vanishingly weak excitonic effect. (b) The case where the Coulombic interaction is much smaller than the Fermi energy ($\varepsilon_F \gg E_B^X$) and where the excitonic effect can be neglected. (c) The case of the system with strong Coulombic interaction ($\varepsilon_F < E_B^X$), which corresponds to the experimental situation reported in this paper.

The commonly used approach has been worked out on the basis of III–V semiconductor heterostructures with rather weak Coulombic interaction and relatively small exciton binding energies. The excitons, being fragile in these III–V structures, are already screened at low electron densities. Also the electron effective masses in III–V QWs are smaller than in II–VI heterostructures, which results in higher Fermi energy at a given electron density ($\varepsilon_F = \pi\hbar^2 n_e/m_e$). As a result, the concentration range for the "intermediate" regimes of exciton screening is very narrow and because of that it has not been studied in detail yet. Schematically, optical transitions for absorption (the edge of absorption corresponds to a resonance in reflectivity spectra) and emission are given in Fig. 6. In the absence of electrons (see Fig. 6a) emission and absorptions transitions have the same energy and no contribution of 2DEG effects into the linewidth is expected. For the case of a doped structure correlation effects in a dense electron gas cause the band gap renormalization and the band gap decreased by a value of $E_{BGR} < \varepsilon_F$ [4, 5]. Then optical transitions to the absorption edge (shown by arrow A) is shifted to higher energies by $\varepsilon_F - E_{BGR}$. Also the emission band exhibits a width of $\varepsilon_F$. Its high energy side coincides with the absorption edge and the low energy side is shifted by $\varepsilon_F$ to smaller energies. It is important for our discussion that both the absorption edge and the high energy side of the emission band correspond to the Fermi energy of the 2DEG. This gives a direct way to measure a value of the band gap renormalization.

When we apply this approach to the ZnSe QWs we meet the following problem. The C-line energy position is at 2.815 eV, and by no means can be associated with a band-to-band transition, which is expected for these structures at energies of about 2.845 eV (i.e. 30 meV higher that the C-line is located). 30 meV is the binding energy of a quasi-two-dimensional exciton in this QW [3] (see also

Fig. 5a). The exciton itself is only slightly modified by the presence of the 2DEG with Fermi energy of only 7.7 meV as the condition $\varepsilon_F < E_B^X$ remains valid. This conclusion is confirmed by the value of the exciton diamagnetic shift in high fields, which is identical to that in undoped QWs. In the case of screened excitons with smaller binding energy the diamagnetic shift should be large. Also we can not expect that the band gap renormalization in the studied ZnSe QW has a value of about 38 meV for presence of a 2DEG with a Fermi energy of 7.7 meV only. Therefore we suggest that the C-line energy does not correspond to the Fermi energy itself, but to the Fermi energy shifted by an energy $E_{eh} \approx E_B^X$, which is the energy of Coulombic interaction of electrons with a photogenerated hole and is about the exciton binding energy. In other words the energy of the C-line $E_C = [E_g + \varepsilon_F - E_{BGR}] - E_{eh}(\approx E_B^X)$, where $E_g$ is the band gap of the empty QW. Schematically this is shown in Fig. 6(c).

## 6. Oscillatory Behavior Of The System Properties

A characteristic feature of modulation-doped QWs in external magnetic fields is the nonmonotonic variation of the system properties with increasing magnetic fields. This is especially pronounced in vicinity of the integer filling factors of a two-dimensional electron gas and was reported for different material systems like GaAs, CdTe, and ZnSe (see e.g. [5]). In the structures studied in this work critical behavior at integer filling factors has been established for the very different characteristics. They are collected in three panels of Fig. 7.

The energy shift of the lowest PL maximum (Fig. 7c) shows an upward cusp at integer filling factors from 1 to 4, and downward convex curves between them [12]. The most pronounced feature at $B = 10.3$ T corresponds to $\nu = 2$. At this magnetic field the linear shift of the PL maxima is changed to the diamagnetic behavior typical for trions. From the magnetic field value corresponding to $\nu = 2$ we derive the 2DEG density $n_e = 5 \times 10^{11}$ cm$^{-2}$ and calculate the expected fields for the set of integer filling factors. These fields are marked by vertical dotted lines in panels (a), (b) and (c). They are in good agreement with estimations done from the linewidth of emission line, which however are much less accurate. This also coincides well with the data on the optical detection resonance spectroscopy with the use of far-infrared radiation performed for the same sample and reported in Ref. [12].

Very pronounced features are observed for the even $\nu = 2, 4, 6$ and 8 for the PL circular polarization degree, which has minima when the Landau level is fully occupied (Fig. 7a). Also PL intensity in sigma minus polarization shows dips at $\nu = 1, 2$ and 4 (Fig. 7b). The integral PL intensity also shows oscillatory behavior (Fig. 7c) which could be explained by a decrease of the radiative recombination rate at integer filling factors. However, this interpretation requires additional experimental support, e.g. time-resolved measurements of the emission decay.

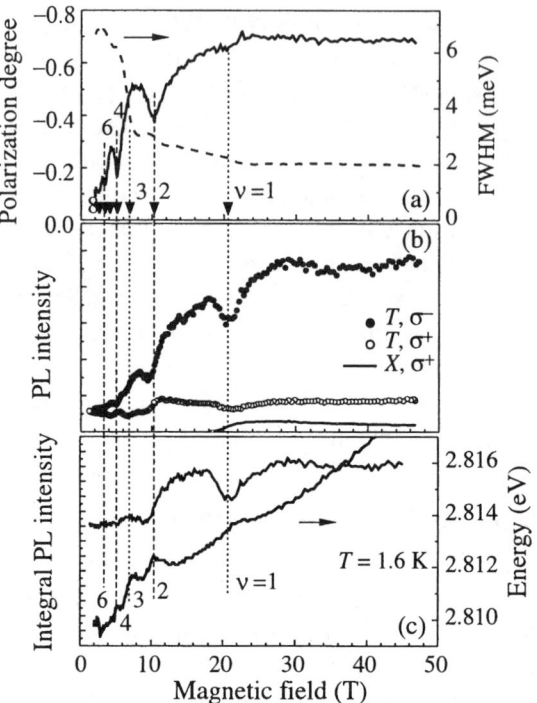

*Figure 7.* Summary of nonmonotonic behavior of the magneto-optical properties of modulation doped QW with $n_e = 5 \times 10^{11}$ cm$^{-2}$: (a) Polarization degree and linewidth of luminescence band; (b) Luminescence intensity for two circular polarization; (c) Integral emission intensity over both polarizations and energy shift of PL maxima detected in $\sigma^-$ polarization. Dotted lines indicate location of even filling factors.

It is worthwhile to note here that the oscillatory behavior has been also detected in experiments with nonresonant heating of a 2DEG by a far-infrared laser (184.3 $\mu$m) [12]. Clear minima in the electron temperature have been found for $\nu = 2$ and 4, i.e. when the electron acceleration inside the same Landau level is prohibited as all the states are occupied.

Finally the linewidth of emission line, plotted in Fig. 7(a) for completeness, shows rather weak nonmonotonic features in vicinity of $\nu = 1$ and 2, where it is already determined by the inhomogeneous broadening. Drastic changes in the linewidth take place between factors 4 and 2, where the electronic system became concentrated on the lowest Landau level. $\nu = 2$ corresponds to the case when at low temperatures all electrons are accumulated at the lowest Landau level.

## 7. Conclusions

We present a set of experimental data on modification of emission and reflectivity spectra of ZnSe-based QW with a dense 2DEG in high magnetic fields. In the considered case the Fermi energy of the 2DEG was four times smaller then that of the exciton binding energy, but exceeds slightly the binding energy of the charged exciton. In this regime the properties of the 2DEG, visualized in optical spectra, are strongly influenced by the Coulombic interaction between electrons and photoexcited holes. We have shown that the problems of exciton screening and band gap renormalization requires further detailing in the intermediate range of electron densities. More experimental data and theoretical efforts are required here.

Modulation-doped ZnSe-based QWs, due to their small inhomogeneous broadening and strong Coulombic interaction, provide a reliable model system to investigate excitonic effects and Coulombic correlation effects in the presence of a dense 2DEG. Technological possibilities for introducing magnetic Mn ions and fabrication of ZnMnSe-based modulation-doped heterostructures give an excellent opportunity to study spin-polarized 2DEG of high density (see e.g. Refs. [21–23]). Also the recent studies of modulation-doped ZnSe QWs under high-density optical excitation shows interesting results and allows to establish the new lasing mechanism involving trion states [24, 25].

## Acknowledgements

The authors are thankful to H.A. Nickel and B.D. McCombe for collaboration at initial stages of this study and to V.P. Kochereshko for useful discussions. The work was supported by the Deutsche Forschungsgemeinschaft (Grant Nos. Os98/6 and SFB 410).

## References

1. Lampert, M.A. (1958) *Phys. Rev. Lett.* **1**, 450.
2. Kheng, K., Cox, R.T., Merle d'Aubigne, Y., Bassani, F., Saminadayar, K., and Tatarenko, S. (1993) *Phys. Rev. Lett.* **71**, 1752.
3. Astakhov, G.V., Yakovlev, D.R., Kochereshko, V.P., Ossau, W., Faschinger, W., Puls, J., Henneberger, F., Crooker, S.A., McCulloch, Q., Wolverson, D., Gippius, N.A., and Waag, A. (2002) *Phys. Rev. B* **65**, 165335.
4. Schmitt-Rink, S., Chemla, D.S., and Miller, D.A.B. (1989) *Advances in Physics* **38**, 89.
5. Perry, C.H., Worlock, J.M., Smith, M.C., and Petrou, A. 1987 *Proc. Int. Conf. on High Magnetic Fields in Semiconductor Physics*, Würzburg, Germany 1986, (Springer, Berlin 1987), Ed. by G. Landwehr, p. 202.
6. Skolnick, M.S., Rorison, J.M., Nash, K.J., Mowbray, D.J., Tapster, P.R., Bass, S.J., and Pitt, A.D. (1987) *Phys. Rev. Lett.* **58**, 2130.
7. Note, that the reference structure has only nominally the same QW width as the doped structure. From the data in high magnetic fields it seems the QW widths differ considerably. At least from the comparison of these two structures we are not able to talk about absolute values in energy renormalization. More studies are required to clarify that point for ZnSe QWs.
8. Astakhov, G.V., Kochereshko, V.P., Yakovlev, D.R., Ossau, W., Nürnberger, J., Faschinger, W., Landwehr, G., Wojtowicz, T., Karczewski, G., and Kossut, J. (2002) *Phys. Rev. B* **65**, 115310.

9. Huard, V., Lovisa, S., Cox, R.T., Saminadayar, K., Potemski, M., Arnoult, A., Cibert, J., Tatarenko, S., and Wasiela, A. (1998) *Physica B* **256-258**, 136.
10. Imanaka, Y., Takamasu, T., Kido, G., Karczewski, G., Wojtowicz, T., and Kossut, J. (2000) *J. Cryst. Growth* **214/215**, 240.
11. Shen, J.X., Oka, Y., Ossau, W., Fischer, F., Waag, A., and Landwehr, G. (1998) *Physica B* **249-251**, 589.
12. Ossau, W., Yakovlev, D.R., Astakhov, G.V., Waag, A., Meinig, C.J., Nickel, H.A., McCombe, B.D., and Crooker, S.A. (2002) *Physica E* **12**, 512.
13. Crooker, S.A., Rickel, D.G., Lyo, S.K., Samarth, N., and Awschalom, D.D. (1999) *Phys. Rev. B* **60**, R2173.
14. Yakovlev, D.R., Astakhov, G.V., Ossau, W., Crooker, S.A., Uchida, K., Miura, N., Waag, A., Gippius, N.A., Sivachenko, A.Yu., and Dzyubenko, A.B. (2001) *Phys. Stat. Sol. (b)* **227**, 353.
15. Gekhtman, D., Cohen, E., Arza Ron, and Pfeiffer, L.N. (1996) *Phys. Rev. B* **54**, 10320.
16. Yoon, H.W., Sturge, M.D. and Pfeiffer, L.N. (1997) *Solid State Commun.* **104**, 287.
17. The respective lines for the reference sample have been shifted to lower energies by 7.5 meV. We believe that the main part of this shift is not caused by the difference in the electron concentration, i.e. by the band-gap renormalization effect, but is due to the technological inconsistency with the QW width for this pair of the structures grown. That is why we will not use these data for evaluation the band-gap renormalization. However, this important point will be addressed in our future studies.
18. Yakovlev, D.R., Kochereshko, V.P., Suris, R.A., Schenk, H., Ossau, W., Waag, A., Landwehr, G., Christianen, P.C.M., and Maan, J.C. (1997) *Phys. Rev. Lett.* **79**, 3974.
19. Ossau, W., Kochereshko, V.P., Astakhov, G.V., Yakovlev, D.R., Landwehr, G., Wojtowicz, T., Karczewski, G., and Kossut, J. (2001) *Physica B* **298**, 315.
20. Kochereshko, V.P., Astakhov, G.V., Yakovlev, D.R., Ossau, W., Faschinger, W., and Landwehr, G. (2002) *Phys. Stat. Sol. (b)* **229**, 543.
21. Keller, D., Astakhov, G.V., Yakovlev, D.R., Crooker, S.A., Hansen, L., and Ossau, W. 2003 *Proc. NATO workshop on Optical Properties of 2D Systems with Interacting Electrons*, June 2002, St. Petersburg, Russia, NATO series book, (this volume).
22. Crooker, S.A., Johnston-Halperin, E., Awschalom, D.D., Knobel, R., and Samarth, N. (2000) *Phys. Rev. B* **61**, R16307.
23. Crooker, S.A., Johnston-Halperin, E., Awschalom, D.D., Knobel, R., and Samarth, N. 2001 *Proc. 25th IC on the Physics of Semiconductors*, Osaka, Japan 2000, (Springer, Berlin 2001), Eds. N. Miura and T. Ando, p. 961.
24. Puls, J., Mikhailov, G.V., Schwertfeger, G.V., Yakovlev, D.R., Henneberger, F., and Faschinger, W. (2001) *Phys. Stat. Sol. (b)* **227**, 331.
25. Puls, J., Mikhailov, G.V., Henneberger, F., Yakovlev, D.R., Waag, A., and Faschinger, W. (2002) *Phys. Rev. Lett.* **89**, 287402.

# MAGNETO-OPTICS OF INHOMOGENEOUS TWO-DIMENSIONAL ELECTRON GAS

P. HAWRYLAK
*Institute for Microstructural Science, National Research Council, Ottawa, K1A OR6 Canada*

Abstract. We review our theoretical understanding of anomalies in the emission from the homogeneous interacting two-dimensional electron gas at integer filling factors, in particular the change in the magnetic field dependence of the emission in the vicinity of filling factor 2 due to the renormalization of the band-gap. We compare these predictions with predictions from a competing model which involves inhomogeneous disordered electron gas leading to re-appearance of exciton and charged exciton-like resonance at filling factor < 2.

Key words: 2DEG, trions, excitons, band-gap renormalization

The photoluminescence from the homogeneous two dimensional electron gas has been intensively studied theoretically [1–6] and experimentally [6–13]. Here we focus on recently observed anomalies at integer filling factors [6–10], and in particular on anomalies in the vicinity of filling factor 2 [5, 6, 11–13]. The anomaly at filling factor 2 is associated with the change of slope of the emission energy dependence on the magnetic field. We have attributed this anomalous dependence to the change of slope of the band-gap renormalization contribution to the emission energy [6]. However, experimental spectra have been interpreted in terms of reappearance of exciton and/or charged exciton at filling factors less then 2 [5, 10–13]. We find it difficult to reconcile the notion of filling factor, which implies homogeneous charge distribution, and a notion of exciton or charged exciton, which implies parts of the sample free of carriers. We find it equally difficult to reconcile the notion of charged exciton, a freely moving object, and carrier localization [12, 13]. It is however possible that long wavelength potential fluctuations in a weakly disordered sample may lead to the reappearance of spectral features which resemble charged exciton. We discuss here a model of inhomogeneous electron gas at filling factor less than 2 which supports the possibility of the reappearance of charged exciton-like exciton recombination.

We start by discussing the emission spectrum in a clean two-dimensional electron gas.

At low densities the spectra are dominated by the negatively charged exciton and free exciton recombination. The negatively charged exciton has three bound states, a radiative singlet, a radiative triplet, and a dark (finite angular momentum) triplet [14, 15]. The dark triplet, first predicted in Ref. [14], is the ground state in the infinite magnetic field, lowest Landau level limit, and hence relevant to phenomena at filling factor less than 2. By contrast, in high-density samples the recombination energy reveals an approximately linear increase with the magnetic field, so called band-to-band recombination, until filling factor two where a significant departure from linear behavior, consistent with excitonic emission, is observed [6, 10–13]. At very low temperature and very low disorder additional discontinuities and line splittings at integer filling factors are observed [8–10], with their understanding developed in Refs [3, 8]. The overall behavior of the energy of the emission line, especially the linear vs. nonlinear dependence on the magnetic field, has been attributed to band-gap renormalization [6]. The energy of interband optical transitions is due to the recombination of an electron from the bottom of the conduction band with a hole in the valence band. The energy of the non-interacting electron-hole pair increases linearly with the magnetic field $B$. However, the energy of the removed electron is strongly renormalized due to interaction with other electrons, and this renormalization scales as $\sqrt{B}$. These two different scales compete in the dependence of the transition energy on the magnetic field. As shown in Ref. [6], the two different dependencies on the magnetic field contribute to the total energy of the quasi-particle at even filling factors in the following way:

$$E_{qp} = \frac{B_\nu}{u^2} - \sqrt{2\pi} \sqrt{\frac{B_\nu}{u^2}} \frac{\Gamma((2/\nu) + 1/2)}{\nu \sqrt{\pi} \Gamma((2/\nu) + 1)}. \tag{1}$$

In Eq. (1) we expressed the magnetic length $l_0$ in terms of the magnetic field $B$ as $(l_0/a_0) = \sqrt{u^2/B}$, with $a_0$ the effective Bohr radius, and $B_\nu = B_1/\nu$ with $B_1$ denoting the magnetic field corresponding to filling factor $\nu = 1$. At low magnetic fields (high filling factors) Eq. (1) can be simplified to arrive at the approximate expression for the energy of the quasi-particle at low magnetic fields (large $\nu$):

$$E_{qp}(B_\nu) = -\sqrt{4B_1/u^2} + \left(\frac{B_\nu}{u^2}\right)\left(1 + \frac{1}{\sqrt{4B_1/u^2}}\right).$$

Hence, if we think of $B_\nu$ as a continuous magnetic field, we obtain a liner dispersion of the quasi-particle energy with the magnetic field up to fields corresponding to the filling factor $\nu = 2$. Despite the fact that Coulomb energy scales as $\sqrt{B}$, e–e interactions lead to a renormalization of a linear $B$ dependence i.e. effective mass. At filling factor 2 and less, the behavior of quasiparticle changes. This is because when we remove the quasi-particle from the spin down lowest Landau level for magnetic fields varying from $B_2$ to $B_1$, the level remains filled in this entire magnetic field range, and there are no other levels filled with electrons with the same spin. Hence for $B_2 < B < B_1$ the quasi-particle energy has only the

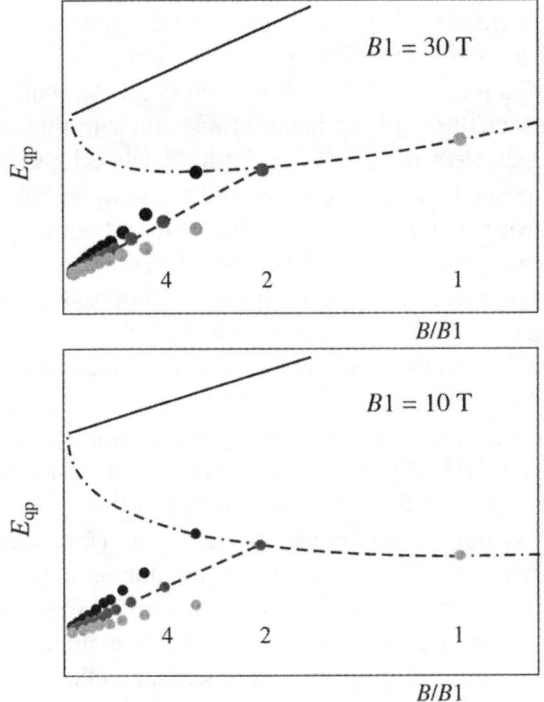

*Figure 1.* (a) Reference emission spectra: Hartree–Fock energies of quasi-particles with spin-up and spin-down as a function of the magnetic field at $T = 0$ for two different densities corresponding to different magnetic fields corresponding to filling factor 1.

contribution proportional to $B$ and exchange contribution proportional to $\sqrt{B}$:

$$E_{qp}(1 < \nu < 2) = \frac{B}{u^2} - \sqrt{2\pi}\sqrt{\frac{B}{u^2}}.$$

The most important fact is that this energy is exactly equal to the recombination energy from a single magneto-exciton. Hence band-to-band recombination and exciton recombination spectra are degenerate for filling factors less than 2. However, the onset of the band-to-band recombination spectrum starts at filling factor 2, and its appearance depends. on carrier concentration. We show in Fig. 1 the calculated quasiparticle energy dependence on magnetic field for two different concentrations corresponding to $B_1 = 10$ T $B_1 = 30$ T. We see that the visibility of the kink in the dispersion depends on concentration, and is much easier to observe at lower concentration. This band-gap renormalized spectra should be treated as "reference emission spectra", from which higher order excitonic (vertex) and correlation effects should be extracted. In these reference spectra, a change in emission, from linear with $B$ to excitonic like $B$ dependence, is expected at filling factor $\sim 2$.

So far we investigated a homogeneous electron gas for which filling factor was well defined. It is possible that the electron gas can effectively screen long wavelength potential fluctuations for filling factors greater than 2 while screening becomes very inefficient for filling factors less than 2. In this scenario, potential maxima might repel electrons while attracting valence holes. The repelling of electrons is only possible when filling factor decreases and empty states, holes, appear. Hence potential maximum is equivalent to the attracting potential for holes, empty states, in the lowest Landau level, which will spatially localize on such potential maxima. Hence there might appear disks of states empty of electrons with valence holes migrating into the center. We therefore study a simple model of a repulsive electronic impurity in the two-dimensional electron gas at filling factor < 2 and an attractive impurity in the valence band. The single particle states $|m\rangle$ are states of the lowest Landau level in a symmetrical gauge and we take a disk of states with "$m < 50$". We take the impurity as extending from "$m = 0$" to "$m = mr$" with single particle dispersion $E(m) = V0(m - mr)$, $(m < mr)$, and $E(m) = 0, m > mr$, and $V0$ as impurity strength. Similar but attractive dispersion is applied to valence holes. For $m > mr$ we assume all conduction band states filled with spin up and down electrons (filling factor 2). We also assume that the impurity potential is an effective potential and hence assume charge neutralizing background for $m > mr$. The model is very similar to the one used in Ref. [3]. The Hamiltonian of many electrons and a single valence hole can be written as:

$$H_0 = \sum_i E_i^e c_i^+ c_i + \frac{1}{2} \sum_{ijkl} \langle ij|V_{ee}|kl\rangle c_i^+ c_j^+ c_k c_l$$
$$+ \sum_i E_i^h h_i^+ h_i - \sum_{ijkl} \langle ij|V_{eh}|kl\rangle c_i^+ h_j^+ h_k c_l. \quad (2)$$

The operators $c_i^+$ ($c_i$), $h_i^+$ ($h_i$) create (annihilate) the electron or valence band hole in the state $|i\rangle$ with the single-particle energy $E_i$. The two-body coulomb matrix elements are $\langle ij|V|kl\rangle$ for electron–electron (ee), hole–hole (hh) and electron–hole (eh) scattering, respectively [3].

We study a situation when additional electron-valence band hole pair has been created optically in the disk. The charge neutral electron-valence hole states in the presence of charge neutral filling factor 2 state can be expanded in terms of excitations of the filling factor 2 droplet, and classified into 3 different spin configurations involving spin of photo-excited electron and spin of electron excited from the filling factor 2 droplet:

$$|X, -1/2\rangle = \left(\sum_m A_m h_m^+ c_{m\downarrow}^+ + \sum_{mpkl} B_{mpkl} h_m^+ c_{p\downarrow}^+ c_{k\downarrow}^+ c_{l\downarrow}\right.$$
$$\left. + \sum_{mpkl} C_{mpkl} h_m^+ c_{p\downarrow}^+ c_{k\uparrow}^+ c_{l\uparrow} + \ldots\right)|\nu = 2\rangle$$

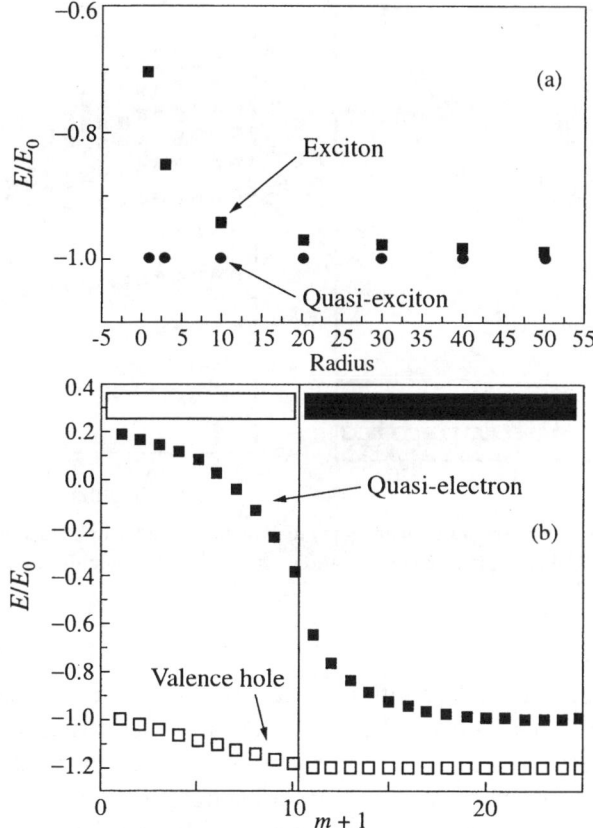

*Figure 2.* Mangneto-exciton binding energy as a function of the radius of the disk in the frozen background of filling factor 2 electrons (a) bare electron in the exciton (b) quasi-electron in the exciton.

$$|X, -3/2\rangle = A_m h_m^+ c_{m\downarrow}^+ B_{mpkl} h_m^+ c_{p\downarrow}^+ c_{k\downarrow}^+ c_{l\uparrow} |\nu = 2\rangle$$
$$|X, +1/2\rangle = A_m h_m^+ c_{m\downarrow}^+ B_{mpkl} h_m^+ c_{p\downarrow}^+ c_{k\uparrow}^+ c_{l\downarrow} |\nu = 2\rangle$$

We shall analyze here only optically active configuration corresponding to spin $S = -1/2$. It consists of a number of contributions: (1) a linear combination of electron-valence hole states with amplitude A, (2) a valence hole, one photo-excited electron and one electron excited from the $\nu = 2$ droplet, and one hole in the $\nu = 2$ droplet, with amplitude B (3) the third term with amplitude C which has the same form as term with amplitude B except that the spin configuration of the two electrons is anti-parallel. The second and the third terms correspond to exciton dressed by charge and spin excitations of the $\nu = 2$ droplet. Charge neutrality requires that they involve two electrons and two holes, i.e. form not a trion but a bi-exciton. This bi-exciton is always mixed with a single exciton.

First let us analyze contribution A, a single exciton in a disk. We first diagonalise

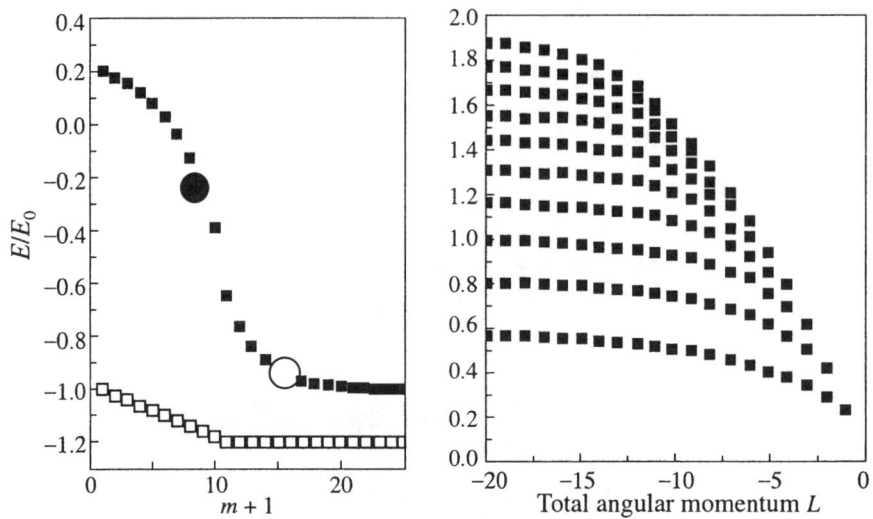

*Figure 3.* Edge excitation spectrum of an empty disk in the filling factor 2 background. Left: single electron-hole pair Right: bands of edge excitations as a function of angular momentum for a disk radius $R = 10$ and $V_0 = 0.1$.

the exciton problem on a disk with finite radius (size of the opening in the $\nu = 2$ state) in the presence of a frozen background of $\nu = 2$ electrons. Figure 2a shows the significant dependence of exciton energy on the size of the disk, implying a very broad emission spectrum. This calculation treated the photo-excited electron as distinguishable from the $\nu = 2$ electrons. This is not the case and the photo-excited electron is a quasiparticle of the $\nu = 2$ state. In Fig. 2b we show the dispersion of the qusiparticle as it moves from the center of the disk, where its energy is only renormalized by the impurity potential (shown in the valence hole dispersion below) to the edge of the $\nu = 2$ droplet where it is equal to the exchange energy $E_0$ of the $\nu = 2$ droplet. When we replace the photo-excited electron by a quasi-electron, the exciton binding energy no longer depends on the disk radius, as shown in Fig. 2a. There is an exact compensation between the loss of states available to form an exciton by the increase of exchange with electrons in the $\nu = 2$ state. Hence in this approximation all impurities (disks), irrespective of their size, emit with the same energy, equal to the free exciton energy and band-to-band emission energy $E_0$. What distinguishes the $\nu = 2$ exciton is the interaction with excitations of the edges of the $\nu = 2$ state.

In Fig. 3 we show the band of electron-hole pair edge excitations of the $\nu = 2$ state. The left panel shows how we construct a single electron–hole excitation for a given angular momentum. The right panel shows the spectrum of all excitations for each total angular momentum. We see the characteristic spectrum of narrow bands. Each band corresponds to different position of quasi-electron and the width is determined primarily by the position of the hole in the $\nu = 2$ state. The width

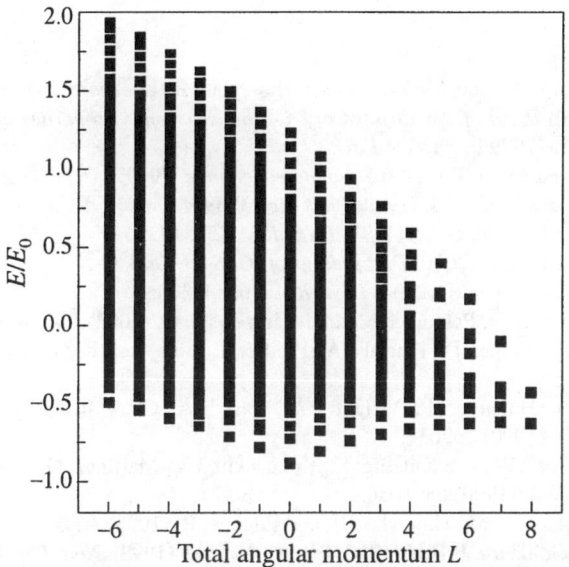

*Figure 4.* Energy spectrum of exciton dressed by charge excitations as a function of angular momentum for parameters of Fig. 3.

of the total band is $\sim E_0$.

In Fig. 4 we show the spectrum of exciton interacting with edge excitations of the form $C$ i.e. with two electrons of opposite spin. We see that for each total angular momentum there is a well separated dressed exciton state and a band of states at higher energies dominated by edge excitation of the $\nu = 2$ state. The exciton energy has a dispersion and the minimum of its energy is lowered in comparison with free exciton by interactions with edge excitations by $\sim 0.05 E_0$. This lowering of energy is typical of "charged exciton" [14, 15]. Hence emission spectrum would result in a characteristic band with energy coinciding with charged exciton energy and a high energy shoulder coinciding with free exciton band. While more extensive calculations are needed, from the cancellation discussed in Fig. 2 we expect the "dressed exciton" emission to weakly depend on the disk size i.e. on filling factor.

In summary, we discussed the characteristic changes in the emission spectrum expected at filling factor $\nu = 2$. Preliminary calculations indicate that in a disordered sample potential fluctuations could produce emission spectrum for $\nu < 2$ which resembles charged and free exciton like recombination bands.

## Acknowledgements

We thank M. Potemski, B. McCombe and I. Bar-Joseph for discussions.

## References

1. For a review and additional references see Hawrylak, P., Lockwood, D., and Sachrajda, A.S. (Editors) (2000) *Proc. of the 13 Int. Conf. on the Electronic Properties of Two-Dimensional Systems*, Ottawa, 1999 in *Physica E* **6**.
2. Katayama, S. and Ando, T. (1989) *Solid State Commun.* **70**, 97; Uenoyama, T. and Sham, L.J. (1989) *Phys. Rev.* **B 39**, 11044.
3. Hawrylak, P. and Potemski, M. (1997) *Phys. Rev.* **B56**, 12386.
4. Asano, K. and Ando, T. (2000) *Proceedings of ICPS 25*, Osaka.
5. Rashba, E.I. and Sturge, M. (2000) *Phys. Rev.* **B63**, 45305.
6. Hawrylak, P., Teran, F., Potemski, M., and Karczewski, G. (2002) *Phys. E* **12**, 495.
7. Goldberg, B.B., Heiman, D., Pinczuk, A., Pfeiffer, L., and West, K. (1990) *Phys. Rev. Lett.* **65**, 641;
   Turberfield, A.J., Haynes, S.R., Wright, P.A., Ford, R.A., Clark, R.G., Ryan, J.F., Harris, J.J., and Foxon, C.T. (1990) **65**, 637;
   Buhmann, H., Joss, W., von Klitzing, K., Kukushkin, I.V., Martinez, G., Plaut, A.S., Ploog, K., and Timofeev, V.B. (1990) **65**, 1056.
8. Gravier, L., Potemski, M., Hawrylak, P., and Etienne, B. (1998) *Phys. Rev. Lett.* **80**, 3344.
9. Manfra, M.J., Goldberg, B.B., Pfeiffer, L., and West, K. (1998) *Phys. Rev.* **B57**, R9467.
10. Finkelstein, G., Shtrikman, H., and Bar-Joseph, I. (1997) *Phys. Rev.* **B56**, 10326.
11. Eytan, G., Yaon, Y., Rappaport, M., Shtrikman, H., and Bar-Joseph, I. (1998) *Phys. Rev. Lett.* **81**, 1666.
12. Yaon, Y., Rappaport, M., Umanski, V., and Bar-Joseph, I. cond.matt0011142.
13. Nickel, H.A., Yeo, T.M., Dzyubenko, A.B., McCombe, B.D., Petrou, A., Sivachenko, A.Yu., Schaff, W., and Umansky, V. (2002) *Phys. Rev. Lett.* **88**, 56081.
14. Wojs, A. and Hawrylak, P. (1995) *Phys. Rev.* **B 51**, 10880.
15. Wojs, A., Quinn, J.J., and Hawrylak, P. (2000) *Phys. Rev.* **B 62**, 4630.

# FILLING FACTOR DEPENDENCE OF OPTICAL SPECTRA FOR CdTe QUANTUM WELLS UNDER MAGNETIC FIELD

R.T. COX, V. HUARD, K. SAMINADAYAR
*Service de Physique des Matériaux et Microstructures,
CEA-Grenoble, France*

C. BOURGOGNON, S. TATARENKO
*Laboratoire de Spectrométrie Physique, Université Joseph Fourier
Grenoble, France*

Abstract. The interband magneto-emission is studied for semiconductor quantum wells containing a high concentration two-dimensional electron system. In first approximation, the field dependence of the transition energies fits with the "fan-diagram" of transitions between conduction band and valence band Landau levels. However, excitonic interactions and electron electron exchange interactions produce marked deviations from linear field dependence. This article presents a method for separating the two types of correction from each other, by measuring splittings between allowed (diagonal) and forbidden (non-diagonal) recombination transitions. Data for modulation-doped CdTe and CdMnTe quantum wells are compared to the first-order theory of the filling factor dependence of the interactions.

Key words: magneto-optics, quantum wells, two-dimensional electron gas, excitons, exchange interactions, CdTe

## 1. Introduction

This article discusses the effects of carrier-carrier interactions on the 2 K, inter-band magneto-optics of semiconductor quantum wells containing a two dimensional electron gas. We will describe a method for distinguishing the separate effects of the electron-valence hole (excitonic) interactions and the electron-electron (exchange) interactions. We suggest that this method could be of general interest for analysing emission spectra from quantum wells having high electron concentrations. Our own data are for 10 nm thick CdTe and $Cd_{1-x}Mn_xTe$ ($x = 0.002$) quantum wells between CdMgZnTe barriers. The structures are grown strained on $Cd_{0.88}Zn_{0.12}Te$ substrates and the barriers are aluminium-doped to give electron concentrations in the well, $n_e$, ranging up to $10^{12}$ cm$^{-2}$.

Attempts to understand the field dependence of the optical absorption and emission usually start from the "fan-diagram" of interactionless energy gaps between valence band and conduction band Landau levels:

$$E_{ij} = E_{\text{gap}} + \left(i + \frac{1}{2}\right)\hbar\omega_{ce} + \left(j + \frac{1}{2}\right)\hbar\omega_{ch} + E_{\text{Zeeman}}, \quad (1)$$

where the parameter $E_{\text{gap}}$ is an interactionless zero-field bandgap.

Here "interactionless" means including all renormalisation effects caused by the $n_e$ electrons, but *excepting* their exchange interaction with the electron that is created or annihilated by the optical transition and excepting any excitonic interactions with the valence hole.

Note that by this definition $E_{\text{gap}}$ in Eq. (1) is *not* the lower edge of the zero field emission band, the lowest energy transition that annihilates a $k = 0$ conduction electron. The emission edge is shifted downwards from $E_{\text{gap}}$ by an amount equal to the exchange interaction between a $k = 0$ electron and the background electrons of the same spin, because annihilating the electron raises the final state energy by that amount, see e.g. Ref. [1]. The emission edge is also shifted downwards from $E_{\text{gap}}$ by the amount of any zero-field excitonic binding $E_b$ (annihilating the valence hole raises the energy by $E_b$).

If the excitonic and exchange corrections gave just a constant energy shift with respect to Eq. (1), independent of the magnetic field, the optical transitions would have a linear field dependence, defined by the cyclotron parameters $\hbar\omega_{ce}$ and $\hbar\omega_{ch}$. This is assuming a two-band model (very reasonable for our samples with 35 meV strain-splitting of the quantum well's valence band) and subtracting the non-linear Zeeman splitting in the particular case of CdMnTe wells.

The interesting physics is in the *deviations* from the simple model of linearly varying transition energies $E_{ij}$. These deviations occur because the exchange and excitonic interactions vary in a complex way as the Landau level filling factor $\nu$ decreases with magnetic field.

Before continuing, we note that for low electron concentration $n_e$, where the situation $\nu < 2$ is reached at quite low fields, the deviations can be so great that the concept of "inter-Landau level" transitions is of no use at all, at least at field strengths where the cyclotron energies are less than the excitonic Rydberg. The principal absorption and emission peaks are then associated with the creation/annihilation of the exciton $X$ and the trion $X^-$. For typical II–VI semiconductor quantum wells ( CdTe, ZnSe, etc.), these transitions are characterised by a quadratic dependence on the field strength $B$, even at fields of order 20 T.

We are interested here in high electron concentrations $n_e$ with electrons filling many Landau levels at fields of a few Teslas. For high $n_e$, a large set of many transitions is observed, which can be assigned to the scheme of inter-Landau level transitions of Eq. (1). We write the transitions $i \leftarrow j$ (absorption) and $i \rightarrow j$ (emission), where $i, j$ are conduction band and valence band Landau level indices respectively. Diagonal transitions $i = j$ are "allowed", and non-diagonal transitions $i \neq j$ are "forbidden".

## 2. Spectra

Figure 1 displays line positions in emission spectra and a gray-scale plot of absorption line intensities for Sample 1, which is a one-side modulation doped, 10 nm thickness CdMnTe (0.2% Mn) quantum well with $n_e = 3.8 \times 10^{11}$ cm$^{-2}$. The labels $LL_{ii}$ identify allowed transitions $i \leftrightarrow i$.

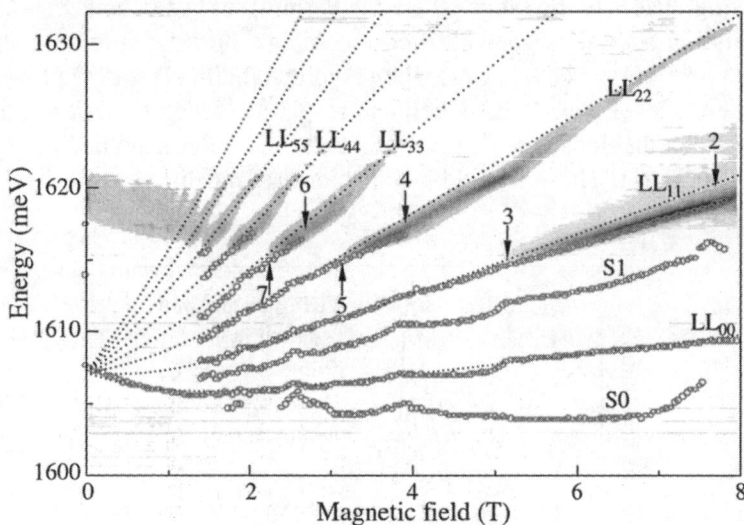

*Figure 1.* Emission peak positions (circles) and absorption spectra (greyscale map) for sample 1, a 10 nm CdMnTe (0.2% Mn) quantum well with $n_e = 3.8 \times 10^{11}$ cm$^{-2}$, in $\sigma^+$ polarisation at 2 K. Integer filling factor values are indicated by arrows. Labels $LL_{ii}$ designate allowed (diagonal) transitions. Emission peak S1 is a forbidden $1 \to 0$ recombination. (Peak S0, not discussed here, is an Anderson–Fano resonance.) The dotted curves are a fit to Eq. (1), taking account of the large non-linear Zeeman effect in Mn doped CdTe.

As the field is swept upwards, allowed (diagonal) absorption transitions $i \leftarrow i$ become visible in $\sigma^+$ polarisation from a threshold filling factor $\nu_{thr} = 2i + 1$. This is the odd integer value of $\nu$ at which spin sublevel $m = -1/2$ of electronic Landau level $i$ begins to empty. (Note that $g_e$ is positive in Cd$_{1-x}$Mn$_x$Te at low fields; in pure CdTe $g_e$ is negative and it is the $m = +1/2$ sublevels that empty first, as in GaAs).

In this paper, we will discuss the emission transitions only. Emission corresponds to transitions from an occupied or partly occupied conduction band level $i$ and, as the magnetic field increases, an emission transition $i \to j$ is seen to *disappear* at the integer value of filling factor where level $i$ empties (at 0 K). This is observed in the 2 K data of Fig. 1, where the number of observed transitions decreases as $\nu$ decreases until, finally, we are left with transitions from the lowest Landau level only.

The dotted lines are a representative fit to the field dependence of Eq. (1) using reduced mass $\mu = 0.080 m_0$ and allowing for the non-linear Zeeman effect associated with the light Mn doping. The origin of this fitted fan diagram falls in the lower edge of the zero-field emission peak and, as explained above, is shifted down with respect to the parameter $E_{\text{gap}}$ of Eq. (1).

Note that there is a good connection between the positions measured in emission (symbols) and in absorption (grayscale data) for allowed transitions $i \leftrightarrow i$. We see a quite accurate overlap in the filling factor range where conduction Landau level $i$ is emptying, that is in the range where the Fermi level lies in level $i$.

In emission at 2 K, higher allowed recombination transitions like $1 \rightarrow 1$, $2 \rightarrow 2$, etc., are quite weak, because the photohole thermalises down towards the lowest valence level $j = 0$. This helps to reveal forbidden emission transitions (non-diagonal transitions $i \neq j$), for example the recombination transition $1 \rightarrow 0$ labelled S1 in Fig. 1. These forbidden transitions are much too weak to be seen in the absorption spectra.

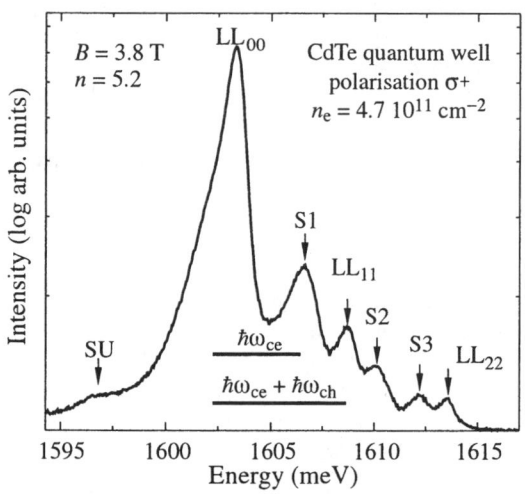

*Figure 2.* Emission spectrum in $\sigma^+$ polarisation on a logarithmic intensity scale for Sample 2, a 10 nm CdTe quantum well with $n_e = 4.7 \times 10^{11}$ cm$^{-2}$, at 2 K, 3.8 T (filling factor $\nu = 5.2$). Three allowed (diagonal) and three forbidden (non-diagonal) transitions $S_1 = 1 \rightarrow 0$, $S_2 = 2 \rightarrow 0$, $S_3 = 2 \rightarrow 1$ are resolved.

Figure 2 shows the 2 K emission spectrum of Sample 2, which is a symmetrically modulation-doped, 10 nm thickness, pure CdTe quantum well with $n_e = 4.7 \times 10^{11}$ cm$^{-2}$. Note that this spectrum is on a *logarithmic* intensity scale. Three forbidden transitions $i \neq j$ ($S_1$, $S_2$, $S_3$) can be seen, along with three allowed transitions $i = j$. Such forbidden transitions were first identified in Ga$_{1-x}$In$_x$As quantum wells with large $x$. They were attributed by Lyo *et al.* [2] to Landau level mixing induced by the strong disorder present in an alloy material. However non-diagonal transitions are increasingly being recognized in electron-doped quantum wells with low disorder, including pure GaAs wells of very high mobility. We will show below that their presence helps greatly in the spectrum analysis, allowing us to distinguish the separate contributions of excitonic and exchange interactions to the recombination energies.

## 3. Analysis

For an initial emitting state with a hole occupying a valence Landau level $j$, the luminescence spectrum is the spectrum of the possible final states of the electron system after removal of one electron, that is the set of one diagonal and many non-diagonal recombination transitions $i \to j$. These transitions leave a hole in conduction band levels ($i = 0, 1, 2, \ldots$) and it is convenient to consider their spectrum as being the spectrum of transitions $i \leftarrow j$ of the hole. The excitonic correction (the binding of the valence hole) is associated with the initial state, whereas the exchange correction can usefully be associated with the *final* state [3]. The latter correction is the exchange interaction between the hole, injected into conduction level $i$, and the $n_e$ background electrons, equal to the negative of the exchange energy of the annihilated electron.

Since the initial state hole is not thermalised completely to 2 K, we have a series of populated initial states, with holes in valence levels $j = 0, 1, 2, \ldots$, and each of these initial states gives rise to its own spectrum of hole transitions $i \leftarrow j$. This multiplicity of spectra makes emission more complicated than absorption.

At first glance, Eq. (1) fits the observed transitions, both emission and absorption, fairly well on average. One reason for this is that excitonic corrections and exchange corrections tend to vary in opposite directions for certain transitions over certain ranges of $\nu$. The two corrections may each be varying considerably, whereas the net recombination energies hardly vary at all. So perhaps the main problem in analysing the optical spectra is separating the two types of correction.

We do not know of any way to extract the absolute values of the excitonic and exchange interactions. However, we will propose that one can extract relative values of the excitonic corrections in different initial states, and also relative value of exchange corrections in different final states, by plotting energy *differences* between successive diagonal and non-diagonal emission peaks.

Also, as already shown by Lemaître et al. [4] who analysed splittings between the successive allowed transitions $i \leftarrow i$ seen in magneto-absorption spectra, measuring differences eliminates the unknown downward shift between the origin of the real fan-diagram and the hypothetical zero-field parameter $E_{\text{gap}}$ in Eq. (1), and also eliminates the Zeeman interaction.

### 3.1. EXCITONIC CORRECTIONS SEEN IN EMISSION

Thus, Fig. 3 displays the energy difference $E_{11} - E_{10}$ between the allowed $1 \to 1$ and forbidden $1 \to 0$ recombination transitions, after subtracting the valence-hole cyclotron energy $\hbar\omega_{ch}$ (with $m_h = 0.25$). The data are for the CdMnTe well (Sample 1) and for the CdTe well (Sample 2). Note that the vertical scale is normalised to $E_c \propto \sqrt{B}$ (which is why the data is increasingly noisy at high $\nu$, low $B$).

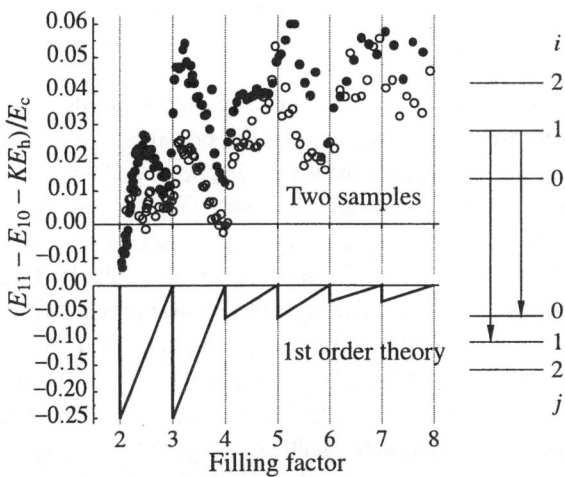

*Figure 3.* Energy difference, normalised to $E_c = \sqrt{\pi/2}e^2/(4\pi\epsilon\epsilon_0 l_B)$, between the allowed $1 \to 1$ and the forbidden $1 \to 0$ recombinations transitions, after subtracting the valence-hole kinetic energy $\hbar\omega_{ch}$ for two samples. Black filled circles are for Sample 1 (CdMnTe well, $\sigma^+$ polarisation, $n_e = 4.7 \times 10^{11}$ cm$^{-2}$) and open circles are for Sample 2 (CdTe well, $\sigma^-$, $n_e = 4.7 \times 10^{11}$ cm$^{-2}$). The lower graph shows the theoretical (first-order) difference between the excitonic corrections for the two transitions.

The final state is the *same* for these two transitions: a hole is left in the $n = 1$ conduction band level. So its self-energy with the $n_e$ electrons does not contribute to the line splitting. The two initial states are different, with the hole in valence levels $j = 1$ and $j = 0$ respectively. So with the valence-hole cyclotron energy $\hbar\omega_{ch}$ subtracted, we have a direct measure of purely excitonic corrections. Or more precisely we have the difference between the excitonic corrections for the same electron system interacting with a $j = 1$ valence hole and with a $j = 0$ valence hole. In Fig. 3, as we decrease $\nu$ progressively by sweeping the field upwards, this difference is seen to zigzag as a function of $\nu$, with jumps at integer values of $\nu$. The energy difference is *positive*, that is the graph shows a $j = 0$ valence hole to be more strongly bound than a $j = 1$ valence hole.

We will attempt to interpret the data using the simplest available theory for the excitonic binding of a hole in valence band Landau level $j$. This is the pure 2-D, infinite field magnetoexciton model. In this theory, there is no binding with filled conduction band Landau levels. So (for an electron system at 0 K) the valence hole can bind only to electrons in the conduction level $i_F$ that lies at the Fermi level, that is the highest occupied level. The excitonic binding is maximum when there is just one electron on this level.

Let $\tau_i$ be the fractional occupation of conduction level $i$. We introduce the effect of phase-space filling of the level $i_F$ following Lemaître et al. [4], who suggested a linear dependence of binding energy on $(1 - \tau_{i_F})$, the fraction of empty states in level $i_F$. The exciton's energy with respect to the energy of a separated electron-hole pair (that is the negative of the excitonic binding energy) is (at 0 K):

$$E_{\text{exciton}} = -\left(1 - \tau_{i_F}\right) C_{i_F j} E_c, \qquad (2)$$

where the coulomb energy parameter $E_c = \sqrt{\pi/2} e^2/(4\pi \epsilon \epsilon_0 l_B)$ with $l_B$ the magnetic length $\sqrt{(\hbar/eB)}$. For CdTe, the parameter $E_c$ is $6.9\sqrt{B}$ meV T$^{-(1/2)}$. The coefficients $C_{ij}$ are rational binary fractions (some values are $C_{00} = 1$, $C_{01} = 1/2$, $C_{11} = 3/4$, $C_{02} = 3/8$, $C_{12} = 7/16$, etc. [5]).

According to Eq. (2), as the highest occupied conduction level $i_F$ empties with increasing field, the binding of the hole in valence band level $j$ increases linearly with $(1 - \tau_{i_F})$, when expressed in units of $E_c$. It reaches a peak value at the field where the level $i_F$ is just reaching the point of emptying completely, at integer $\nu$ plus $\epsilon$ (where $\epsilon \approx 0$), because the valence hole can then correlate totally with one electron on level $i_F$. With a small further increase in field, the excitonic binding drops immediately to zero, at integer $\nu$ minus $\epsilon$. Then it increases linearly to a new peak at integer $\nu$ minus 1, and so on.

Note that with a valence hole in, for example, spin sublevel $j = 1, m$ where $m$ is the spin quantum number, and the Fermi level dropping down progressively through conduction levels $i = \ldots, 3, 2, 1$ with increasing field, the initial state of the recombination is alternately a spin-forbidden and a spin-allowed, *indirect* exciton, until $\nu < 4$ when it becomes the $i = 1, j = 1$ direct exciton. And similarly for a hole in valence band level $j = 0, m$. In our measurement (Fig. 3), this varying binding of the valence holes with electrons at the Fermi level is being monitored by spin-allowed recombinations of the electrons on the second lowest conduction band level $i = 1$. More exactly, we are monitoring the *difference* between the series of excitonic states $i_F, 1$ and $i_F, 0$ (with $i_F = \ldots, 3, 2, 1$), as a function of $\nu$.

Equation (2) gives the excitonic correction to the transition energy, which is negative, that is the transition is red-shifted (excitonic binding lowers the initial state). The splitting between the two transitions considered here is then:

$$E_{11} - E_{10} = \hbar \omega_{ch} - \left(1 - \tau_{i_F}\right) \left(C_{i_F 1} - C_{i_F 0}\right) E_c, \qquad (3)$$

over the ranges where the Fermi level is in a spin sublevel $m$. Values of $(C_{i1} - C_{i0})$ are $+1/32, +1/16, +1/4$ for $i = 3, 2, 1$.

The theoretical contribution of the excitonic corrections to the splitting, to be compared with $E_{11} - E_{10} - \hbar \omega_{ch}$, is plotted in the lower part of Fig. 3. Reading the Figure from right to left (increasing field, decreasing $\nu$), this quantity varies in zigzags of increasing amplitude as the Fermi level drops down through the Landau levels. Its sign is *negative*: the magnetoexciton theory predicts a $j = 1$ valence hole to be more strongly bound than a $j = 0$ valence hole for all $\nu > 2$.

There is almost total disagreement between predicted and experimental splittings in Fig. 3. The measured splittings have opposite sign to the predicted splittings and much smaller magnitude. The zigzags in the data and in the theoretical graph do appear to be correlated, but they are in anti-phase. The amplitude measured for the zigzags is about $0.04E_c$ instead of $0.25E_c$ at low fillings.

Considerable discrepancy in the size of the excitonic corrections is expected due to the finite well width of 10 nm, since Eq. (3) is derived for pure 2D. Also, screening reduces excitonic binding energies, even if its effect is usually considered to be weak in quasi-2D. There is also the unknown effect of disorder. One might ask whether these large effects could act in such a way as to weaken the binding more effectively for a $j = 1$ hole than for a $j = 0$ hole but this does not seem likely.

In summary of this section, we can say only that the data confirm that the valence hole in levels $j = 0$ and $j = 1$ is bound, even at a high filling factor like $\nu = 6$, because we can observe a binding energy difference for the two hole states. But the simple theory used here appears totally inadequate for interpreting the data.

### 3.2. EXCHANGE CORRECTION IN EMISSION

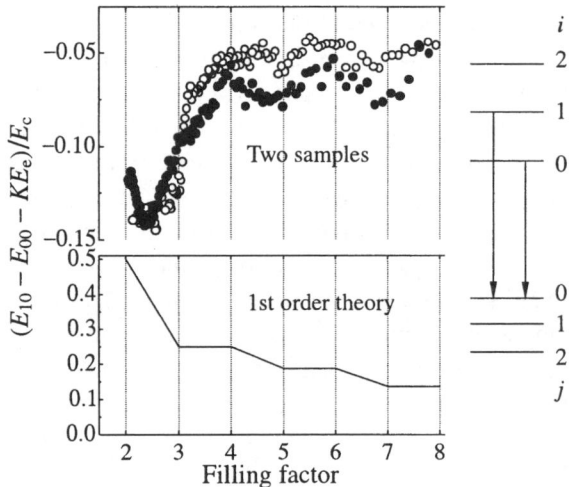

*Figure 4.* Energy difference, normalised to $E_c$, between the forbidden $1 \to 0$ and allowed $0 \to 0$ recombination transitions, after subtracting the electron kinetic energy $\hbar\omega_{ce}$, for Sample 1 (CdMnTe well, $\sigma^+$ polarisation, black filled circles) and Sample 2 (CdTe well, $\sigma^-$, open circles). The lower graph shows the theoretical (first-order) difference between the final state exchange corrections for the two transitions.

Similarly, Fig. 4 displays the energy difference between the two recombination transitions $1 \to 0$ and $0 \to 0$, again for Sample 1 (CdMnTe well, black filled circles) and Sample 2 ( CdTe well, open circles). The electron cyclotron energy $\hbar\omega_{ce}$ is subtracted ($m_e = 0.105m_0$). Here, it is the *initial state* of the recombination which is the same for the two transitions: a hole in the $j = 0$ valence band Landau level interacting with the system of conduction electrons. The final states for the two transitions differ in that there is a hole in the $i = 1$ and $i = 0$ conduction band levels respectively.

So (after subtraction of $\hbar\omega_{ce}$) the splitting between the forbidden and the allowed transitions is just the difference in the self-energy of the conduction band hole in the two final states. The initial state excitonic binding is eliminated in the difference. The observed difference is *negative* (data points in Fig. 4).

We again attempt to interpret the data in a simple first-order theory. The first-order estimate of the exchange interaction of a 2D electron of spin $m$ in conduction Landau level $i$ with a background of 2D electrons of the same spin $m$ on Landau levels $n$ is:

$$E_{\text{exchange}} = -\sum_n \tau_{n,m} C_{in} E_c, \qquad (4)$$

where $\tau_{n,m}$ is the filling fraction of Landau sublevel $n, m$. The numerical coefficients $C_{in}$ are identical to the coefficients $C_{ij}$ that appear in the excitonic binding, see Ref. [6]. The sign of the exchange interaction, being negative for an electron, is positive for the final state conduction hole, and this lowers (red-shifts) the transition energy (as did the initial state excitonic interaction).

From Eq. (4), as the Fermi level drops down through the higher conduction levels with increasing field, the magnitude of the exchange interaction of a hole in spin sublevel $m$ of the lowest conduction level ($i = 0$) decreases in a series of linear steps with intervening plateaux, in units of $E_c$. (Note: This decrease is surprisingly slight in meV units, see formulae in Ref. [1].) Similarly for a hole in conduction level $i = 1$. But the *difference* between the first-order exchange corrections actually *increases* in magnitude with field. The splitting $E_{10} - E_{00}$ is (for 0 K):

$$E_{10} - E_{00} = \hbar\omega_{ce} - \sum_n \tau_{n,m} (C_{1n} - C_{0n}) E_c. \qquad (5)$$

Some values of $\sum_n (C_{1n} - C_{0n})$ are: $-5/32$, $-3/16$, $-1/4$ and $-1/2$ at $\nu = 8, 6, 4, 2$ respectively. The splitting is thus predicted to be *positive* (after subtraction of $\hbar\omega_{ce}$): the theory gives the strongest exchange interaction for the lowest Landau level $i = 0$. This predicted splitting is plotted in the lower part of Fig. 4.

Again, there is near total disagreement between theory and experiment. Although the experimental splitting in the upper part of the Figure does increase in magnitude with decreasing $\nu$, the variation is in antiphase with the theory: the relative signs are wrong. Also, as for the excitonic splittings, the magnitude of the exchange splitting is strongly reduced compared to theoretical values. The largest measured splitting, about $-0.15 E_c$, is much smaller in magnitude than the expected $+0.5 E_c$.

Some of the discrepancy in order of magnitude certainly comes from the intermediate dimensionality and effects of disorder. But the sign error has no obvious explanation: from the data, one would conclude that the exchange correction is greater for a hole injected into the $i = 1$ level than for a hole injected into the $i = 0$ level.

## 4. Conclusion

In summary, we have demonstrated a method for separating exchange and excitonic corrections to the magneto-emission spectra, by measuring splittings between adjacent allowed (direct) and forbidden (indirect) transitions. Depending on which spectrum splittings are examined, this gives the difference between the excitonic binding for holes in two adjacent valence band Landau levels $j$ and $j + 1$, or it gives the difference between the exchange interaction of holes injected in adjacent conduction band Landau levels $i$ and $i + 1$.

The results are very surprising, being contrary to simple theoretical predictions. There is agreement on one point only: the experiments show the magnitude of both the excitonic and exchange splittings increasing with decreasing filling factor. But the observed energy differences are very small, of order a factor of 5 smaller than expected. And the *sign* of the energy differences disagrees with the theoretical predictions, for both the excitonic and the exchange effects.

The theory used is for pure 2D and may be overestimating interactions by a factor of perhaps two for our quasi-2D samples (the binding energy of the 1s zero-field exciton is only about 2 Rydbergs, not 4 Rydbergs, in undoped 10 nm CdTe wells). Also screening will reduce all interactions. But screening is supposed to be weak in quasi-2D and, moreover, the discrepancies persist near integer fillings, where the electron system is in principle incompressible and screening is supposed to disappear. Another obvious factor is disorder, but while structural disorder is commonly considered to reduce exchange interactions, it is less obvious that it will reduce excitonic effects. Perhaps electrostatic disorder needs to be considered, because it tends to separate conduction electrons and valence holes laterally. But it is not evident how any of these factors could change the *signs* of the energy differences between different initial state hole configurations, or between different final state hole configurations.

In conclusion, these measurements of the splittings between allowed and forbidden transitions suggest that the effects of excitonic and exchange interactions are not yet well understood for modulation-doped quantum wells.

## References

1. Hawrylak, P., Teran, F., Potemski, M., and Karczewski, G. (2002) *Physica E* **112**, 495.
2. Lyo, S.K., Jones, E.D., and Klem, J.F. (1988) *Phys. Rev. Lett.* **61**, 2265.
3. Hawrylak, P. and Potemsk, M. (1997) *Phys. Rev. B* **56**, 12386.
4. Lemaître, A., Testelin, C., Rigaux, C. Wojtowicz, T., and Karczewski, G. (2000) *Phys. Rev. B* **62**, 5059.
5. Macdonald, A.H and Ritchie, D.S. (1986) em Phys. Rev. B **33**, 8336.
6. MacDonald, A.H., Oji, H.C.A., and Liu K.L. (1986) *Phys. Rev. B* **34**, 2681.

# CLUSTER MEANFIELD APPROXIMATION FOR THE OPTICAL RESPONSE OF WEAKLY DOPED SEMICONDUCTOR QUANTUM WELLS

FRANZ X. BRONOLD
*Institut für Theoretische Physik, Otto-von-Guericke-Universität Magdeburg, D-39016 Magdeburg, Germany*
*Physikalisches Institut, Universität Bayreuth, D-95440 Bayreuth, Germany*

Abstract. The calculation of the optical properties of doped semiconductor quantum wells is an intricate many-body problem because of the dynamical response of the excess carriers to the photogenerated valence band hole. At low densities, however, where the main effect of the dynamical response is the formation of trions, a simple cluster meanfield approximation can be effectively employed to calculate the optical susceptibility.

Key words: dynamical response of excess carriers, cluster meanfield approximation, trions

## 1. Introduction

The interaction of an optically generated exciton with excess carriers in the conduction band has been the subject of extensive experimental and theoretical investigations. Especially the high density regime attracted much attention and is qualitatively understood [1–3]: Dynamical screening and Pauli blocking prohibit the formation of bound *eh* [4] states, the exciton is therefore unstable and its spectral weight distributed among *eh* scattering states. Residual correlations between the valence band (VB) hole and conduction band (CB) electrons are still important, however, and, depending on the CB electron to VB hole mass ratio, give rise to a more or less pronounced Fermi edge singularity [5], for instance, in optical absorption spectra.

With the seminal experiment of Kheng et al. [6], which showed that in a weakly n-doped semiconductor quantum well (QW), photogenerated *eh* pairs give rise to trions, i.e., bound *eeh* states comprising two CB electrons and one VB hole, the center of interested shifted to low densities. As a result of numerous theoretical and experimental studies [7–23], the importance of *eeh* states in the low density regime is now unambiguously established and a qualitative understanding of the low density regime is also starting to emerge.

Both the Fermi edge singularity and the trions are consequences of the dynamical response of the CB electrons to the sudden appearance of the photogenerated VB hole. A typical lowest order process is depicted in Fig. [1]. The photogenerated VB hole scatters from one momentum state to another and simultaneously excites a virtual $e\bar{e}$ pair (recall our notation [4]) to compensate for the momentum transfer. At low densities, the Coulomb interaction is strong enough to correlate the VB hole with two CB electrons (the photoexcited CB electron and the CB electron from the $e\bar{e}$ pair) and a (negatively charged) trion appears. At higher CB electron densities, an *arbitrary* number of $e\bar{e}$ pairs is excited, giving rise to, among others, the screening of the Coulomb interaction. At some density, the Coulomb interaction is then too weak to support trions and/or excitons, but the residual correlations might produce a Fermi edge singularity.

The close connection between trion formation at low densities and the appearance of the Fermi edge singularity at high densities has been verified experimentaly [20–22] but, so far, a unified theoretical description of the full density dependence of the optical susceptibility, based, e.g., on the qualitative considerations of the previous paragraph, is still missing. To construct such a theory is a rather formidable task. Here, we focus therefore on the low density regime, where the cluster meanfield approximation (CMFA) [24] can be effectively used to calculate the optical susceptibility [12]. In contrast to early theoretical studies of trion states [7–11], the CMFA describes trion and exciton states simultaneously. Moreover, it explicitly accounts for a (low) density of excess carriers. [In this sense it is closely related to the density matrix method of Esser et al. [14] and the exciton-electron T-matrix model of Suris et al. [15].] The high density regime, however, cannot be systematically reached, despite the earlier optimistic assessment of the author.

## 2. Cluster Meanfield Approximation (CMFA)

### 2.1. OPTICAL SUSCEPTIBILITY

We consider for simplicity an idealized strictly two-dimensional quantum well with two isotropic parabolic bands. With an appropriate choice of single particle states, our formalism can be of course also applied to more realistic QW models, which include, e.g., valence band mixing and finite band off-sets. We assume a single hole in the VB and a low concentration $n$ of electrons in the CB. The charge carriers are coupled through the Coulomb interaction and the total Hamiltonian is given by

$$H = H_0 + V$$

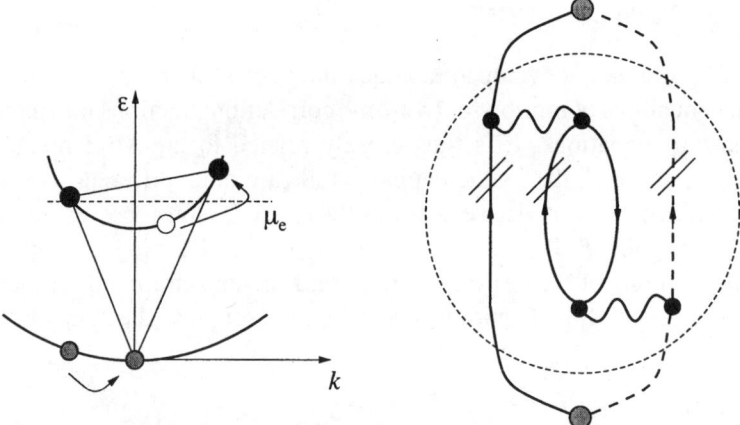

*Figure 1.* The left panel illustrates a typical dynamical response of the CB electrons to the photogenerated VB hole in an idealized n-doped semiconductor ($\mu_e$ is the chemical potential for the CB electrons). The right panel shows the corresponding Feynman diagram. The dashed forward running line denotes a VB hole, whereas the forward and backward running solid lines depict, respectively, a CB electron and CB hole. Multiple scattering, indicated by the thin stripes attached to the participating lines, correlates the intermediate *eeh* cluster to a trion.

$$= \sum_{\vec{k}\sigma}\left[\epsilon_c(k)a^{\dagger}_{\vec{k}\sigma}a_{\vec{k}\sigma} + \epsilon_v(k)b^{\dagger}_{\vec{k}\sigma}b_{\vec{k}\sigma}\right]$$

$$+ \frac{1}{2}\sum_{\vec{k}\vec{p}\vec{q}\sigma\tau} v(q)\left[a^{\dagger}_{\vec{k}+\vec{q}\sigma}a^{\dagger}_{\vec{p}-\vec{q}\tau}a_{\vec{p}\tau}a_{\vec{k}\sigma} - 2a^{\dagger}_{\vec{k}+\vec{q}\sigma}b^{\dagger}_{\vec{p}-\vec{q}\tau}b_{\vec{p}\tau}a_{\vec{k}\sigma}\right], \quad (1)$$

where $\epsilon_c(k) = E_g + k^2/2m_e$ and $\epsilon_v(k) = k^2/2m_h$ are the CB and VB dispersions ($\hbar = 1$), respectively, and $V(q) = 4\pi e^2/\epsilon_0 q$ is the Coulomb interaction in standard notation.

If the momentum dependence of the dipole matrix element $r_\gamma$ is ignored, linear response theory gives for the optical susceptibility ['1' stands for $\vec{k}_1$]

$$\chi(\omega) = 2|r_\gamma|^2 \sum_{12}\langle\langle b_{-1}a_1; a_2^{\dagger}b^{\dagger}_{-2}\rangle\rangle_\omega \quad (2)$$

where $\langle\langle b_{-1}a_1; a_2^{\dagger}b^{\dagger}_{-2}\rangle\rangle_\omega$ denotes the Fourier transform of the two-time correlation function

$$\langle\langle (b_{-1}a_1)(t); (a_2^{\dagger}b^{\dagger}_{-2})(0)\rangle\rangle = -i\Theta(t)\langle\left[(b_{-1}a_1)(t), \left(a_2^{\dagger}b^{\dagger}_{-2}\right)(0)\right]\rangle. \quad (3)$$

Here, the thermodynamic expectation value is defined by $\langle\ldots\rangle = Z^{-1}Tr(e^{-\beta\mathcal{H}}\ldots)$, with $Z = Tre^{-\beta\mathcal{H}}$, and $\mathcal{H} = H - \mu_e N_e$. The imaginary part of $\chi(\omega)$ determines, apart from a constant factor, the optical absorption $I_{abs}(\omega) \sim -Im\chi(\omega)$. The overall factor of two in front of the sum on the rhs of Eq. (2) comes from the spin.

## 2.2. DYSON-TYPE EQUATIONS

The CMFA, originally developed to attack the nuclear many-body problem, rearranges the equations of motion for two-time correlation functions into an hierarchy of Dyson-type equations. It is thus closely related to the Mori memory function method, although the mathematical details are quite different. Using CMFA techniques [24], we can derive a formally exact Dyson-type equation for the $eh$ pair propagator $P(12; \omega) = \langle\langle b_{-1}a_1; a_2^{\dagger}b_{-2}^{\dagger}\rangle\rangle_\omega$. Keeping in mind that the spin configurations of the various correlation functions are fixed, because of the spin-independence of the Coulomb scattering, we suppress the spin variables and write:

$$[\omega + i\eta - Z(1)] P(12; \omega) = N(1)\{\delta_{12} + \sum_3 [M^{st}(13) + \delta M(13; \omega)] P(32; \omega), \quad (4)$$

with

$$M^{st}(12) = N^{-1}(1)N^{-1}(2)\langle[[b_{-1}a_1, V], a_2^{\dagger}b_{-2}^{\dagger}]\rangle \quad (5)$$

$$\delta M(12; \omega) = N^{-1}(1)N^{-1}(2)\langle\langle[b_{-1}a_1, V]; [V, a_2^{\dagger}b_{-2}^{\dagger}]\rangle\rangle_\omega \quad (6)$$

$$N(1) = \langle[b_{-1}a_1, a_1^{\dagger}b_{-1}^{\dagger}]\rangle \quad (7)$$

$$Z(1) = \epsilon_c(1) + \epsilon_v(-1). \quad (8)$$

Applying the CMFA technique to the correlation function defining $\delta M(12; \omega)$, and so on, we obtain an hierarchy of Dyson-type equations for correlation functions with an increasing number of particles. Clearly this set of equations has to be truncated and it is the truncation which restricts the CMFA to low CB electron densities.

We now briefly recall the truncation procedure adopted in Ref. [12]. The vanishing VB hole concentration implies $\langle b_1^{\dagger}b_1\rangle = 0$, i.e., $N(1) = 1 - \langle a_1^{\dagger}a_1\rangle$. By assumption, the CB electron density is also small, i.e., $N(1) \simeq 1$. Thus, in leading order in the CB electron density, we ignore Pauli blocking. In the low density regime, we also neglect screening and single-particle selfenergy corrections and Eq. (4) reduces to

$$P(12; \omega) = P^{(0)}(12; \omega) + \sum_{34} P^{(0)}(13; \omega) [M^{st}(34) + \delta M(34; \omega)] P(42; \omega), (9)$$

with a bare $eh$ pair propagator, $P^{(0)}(12; \omega) = \delta_{12}[\omega + i\eta - Z(1)]^{-1}$, and a static part of the $eh$ pair selfenergy $M^{st}(45) = -v(4-5)$. The dynamical response of the CB electrons, i.e., the creation of virtual $e\bar{e}$ pairs, is encoded in the dynamical (resonating) part of the $eh$ pair selfenergy, which, in the low density regime, becomes

$$\begin{aligned}\delta M(12;\omega) = \sum_{4567} v(5)v(7)\big(&R(1,4,4+5,5-1|2,6,6+7,7-2;\omega)\\ -\,&R(1-5,4,4-5,-1|2,6,6+7,7-2;\omega)\\ -\,&R(1,4,4+5,5-1|2-7,6,6-7,-2;\omega)\\ +\,&R(1-5,4,4-5,-1|2-7,6,6-7,-2;\omega)\big).\end{aligned} \quad (10)$$

We defined an eightpoint $ee\bar{e}h$ function $R(1235|5678;\omega) = \langle\langle b_4 a_3^\dagger a_2 a_1; a_5^\dagger a_6^\dagger a_7 b_8^\dagger\rangle\rangle_\omega$, for which, as indicated above, we again derive a Dyson-type equation, symbolically written as $R(\omega) = R^{(0)}(\omega) + R^{(0)}(\omega)[K^{st} + \delta K(\omega)]R(\omega)$. [Assuming that the spin configuration of the photogenerated $eh$ pair is $(\uparrow\downarrow)$, the spin configuration of the $ee\bar{e}h$ cluster defining $R$ is $(\uparrow \sigma\sigma \downarrow)$; $\sigma$ is then summed over in Eq. (10).] In the dilute limit, we need $\delta M(12;\omega)$ only to $o(n)$. Using an $\bar{e}$-line expansion for $\delta M(12;\omega)$, we can show (i) that the term $\delta K(\omega)$ gives rise to $o(n^2)$ contributions to $\delta M(12;\omega)$ and is thus negligible at low densities, and (ii) that, in leading order in the CB electron density, the eightpoint $ee\bar{e}h$ function appearing in Eq. (10) in fact factorizes into

$$R(1234|5678;\omega) = \delta_{37} f_e(3) G_3(12|56;\omega + \epsilon_c(3)), \quad (11)$$

with $f_e(3) = \langle a_3^\dagger a_3 \rangle$ and a sixpoint $eeh$ function $G_5(12|34;\omega) = \langle\langle b_{5-1-2} a_2 a_1; a_3^\dagger a_4^\dagger b_{5-3-4}^\dagger\rangle\rangle_\omega$, which satisfies

$$[\omega + i\eta - Z_5(12)]\,G_5(12|34;\omega) = I(12|34) + \sum_{67} \mathcal{V}_5(12|67) G_5(67|34;\omega), \quad (12)$$

with

$$\mathcal{V}_5(12|67) = v(7-2)\delta_{6,1+2-7} - v(1-6)\delta_{2,7} - v(2-7)\delta_{6,1}, \quad (13)$$

and

$$Z_5(12) = \epsilon_c(1) + \epsilon_c(2) + \epsilon_v(5-1-2). \quad (14)$$

Thus, at low densities, the hierarchy of Dyson-type equations can be truncated at the level of a sixpoint $eeh$ function and, in leading order in the CB electron density, the optically generated $eh$ pair is only coupled to an $eeh$ cluster. The groundstate of the $eeh$ cluster has antiparallel CB electron spin. Focusing on the $eeh$ groundstate, we fix therefore the spin configuration of the $eeh$ cluster defining $G$ to $(\uparrow\downarrow\downarrow)$. For this spin configuration, $I(12|34) = \delta_{13}\delta_{24}$, and the spin eventually leads to the overall factor of two in Eq. (2) for the optical susceptibility.

Using the spectral representation for the sixpoint $eeh$ function,

$$G_5(12|34; \omega) = \sum_n \frac{\Psi_{5n}(12)\Psi_{5n}^*(34)}{\omega + i\eta - \Omega_n(5)}, \quad (15)$$

we finally obtain for the dynamic part of the $eh$ pair selfenergy

$$\delta M(34; \omega) = \sum_{5n} f_e(5) \frac{\Xi_{5n}(3)\Xi_{5n}^*(4)}{\omega + i\eta + \epsilon_c(5) - \Omega_n(5)}, \quad (16)$$

where all multiple scattering events within the $eeh$ cluster are now encoded in a vertex function

$$\Xi_{5n}(3) = \sum_7 v(5-7)[\Psi_{5n}(37) - \Psi_{5n}(7, 3+5-7)], \quad (17)$$

given in terms of (normalized and complete) momentum-space $eeh$ wavefunctions $\Psi_{5n}(37)$, which satisfy an $eeh$ momentum-space Schrödinger equation:

$$[\Omega_n(5) - Z_5(12)]\Psi_{5n}(12) - \sum_{34} \mathcal{V}_5(12|34)\Psi_{5n}(34) = 0. \quad (18)$$

Here, 'n' and '5' depict, respectively, internal quantum numbers and the (center-of-mass) momentum of the (propagating) bound $eeh$ cluster.

To calculate $P(12; \omega)$ it is advantageous to split Eq. (9) into two coupled integral equations with, respectively, $M^{st}(=-v)$ and $\delta M$ as kernels:

$$P^{st}(12; \omega) = P^{(0)}(12; \omega) - \sum_{34} P^{(0)}(13; \omega)v(3-4)P^{st}(42; \omega) \quad (19)$$

$$P(12; \omega) = P^{st}(12; \omega) + \sum_{34} P^{st}(13; \omega)\delta M(34)P(42; \omega) \quad (20)$$

The first integral equation is readily solved using again a spectral representation

$$P^{st}(12; \omega) = \sum_\nu \frac{\Phi_\nu(1)\Phi_\nu^*(2)}{\omega + i\eta - E_\nu}, \quad (21)$$

where the (normalized and complete) momentum-space exciton wave functions $\Phi_\nu(1)$ obey an $eh$ momentum-space Schrödinger equation (Wannier equation)

$$[E_\nu - Z(1)]\Phi_\nu(1) + \sum_2 v(1-2)\Phi_\nu(2) = 0. \quad (22)$$

The solutions of this equation are analytically known. [25]

The second integral equation cannot be solved exactly. In order to avoid heavy numerics, we employ for its solution a simple approximation scheme based on the

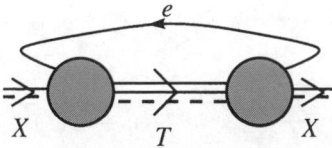

*Figure 2.* Graphical illustration (not a Feyman diagram) of the dynamical part of the exciton selfenergy $\delta M_X(\omega)$. The full circles denote the exciton-trion coupling constants $u_{5T}^X$ and $[u_{5T}^X]^*$.

"exciton representation", i.e., on a bilinear expansion in terms of the solutions of Eq. (22). Writing $P(12; \omega) = \sum_{\nu,\mu} P_{\nu\mu}(\omega)\Phi_\nu(1)^*\Phi_\mu(2)$, we transform Eq. (20) into an infinite set of algebraic equations for the expansion coefficients,

$$[\omega + i\eta - E_\nu] P_{\nu\mu}(\omega) = \delta_{\nu\mu} + \sum_\lambda \delta M_{\nu\lambda}(\omega) P_{\lambda\mu}(\omega), \quad (23)$$

where

$$\delta M_{\nu\lambda}(\omega) = \sum_{5n} f_e(5) \frac{u_{5n}^\nu \left[u_{5n}^\lambda\right]^*}{\omega + i\eta + \epsilon_c(5) - \Omega_n(5)}, \quad (24)$$

and $u_{5n}^\nu = \sum_1 \Phi_\nu^*(1)\Xi_{5n}(1)$. In the vicinity of the exciton resonance, i.e., for energies $\omega$ close to the exciton groundstate energy $E_X$, we expect the exciton and trion groundstate to be dominant. Thus, we solve Eq. (23) by employing one-pole approximations: First, we keep only the term corresponding to the exciton ground state $\nu = X$ and neglect the coupling to excited exciton states $\nu \neq X$. Next, assuming that the main modification of the exciton ground state comes from the groundstate of the *eeh* cluster, i.e., from the trion denoted by $n = T$, we restrict the sum in Eq. (24) to the term $n = T$.

Let us emphasize, the one-pole approximations are by no means mandatory and various improvements are possible. For example, if the exciton binding energy is small, i.e., if the exciton groundstate is not well separated from the first excited state, the first excited state has to be kept. This gives rise to two coupled algebraic equations (two-pole approximation). It is even conceivable to treat all *eh* scattering states in a similar way, introducing an "effective" excited state, which carries the spectral weight of the whole continuum. Excited trion states can be also taken into account by simply extending the sum in Eq. (24) to the respective states.

The theoretical work by Esser et al. [14] indicates that these refinements are indeed necessary to obtain good quantitative agreement between theory and experiment. For the purpose of demonstration, however, we neglect here scattering states and stick to the simple one-pole approximation described above. The optical susceptibility in the vicinity of the exciton resonance is then given by

$$\chi(\omega) \simeq \frac{2|s_X|^2}{\omega + i\eta - E_X - \delta M_X(\omega)}, \quad (25)$$

with an exciton selfenergy (see Fig. 2)

$$\delta M_X(\omega) = \sum_5 f_e(5) \frac{|u_{5T}^X|^2}{\omega + i\eta + \epsilon_c(5) - \Omega_T(5)}, \quad (26)$$

and an exciton oscillator strength $s_X = r_\gamma \sum_1 \Phi_X(1)$. Note, the one-pole approximation is conceptually close to the T-matrix model considered by Suris et al. in their analysis of absorption and reflection spectra of modulation-doped quantum wells. [15]

To proceed further, we assume that the CB electron density is such that the Fermi function in Eq. (26) can be replaced by a Boltzmann function parametrized by the CB electron density $n$. Measuring energies in units of the 2D exciton binding energy $4R$ ($R$ is the 3D exciton Rydberg) and length in units of the 3D Bohr radius $a_B$, we define $\tilde{\chi}(\tilde{\omega}) \equiv \chi(4R\tilde{\omega} + E_g)/4R$, with $\tilde{\omega} = [\omega - E_g]/4R$, and rewrite the optical susceptibility into dimensionless form

$$\tilde{\chi}(\tilde{\omega}) = \frac{|s_X|^2}{8R^2} \frac{1}{\tilde{\omega} + i\tilde{\eta} + 1 - \delta \tilde{M}_X(\tilde{\omega})}. \quad (27)$$

Here, $\tilde{\eta} = \eta/4R$ and

$$\delta \tilde{M}_X(\tilde{\omega}) \equiv \delta M_X(4R\tilde{\omega} + E_g)/4R \quad (28)$$

$$= \tilde{\beta} \frac{M_T}{M_X} (na_B^2) |u_T^X|^2 I(\tilde{\omega}), \quad (29)$$

where $\tilde{\beta} = 4\beta R$, $M_X$, and $M_T$ are, respectively, the thermal energy measured in units of $4R$, the exciton mass, and the trion mass. The exponential integral,

$$I(\tilde{\omega}) = \int_0^\infty dy \frac{e^{-\tilde{\beta} \frac{M_T}{M_X} y}}{\tilde{\omega} + i\tilde{\eta} - \tilde{\epsilon}_T + y}, \quad (30)$$

where $\tilde{\epsilon}_T = [\Omega_T(0) - 2E_g]/4R$ is the energy of a trion at rest, i.e., the *internal* trion energy, which is the binding energy of *two* CB electrons to one VB hole. The dimensionless exciton-trion coupling constant $u_T^X$ finally is given by an integral involving the wave function of the trion groundstate, the Coulomb interaction, and the wavefunction of the exciton groundstate. For simplicity we neglected the weak dependence of $u_T^X$ on the (center-of-mass) momentum of the trion. Note, for vanishing CB electron density $n$, $\delta \tilde{M}_X(\tilde{\omega}) \to 0$ and $\tilde{\chi}(\tilde{\omega})$ describes a single exciton resonance.

To obtain the exciton-trion coupling constant $u_T^X$ and the internal trion energy $\tilde{\epsilon}_T$, we employ the variational technique originally used by Stébé and coworkers [9, 10] to calculate various properties of an isolated trion. For that purpose we transform Eq. (18) to real space (to solve the *eeh* problem we use atomic units, which are, however, at the end of the calculation, translated into the excitonic units adopted above)

$$[T + V]\Psi_T(\vec{r}_{e1}\vec{r}_{e2}\vec{r}_h) = \epsilon_T \Psi_T(\vec{r}_{e1}\vec{r}_{e2}\vec{r}_h),$$

with kinetic and potential energies ($\hbar = 1$),

$$T = -\frac{1}{2m_e}(\Delta_{e1} + \Delta_{e2}) - \frac{1}{2m_h}\Delta_h,$$

$$V = -\frac{e^2}{\epsilon_0}\left(\frac{1}{|\vec{r}_{e1} - \vec{r}_h|} + \frac{1}{|\vec{r}_{e2} - \vec{r}_h|} - \frac{1}{|\vec{r}_{e1} - \vec{r}_{e2}|}\right),$$

separate away the center-of-mass motion, which is simply a plane wave, and make a variational Hylleraas Ansatz for the internal part of the trion groundstate wavefunction,

$$\Psi(s, t, u) = \sum_{lnm} c_{lnm} |lnm\rangle, \tag{31}$$

where $|lnm\rangle = e^{-\frac{s}{2}}s^l t^m u^n$ and $u$, $s$, and $t$ are ellipitical coordinates for the internal motion of the *eeh* cluster. Note, since the trion groundstate is a singlet, $\Psi(s, t, u)$ has to be an even function in $t$, i.e., $m$ has to be even. The expansion coefficients are determined by a Ritz variational principle, i.e., by minimizing the energy functional $E[\vec{c}, \epsilon_T] = \langle\Psi|T + V|\Psi\rangle/\langle\Psi|\Psi\rangle$. The variation yields an eigenvalue problem which is then iteratively solved. The largest eigenvalue is related to $\epsilon_T$ ($\to \tilde{\epsilon}_T$) and the associated eigenfunction gives the expansion coefficients $c_{lnm}$. For more details we refer the readers to Refs. [9, 10] and [26]. With the Hylleraas variational function, the exciton-trion coupling constant becomes

$$u_T^X = \sqrt{\frac{2}{N\pi}}\frac{1}{k(1+\sigma)} \int_0^\infty ds \int_0^s du \int_0^u dt \frac{2\pi(s^2 - t^2)u}{\sqrt{(u^2 - t^2)(s^2 - u^2)}}$$

$$\times e^{-\frac{s+t}{k(1+\sigma)}}\left[\frac{2}{s-t} - \frac{1}{u}\right]\Psi(s, t, u), \tag{32}$$

where $k$ is the effective charge of a CB electron, $N$ is the norm of $\Psi$, and $\sigma = m_e/m_h$.

## 2.3. SOME NUMERICAL RESULTS

Using a 2D 22-term Hylleraas wavefunction, we calculated in Ref. [12] the optical absorption along the lines explained in the previous subsections. For a mass ratio $\sigma = m_e/m_h = 0.146$, corresponding to GaAs, we found $u_T^X \simeq 1.0896$ and $\tilde{\epsilon}_T = -1.1151$. ($u_T^X$ was obtained by Gaussian quadrature.) The binding energy (in units of $4R$) for *one* electron is therefore $\tilde{W}_T = 0.1151$ in agreement with the binding energies obtained by other means. [7]

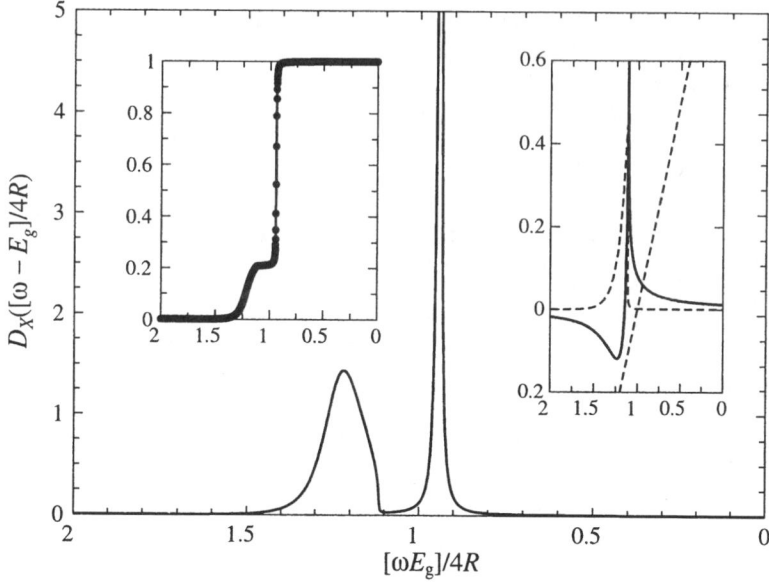

*Figure 3.* The spectral function $D_X(\tilde{\omega})$ for $na_B^2 = 0.008$ and $4R\beta = 10$. The effective mass ratio $\sigma = 0.146$, corresponding to GaAs, and $\tilde{\eta} = 0.001$. The left inset depicts the integrated spectral weight $C_X(\tilde{\omega})$, whereas the right inset displays $\tilde{\Gamma}_X(\tilde{\omega})$ (dot-dashed line) and $\tilde{\Delta}_X(\tilde{\omega})$ (solid line). The energy for which $\tilde{\Delta}_X(\tilde{\omega})$ crosses $\tilde{\omega} + 1$ (dotted line) defines the exciton line at $\tilde{\omega} = -0.94$. From Ref. [12].

The optical absorption in the vicinity of the exciton resonance is related to the exciton spectral function

$$D_X(\tilde{\omega}) = -Im\tilde{\chi}(\tilde{\omega}), \qquad (33)$$

shown in Fig. 3 for $na_B^2 = 0.008$ and $\tilde{\beta} = 10$. The overall structure is in qualitative agreement with experiments, despite the simplicity of the model. Below a narrow peak at $\tilde{\omega} = -0.94$, we see a broad absorption band with a sharp high energy edge at $\tilde{\omega} = -1.1151$ and a low energy tail. The narrow peak is the exciton resonance whereas the broad band corresponds to the absorption due to a trion. As shown in Ref. [12], the sharp high energy edge of the absorption band comes from the pole of the exciton selfenergy at $\tilde{\epsilon}_T = -1.1151$, i.e., a photon in resonance with the high energy edge creates a trion with momentum zero. The band has a tail at the low energy side because of the recoil of the trion, i.e., the tail comprises trion states with finite momentum. The right inset of Fig. 3 displays the real and imaginary parts of the exciton selfenergy, $\delta \tilde{M}_X(\tilde{\omega}) = \tilde{\Delta}_X(\tilde{\omega}) - i\tilde{\Gamma}_X(\tilde{\omega})$. As can be seen, trion states with small momenta (high energy edge) are strongly damped. The maximum of the absorption band does therefore not correspond to a trion with zero momentum (as in the case of vanishing CB electron concentration [11]) but to a trion state with a finite momentum. The left inset depicts the integrated spectral weight up to an energy $\tilde{\omega}$ defined in terms of the cumulant $C_X(\tilde{\omega}) = \int_{-\infty}^{\tilde{\omega}} d\tilde{E} D_X(\tilde{E})$. As a consequence of the one-pole approximation, the total spectral weight associated with the trion band and the exciton line adds up to one. If scattering states had been taken into account, they would also carry some spectral weight, i.e., in a complete theory, the combined spectral weight of trion and exciton has to be of course less then one. Our exploratory calculation neglected scattering states, because, at low enough CB densities, in particular, within the range of validity of our theory, scattering states are expected to carry almost no spectral weight. Indeed, the more refined calculations of Esser et al. [14] seem to verify this assessment. The spectral weight, which can be associated with scattering states, is less then 5% of the spectral weight of the exciton groundstate.

## 3. Conclusions

We presented a detailed description of the CMFA for the optical response of a weakly n-doped idealized semiconductor QW. The CMFA, tailormade for low CB electron densities, reduces the calculation of the optical susceptibility to the solution of the Schrödinger equations for, respectively, a single *eh* pair and a single *eeh* cluster. Excess carriers are only treated as a reservoir for bound state formation, which does not, in leading order in the CB electron density, modify the properties of the *eh* and *eeh* states. We derived the CMFA for an idealized QW model, but it is straightforward to generalize the formalism to more realistic models. Furthermore, relaxing the one-pole approximations, which we solely used for illustration, would allow to take exciton and trion continuum states into account. At intermediate-to-high CB electron density, many-body medium corrections, such as Pauli blocking, screening, and single particle selfenergy corrections, substantially modify *eh* and *eeh* states. Within the CMFA it is however not clear how to systematically account for these effects.

## References

1. Haug, H. and Schmitt-Rink, S. (1984), Electron theory of the optical properties of laser-excited semiconductors, *Prog. Quant. Electr.* **9**, 3–100.
2. Schmitt-Rink, S., Chemla, D.S., and Miller, D.A.B. (1989), Linear and nonlinear optical properties of semiconductor quantum wells, *Adv. Phys.* **38**, 89–188.
3. Zimmermann, R. (1988), *Many-Particle Theory of Highly Excited Semiconductors*, Teubner Texte für Physik, Bd. 18, Teubner Verlagsgesellschaft, Leibzig.
4. $e, \bar{e}$, and $h$ denote, respectively, CB electrons, CB holes, and VB holes; a trion, e.g., is therefore a *eeh* cluster.
5. For a recent review see, e.g., Brum, J.A. and Hawrylak, P. (1997), Fermi egde singularity in the optical properties of two-dimensional electron gas, *Comments Cond. Mat. Phys.* **18**, 135–161.
6. Kheng, K., Cox, R.T., Merle d'Aubigné, Y., Bassani, F., Saminadayar, K., and Tatarenko, S. (1993), Observation of negatively charged excitons $X^-$ in semiconductor quantum wells, *Phys. Rev. Lett.* **71**, 1752–1755.

7. Usukura, J., Suzuki, Y., and Varga, K. (1999), Stability of two- and three-dimensional excitonic complexes, *Phys. Rev. B* **59**, 5652–5661.
8. Ruan, W.J., Chan, K.S., Ho, H.P., Zhang, R.Q., and Pun, E.Y.P. (1999), Hyperspherical approach for charged excitons in quantum wells, *Phys. Rev. B* **60**, 5714–5720.
9. Stébé, B. and Ainane, A. (1989), Ground-state energy and optical absorption of excitonic trions in two-dimensional semiconductors, *Superlattices and Microstructures* **5**, 545–548.
10. Stébé, B., Munschy, G., Stauffer, L., Dujardin, F., and Murat, J. (1997), Excitonic trion $X^-$ in semiconductor quantum wells, *Phys. Rev. B* **56**, 12454–12461.
11. Stébé, B., Feddi, E., Ainane, A., and Dujardin, F. (1998), Optical and magneto-optical absorption of negatively charged excitons in three- and two-dimensional semiconductors, *Phys. Rev. B* **58**, 9926–9932.
12. Bronold, F. X. (2000), Optical absorption of a weakly n-doped semiconductor quantum well, *Phys. Rev. B* **61**, 12620–12623.
13. Esser, A., Runge, E., Zimmermann, R., and Langbein, W. (2000), Photoluminescence and radiative lifetime of trions in GaAs quantum wells, *Phys. Rev. B* **62**, 8232–8239.
14. Esser, A., Zimmermann, R., and Runge, E. (2001), Theory of trion spectra in semiconductor nanostructures, *phys. stat. sol. (b)* **227**, 317–330.
15. Suris, R.A., Kochereshko, V.P., Astakhov, G.V., Yakovlev, D.R., Ossau, W., Nürnberger, J., Faschinger, W., Landwehr, G., Wojtowicz, T., Karczewski, G., and Kossut, J. (2001), Excitons and Trions modified by interaction with a two-dimensional electron gas, *phys. stat. sol. (b)* **227**, 343–353.
16. Lovisa, S., Cox, R.T., Magnea, N., and Saminadayar, K. (1997), Filling-factor dependence of the negatively-charged-exciton absorption in a CdTe quantum well, *Phys. Rev. B* **56**, R12787–R12790.
17. Siviniant, J., Scalbert, D., Kavokin, A.V., Coquillat, D., Lascaray, J.-P. (1999), Chemical equilibrium between excitons, electrons, and negatively charged excitons in semiconductor quantum wells, *Phys. Rev. B* **59**, 1602–1604.
18. Eytan, G., Yayon, Y., Rappaport, M., Shtrikman, H., and Bar-Joseph, I. (1998), Near-Field spectroscopy of gated electron gas: A direct evidence for electron localization, *Phys. Rev. Lett.* **81**, 1666–1669.
19. Manassen, A., Cohen, E., Arza Ron, Linder, E., and Pfeiffer, L.N. (1996), Exciton and trion spectral line shape in the presence of an electron gas in GaAs/AlAs quantum wells, *Phys. Rev. B* **54**, 10609–10613.
20. Shields, A.J., Pepper, M., Ritchie, D.A., and Simmons, M.Y. (1995), Influence of excess electrons and magnetic fields on Mott-Wannier excitons in GaAs quantum wells, *Adv. Phys.* **44**, 47–72.
21. Brown, S.A., Young, Jeff F., Brum, J.A., Hawrylak, P., and Wasilewski, Z. (1996), Evolution of the interband absorption threshold with the density of a two-dimensional electron-gas, *Phys. Rev. B* **54**, R11082–R11085.
22. Yusa, G., Shtrikman, H., and Bar-Joseph, I. (2000), Onset of exciton absorption in modulation-doped GaAs quantum wells, *Phys. Rev. B* **62**, 15390–15393.
23. For a recent comprehensive overview see, e.g., Zimmermann, R. (Ed.) (2001), Proceedings of the *Miniworkshop on Trion Physics*, *phys. stat. sol. (b)* **227**.
24. See, e.g., Dukelsky, J., Röpke, G., and Schuck, P. (1998), Generalized Brückner-Hartree-Fock theory and self-consistent RPA, *Nucl. Phys. A* **628**, 17–40.
25. Haug, H. and Koch, S.W. (1990), *Quantum Theory of the Optical and Electronic Properties of Semiconductors*, World Scientific, Singapore.
26. Bethe, H.A. and Salpeter, E.E. (1957), *Quantum Mechanics of one- and two-electron atoms*, Springer Verlag, Berlin.

# D'YAKONOV–PEREL'S SPIN RELAXATION UNDER ELECTRON–ELECTRON COLLISIONS IN QWs

M.M. GLAZOV AND E.L. IVCHENKO
*Ioffe Physico-Technical Institute, 194021, St Petersburg, Russia*

Key words: quantum wells, spin relaxation, electron-electron interaction, Boltzmann equation

## 1. Introduction

In recent years the electron spin relaxation processes have received much attention in connection with various spintronics applications. For free electrons in quantum well structures the following four mechanisms of spin decoherence are most important (see [1, 2] and references therein): the Elliot–Yafet, D'yakonov–Perel', Bir–Aronov–Pikus mechanisms and electron spin-flip scattering by paramagnetic centers. This paper is devoted to the D'yakonov–Perel' spin relaxation (DPSR) in which case the spin relaxation time, $\tau_s$, is related to the spin splitting and given by [3]

$$\tau_s^{-1} \propto \langle \Omega_k^2 \rangle \tau . \tag{1}$$

Here $\Omega_k$ is the effective Larmor frequency describing the splitting of the electron spin dispersion branches, the angle brackets mean averaging over the electron energy distribution and $\tau$ is a microscopic electron relaxation time. In a two-dimensional (2D) system laking a center of a symmetry, e.g. quantum well or heterojunction, the frequency $\Omega_k$ is linear in $k$. In this case the time $\tau$ was usually identified with the momentum relaxation time that determines the electron mobility [4–8]. In [9] we have shown that the inverse time $\tau^{-1}$ is determined not only by the momentum scattering rate but contains also an independent contribution from electron–electron collisions which make no effect on the electron mobility. Really, electron–electron collisions change the direction of $k$ and $\Omega_k$ and, therefore, they control the Dyakonov–Perel' spin relaxation exactly in the same way as any other scattering processes do.

The paper is organized as follows. Section 2 contains the discussion of mechanisms of the spin splitting of the electron subbands in quantum wells and electron spin decoherence due to spin splitting, Sec. 3 describes the D'yakonov–Perel' spin relaxation mechanism in terms of the kinetic theory, in Sec. 4 we analyze the electron–electron collision integral for the spin-polarized electron gas, and Sec. 5 describes the solution of the kinetic equation with the account of electron–electron collisions.

## 2. Spin Splitting of Electron Subbands

In the parabolic approximation the effective electron Hamiltonian in the zinc-blende-based (001)-grown quantum well (QW) can be written as

$$\mathcal{H} = \frac{\hbar^2 k^2}{2m} + \beta_1(\sigma_y k_y - \sigma_x k_x) + \beta_2(\sigma_x k_y - \sigma_y k_x), \quad (2)$$

where $x \| [100]$, $y \| [010]$, $\beta_{1,2}$ are constants, $m$ is the in-plane effective mass, $\sigma_l$ are the Pauli matrices and $k_n$ are the components of the 2D electron wavevector, $\boldsymbol{k}$. The terms with higher powers in $\boldsymbol{k}$ are not considered here. In the symmetrical QWs (the $D_{2d}$ point group) the linear-$\boldsymbol{k}$ spin-dependent term proportional to $\beta_1$ is allowed only, it is called the bulk-inversion asymmetry (BIA) term. In asymmetrical QW structures characterized by the $C_{2v}$ point-group symmetry, there exists another spin-dependent contribution proportional to $\beta_2$ and called the structure-inversion asymmetry (SIA) term or the Rashba term [10, 11] (see also [12] and references therein). The structure asymmetry can be related with non-equivalent normal and inverted interfaces, external or built-in electric fields, compositionally stepped QWs etc.

It is convenient to introduce the Cartesian coordinates $x' \| [1\bar{1}0]$, $y' \| [110]$, $z \| [001]$ which allow to write a sum of the BIA and SIA terms in the form

$$\mathcal{H}_{c1}(\boldsymbol{k}) = \left(\beta_- \sigma_{x'} k_{y'} + \beta_+ \sigma_{y'} k_{x'}\right)/2, \quad (3)$$

where $\beta_\pm = 2(\beta_2 \pm \beta_1)$. The effective Larmor frequency $\boldsymbol{\Omega}_k$ is defined by $\mathcal{H}_{c1} = (\hbar/2)\boldsymbol{\Omega}_k \cdot \boldsymbol{\sigma}$, its components are

$$\Omega_{k,x'} = \beta_- k_{y'}/\hbar, \quad \Omega_{k,y'} = \beta_+ k_{x'}/\hbar, \quad \Omega_{k,z} = 0. \quad (4)$$

The spin splitting at $\boldsymbol{k} = (k_{x'}, k_{y'})$ is $\hbar \Omega_k = \sqrt{\beta_+^2 k_{x'}^2 + \beta_-^2 k_{y'}^2}$. If only one of the linear-$\boldsymbol{k}$ terms, BIA or SIA, is nonzero then $\beta_-^2 = \beta_+^2 \equiv \beta^2$ and the splitting $\hbar \Omega_k = \beta k$ is angular independent.

Consider an electron gas occupying the lowest conduction subband $e1$ and assume that, at the moment $t = 0$, the electrons are spin-polarized in the same direction along, say, the growth axis $z$. Owing to the spin-splitting of the electron subband the electron spin in the state with the wavevector $\boldsymbol{k}$ precesses around the axis $\boldsymbol{\Omega}_k$ which lies in the interface plane, see Eq. (4). In case of the large splitting, $|\boldsymbol{\Omega}_k| \gg 1/\tau$, the spin of electron in the state $\boldsymbol{k}$ will oscillate with time as $s_z(t) = s_z(0) \cos \Omega_k t$. In the case of $\Omega_k$ being isotropic in the QW plane, the spin polarized electrons which occupy the circle of the fixed radius in the $k$ space show the similar oscillatory behavior for their $s_z$ component. However, if $\beta_-^2 \neq \beta_+^2$ and/or the electrons occupy states with different values of $|k|$, the scatter in $\Omega_k$ results in a fast non-exponential decay of $s_z$.

D'yakonov and Perel' [3] were the first to show that the processes of electron-momentum scattering result in the slowing off the spin decoherence caused by the spin splitting. In the collision-dominated ("motional-narrowing") limit $|\Omega_k| \ll 1/\tau$, this results in an exponential decay of the average spin: $\bar{s}_z(t) = \bar{s}_z(0) \exp(-t/\tau_s)$, where $\tau_s$ is given by Eq. (1). The dimensionless coefficient in Eq. (1) can be obtained from the solution of kinetic equation [3, 4]. This approach is valid as long as the energy relaxation time $\tau_\varepsilon$ is short as compared to $\tau_s$.

## 3. DPSR in Terms of the Kinetic Theory

In the frame of kinetic theory, the electron distribution in the wavevector and spin spaces is described by a $2 \times 2$ spin-density matrix

$$\hat{\rho}_k = f_k + s_k \cdot \sigma. \tag{5}$$

Here $f_k = \text{Tr}(\hat{\rho}_k/2)$ is the average occupation of the two spin states with wavevector $k$, or distribution function of electrons in the $k$-space, and the average spin in the $k$ state is $s_k = \text{Tr}[\hat{\rho}_k(\sigma/2)]$. If we neglect the spin splitting then, for arbitrary degeneracy of an electron gas with non-equilibrium spin-state occupation but equilibrium energy distribution within each spin branch, the electron spin-density matrix can be presented as

$$\hat{\rho}_k^0 = \left\{ \exp\left[\frac{E_k - \bar{\mu} - \tilde{\mu}(\sigma o_s)}{k_B T}\right] + 1 \right\}^{-1}, \tag{6}$$

where $E_k = \hbar^2 k^2/2m$, $k_B$ is the Boltzmann constant, $T$ is the temperature, $o_s$ is the unit vector in the spin polarization direction, $\mu_\pm = \bar{\mu} \pm \tilde{\mu}$ are the effective Fermi energies for electrons with the spin component $1/2$ or $-1/2$ along $o_s$ so that the energy distribution functions of electrons with the spin $\pm 1/2$ are given respectively by

$$f_{k,\pm} = \left[\exp\left(\frac{E_k - \mu_\pm}{k_B T}\right) + 1\right]^{-1}.$$

Note that Eq. (6) can be rewritten in the equivalent form [3]

$$\hat{\rho}_k^0 \equiv f_k^0 + s_k^0 \cdot \sigma = \frac{1}{2}\left[f_{k,+} + f_{k,-} + (f_{k,+} - f_{k,-})(\sigma o_s)\right].$$

The densities $n_\pm$ of 2D electrons with a particular spin can be related with the effective Fermi energies by

$$n_\pm = \frac{m}{2\pi\hbar^2} k_B T \ln\left(1 + e^{\mu_\pm/k_B T}\right). \tag{7}$$

If the spin splitting is non-zero but small compared to $\hbar/\tau$, the distribution function $\mathrm{Tr}[\rho_k/2] = f_k^0$ does not change, whereas the spin vector obtains a correction $\delta s_k = s_k - s_k^0$ proportional to the spin splitting. Therefore, the spin-density matrix may be presented as

$$\hat{\rho}_k = \hat{\rho}_k^0 + \delta s_k \cdot \sigma . \tag{8}$$

The quantum kinetic equation for the spin-density matrix taking into account the electron–electron collisions has the form

$$\frac{\partial \hat{\rho}_k}{\partial t} + \frac{i}{\hbar}[\mathcal{H}_{c1}(k), \hat{\rho}_k] + \hat{Q}_k\{\hat{\rho}\} = 0 , \tag{9}$$

where $[P, R] = PR - RP$, $\mathcal{H}_{c1}(k)$ is the linear-$k$ Hamiltonian (3), and the third term in the left-hand side is the collision integral or the scattering rate, in this equation it is a 2×2 matrix. It follows from Eq. (9) that the pseudovector kinetic equation for $s_k$ can be written as

$$\frac{ds_k}{dt} + \Omega_k \times s_k + Q_k\{s\} = 0 , \tag{10}$$

where $Q_k\{s, f\} = (1/2)\mathrm{Tr}(\sigma \hat{Q}_k\{\rho\})$. In the following we ignore spin flips under scattering. Then, say, for the elastic scattering one has

$$Q_k\{F\} = \sum_{k'} W_{k'k}(F_k - F_{k'}) ,$$

where $W_{k'k}$ is the probability rate for the electron transition from the state $k$ to $k'$. The collision integral for electron–electron scattering is considered in the next section. Here it suffices to note that, for the distribution $\hat{\rho}_k^0$, the collision integral vanishes identically. This integral also vanishes after the summation over $k$ which allows, in particular, to derive from Eq. (10) the following equation of balance for the total average spin $S_0 = o_s(n_+ - n_-)$:

$$\frac{dS_0}{dt} + \sum_k \Omega_k \times \delta s_k = 0 . \tag{11}$$

The angular dependence of the non-equilibrium correction $\delta s_k$ is the linear combination of $\cos\Phi_k = k_{x'}/k$ and $\sin\Phi_k = k_{y'}/k$, where $\Phi_k$ is the angle between $k$ and the axis $x'$. Retaining in the kinetic equation (10) terms proportional to the first angular harmonics we obtain the equation for $\delta s_k$ with the inhomogeneous term linear in $S_0$. Then one can substitute the solution in the second term of Eq. (11). The final result is that the tensor of inverse spin relaxation times, $1/\tau^s_{\alpha\beta}$, is diagonal in the coordinate system $x'$, $y'$, $z$ and given by [4, 13]

$$\frac{1}{\tau^s_{x'x'}} = \frac{1}{2}\left(\frac{\beta_+}{\hbar}\right)^2 \langle k^2 \tau_p \rangle , \tag{12}$$

$$\frac{1}{\tau^s_{y'y'}} = \frac{1}{2}\left(\frac{\beta_-}{\hbar}\right)^2 \langle k^2 \tau_p \rangle,$$

$$\frac{1}{\tau^s_{zz}} = \frac{1}{\tau^s_{x'x'}} + \frac{1}{\tau^s_{y'y'}},$$

where $\tau_p$ is the momentum relaxation time. If among the two contributions, BIA and SIA, to the spin splitting one is dominant and $|\beta_+| = |\beta_-|$, the spin relaxation times are interconnected by [4]

$$\tau^s_{x'x'} = \tau^s_{y'y'} = 2\tau^s_{zz}.$$

Interplay between the BIA and SIA contributions can lead to a giant spin relaxation anisotropy [13]. In particular, if these contributions coincide, $\beta_1 = \beta_2$, so that $\beta_- = 0$ one has $\tau^s_{x'x'} = \tau^s_{zz}$ and $\tau^s_{y'y'} = \infty$. In the case $\beta_1 = -\beta_2$ the coefficient $\beta_+ = 0$, the time $\tau^s_{x'x'}$ is infinite and $\tau^s_{y'y'}$ coincides with $\tau^s_{zz}$.

## 4. Electron–Electron Collisions in QWs

Here we will write the electron–electron collision term $\hat{Q}_k\{\hat{\rho}\}$ in the general case of arbitrary spin-density matrix $\hat{\rho}_k$ (in particular, arbitrary degeneracy and arbitrary distribution of spin in the $k$-space). For this purpose we remind that the matrix element of the Coulomb scattering $k, s_k + k', s_{k'} \to p, s_p + p', s_{p'}$ is given by

$$M(p, s_p; p', s_{p'}|k, s_k; k', s_{k'}) = V_{k-p}\, \delta_{s_p,s_k}\delta_{s_{p'},s_{k'}} - V_{k-p'}\, \delta_{s_p,s_{k'}}\delta_{s_{p'},s_k}, \quad (13)$$

where $s_k, s_{k'}... = \pm 1/2$, $V_q$ is a Fourier transform of the 2D Coulomb potential of the electron–electron interaction

$$V_q = \frac{2\pi e^2}{\varkappa q \Sigma}, \quad (14)$$

$e$ is the elementary charge, $\varkappa$ is the dielectric constant, and $\Sigma$ is the sample area in the interface plane; in the following we set $\Sigma = 1$. Equation (13) takes into account both the direct and exchange Coulomb interaction.

In order to present $\hat{Q}_k\{\hat{\rho}\}$ in a compact form we introduce the 2×2 unit matrix $I^{(1)}$ and Pauli matrices $\sigma^{(1)}_\alpha$ ($\alpha = x, y, z$) for the spin coordinates $s_p, s_k$ and a similar set of four matrices, $I^{(2)}$ and $\sigma^{(2)}_\alpha$, for the spin coordinates $s_{p'}, s_{k'}$. One can check that Eq. (13) allows the following matrix representation

$$\hat{M} = A\, I^{(1)} I^{(2)} + B\, \sigma^{(1)} \cdot \sigma^{(2)}, \quad (15)$$

$$A = V_{k-p} - \frac{1}{2} V_{k-p'}, \quad B = -\frac{1}{2} V_{k-p'}. \quad (16)$$

Now the collision term for the electron spin-density matrix can be presented as

$$\hat{Q}_k\{\rho\} = \frac{\pi}{2\hbar} \sum_{k'pp'} \delta_{k+k',\,p+p'}\, \delta(E_k + E_{k'} - E_p - E_{p'})\, \text{Tr}_2 G(p, p'; k, k'), \quad (17)$$

$$\begin{aligned} G(p, p'; k, k') =\ & \hat{M}\left(I^{(1)} - \hat{\rho}_p^{(1)}\right)\left(I^{(2)} - \hat{\rho}_{p'}^{(2)}\right)\hat{M}\hat{\rho}_k^{(1)}\hat{\rho}_{k'}^{(2)} \\ & + \hat{\rho}_k^{(1)}\hat{\rho}_{k'}^{(2)}\hat{M}\left(I^{(1)} - \hat{\rho}_p^{(1)}\right)\left(I^{(2)} - \hat{\rho}_{p'}^{(2)}\right)\hat{M} \\ & - \hat{M}\hat{\rho}_p^{(1)}\hat{\rho}_{p'}^{(2)}\hat{M}\left(I^{(1)} - \hat{\rho}_k^{(1)}\right)\left(I^{(2)} - \hat{\rho}_{k'}^{(2)}\right) \\ & - \left(I^{(1)} - \hat{\rho}_k^{(1)}\right)\left(I^{(2)} - \hat{\rho}_{k'}^{(2)}\right)\hat{M}\hat{\rho}_p^{(1)}\hat{\rho}_{p'}^{(2)}\hat{M}. \end{aligned} \quad (18)$$

Here the spin-density matrices $\hat{\rho}^{(1)}(k) = I^{(1)} f_k + \sigma^{(1)} \cdot s_k$, $\hat{\rho}^{(2)}(k') = I^{(2)} f_{k'} + \sigma^{(2)} \cdot s_{k'}$ etc., the symbol $\text{Tr}_2$ means the trace over the spin variable 2. After the trace is found the index 1 in $\text{Tr}_2\, G(p, p'; k, k')$ can be omitted. In order to derive Eqs. (17), (18) we used the standard diagram technique.

Instead of equation (9) for the spin-density matrix one can use a scalar equation for the distribution function $f_k$ in the form

$$\frac{df_k}{dt} + Q_k\{f, s\} = 0 \quad (19)$$

and a equation for the spin-distribution vectorfunction as

$$\frac{ds_k}{dt} + \Omega_k \times s_k + Q_k\{s, f\} = 0, \quad (20)$$

where

$$\begin{aligned} Q_k\{f,s\} &= \frac{1}{2}\text{Tr}_1[\hat{Q}_k\{\hat{\rho}\}] \\ &= \frac{\pi}{4\hbar} \sum_{k'pp'} \delta_{k+k',p+p'}\delta(E_k + E_{k'} - E_p - E_{p'})\text{Tr}_1\text{Tr}_2[G(p, p'; k, k')]. \end{aligned} \quad (21)$$

$$\begin{aligned} Q_k\{s,f\} &= \frac{1}{2}\text{Tr}_1[\sigma^{(1)}\hat{Q}_k\{\hat{\rho}\}] \\ &= \frac{\pi}{4\hbar} \sum_{kpp'} \delta_{k+k',p+p'}\delta(E_k + E_{k'} - E_p - E_{p'})\text{Tr}_1\text{Tr}_2[\sigma^{(1)}G(p, p'; k, k')]. \end{aligned} \quad (22)$$

For the analysis of the general equations (9), (17), (21), (22) we consider below few particular cases.

*Spin-unpolarized electrons.* In this case $s_k \equiv 0$ and the spin-density matrix reduces to a product of the unit $2\times 2$ matrix and the distribution function $f_k$. Taking into account that

$$\hat{M}^2 = (A^2 + 3B^2)\, I^{(1)} I^{(2)} + 2B(A - B)\, \sigma^{(1)} \cdot \sigma^{(2)}$$

we come to the conventional collision term

$$Q_k\{f\} = \frac{2\pi}{\hbar} \sum_{k'pp'} \delta_{k+k',p+p'} \delta \left(E_k + E_{k'} - E_p - E_{p'}\right) \left(A^2 + 3B^2\right) \quad (23)$$
$$\times \left[f_k f_{k'} \left(1 - f_p\right) \left(1 - f_{p'}\right) - f_p f_{p'} \left(1 - f_k\right) \left(1 - f_{k'}\right)\right].$$

Note that according to Eq. (16) one has

$$A^2 + 3B^2 = V_{k-p}^2 + V_{k-p'}^2 - V_{k-p} V_{k-p'}$$

which is one-forth of the function $R$ introduced in Eq. (2.4b) in [16] and the above collision term agrees with the equation (2.4a) in the cited paper. It is worth to note that in the sum (23) $V_{k-p}^2 + V_{k-p'}^2$ can be replaced by $2V_{k-p}^2$.

*Electrons polarized along the same axis.* By using the coordinate system with $z$ parallel to the electron spin polarization one has $s_{k,x} = s_{k,y} \equiv 0$ and the spin-density matrix is a diagonal matrix with the diagonal components $f_{k,s}$ ($s = \pm 1/2$). It follows then that the products $\hat{\rho}^{(1)}(k)\hat{\rho}^{(2)}(k')$ and $[I^{(1)} - \hat{\rho}^{(1)}(p)][I^{(2)} - \hat{\rho}^{(2)}(p')]$ are diagonal as well. We can take into account the spin conservation

$$M(p, s_3; p', s_4 | k, s_1; k', s_2) \propto \delta_{s_3+s_4, s_1+s_2}$$

and use the identity

$$M(p, s_3; p', s_4 | k, s_1; k', s_2) M(p, s_3; p', s_4 | k, -s_1; k', s_2) = 0.$$

This allows, in agreement with [17], rewrite the collision term for $f_{k,s}$ as

$$\frac{2\pi}{\hbar} \sum_{k'pp'} \sum_{s's_1s_2} \delta_{k+k', p+p'} \delta(E_k + E_{k'} - E_p - E_{p'}) W(p, s_1; p', s_2 | k, s; k', s')$$

$$\times [f_{k,s} f_{k',s'} (1 - f_{p,s_1})(1 - f_{p',s_2}) - f_{p,s_1} f_{p',s_2} (1 - f_{k,s})(1 - f_{k',s'})].$$

Here

$$W(p, s; p', s | k, s; k', s) = (A + B)^2 = (V_{k-p} - V_{k-p'})^2,$$
$$W(p, s; p', -s | k, s; k', -s) = (A - B)^2 = V_{k-p}^2,$$
$$W(p, -s; p', s | k, s; k', -s) = (2B)^2 = V_{k-p'}^2,$$

and other values of $W$ with $s_1 + s_2 \neq s + s'$ are zero.

*Low electron polarization.* If the average electron spin $s_k$ is small as compared with the occupation probability $f_k$ then, in the equation for $f_k$, one can ignore the spin polarization at all and use Eq. (23), while, in the equation for $s_k$, one can retain in the collision term only the contribution linear in $s_k$. The linearized collision term is given by

$$Q_k\{s, f\} = \frac{2\pi}{\hbar} \sum_{k',p,p'} \delta_{k+k',p+p'} \delta(E_k + E_{k'} - E_p - E_{p'}) \quad (24)$$
$$\times \left[ (V_{k-p}^2 + V_{k-p'}^2 - V_{k-p}V_{k-p'}) s_k F(k'; p, p') \right.$$
$$- V_{k-p}V_{k-p'} s_{k'} F(k; p, p') - V_{k-p}(V_{k-p} - V_{k-p'}) s_p F(p'; k, k')$$
$$\left. - V_{k-p'}(V_{k-p'} - V_{k-p}) s_{p'} F(p; k, k') \right],$$

where

$$F(k_1; k_2, k_3) = f_{k_1}(1 - f_{k_2})(1 - f_{k_3}) + (1 - f_{k_1}) f_{k_2} f_{k_3} \quad (25)$$
$$= f_{k_1}(1 - f_{k_2} - f_{k_3}) + f_{k_2} f_{k_3}.$$

Equation (24) can be transformed to

$$Q_k\{s,f\} = \frac{2\pi}{\hbar} \sum_{k',p,p'} \delta_{k+k',p+p'} \delta(E_k + E_{k'} - E_p - E_{p'}) \quad (26)$$
$$\times \left\{ 2V_{k-p}^2 \left[ s_k F(k'; p, p') - s_p F(p'; k, k') \right] \right.$$
$$\left. - V_{k-p}V_{k-p'} \left[ s_k F(k'; p, p') + s_{k'} F(k'; p, p') - 2 s_p F(p'; k, k') \right] \right\}.$$

Here the term proportional to $2V_{k-p}^2$ is due to the direct Coulomb interaction whereas the term proportional to $V_{k-p}V_{k-p'}$ comes from the exchange interaction.

*Non-degenerate 2D electron gas.* In this case the function $F(k_1; k_2, k_3)$ reduces to $f_{k_1}$ and the collision terms are as follows

$$Q_k\{f,s\} = \frac{2\pi}{\hbar} \sum_{k'pp'} \delta_{k+k',p+p'} \delta(E_k + E_{k'} - E_p - E_{p'}) \quad (27)$$
$$\times [(2V_{k-p}^2 - V_{k-p}V_{k-p'})(f_k f_{k'} - f_p f_{p'}) - V_{k-p}V_{k-p'}(s_k s_{k'} - s_p s_{p'})],$$

$$Q_k\{s, f\} = \frac{2\pi}{\hbar} \sum_{k'pp'} \delta_{k+k',p+p'} \delta(E_k + E_{k'} - E_p - E_{p'}) \quad (28)$$
$$\times [2V_{k-p}^2 (s_k f_{k'} - s_p f_{p'}) - V_{k-p}V_{k-p'}(s_k f_{k'} + s_{k'} f_k - 2 s_p f_{p'})].$$

Neglecting the exchange interaction given by the term proportional to the $V_{k-p}V_{k-p'}$ the scattering rate $Q_k\{f, s\}$ is independent on the electron spin distribution. Both $Q_k\{f, s\}$ and $Q_k\{s, f\}$ take a simple form

$$Q_k\{f,s\} = \sum_{k'pp'} W_{pp',kk'}(f_k f_{k'} - f_p f_{p'}), \quad Q_k\{s, f^0\} \quad (29)$$
$$= \sum_{k'pp'} W_{pp',kk'}(s_k f_{k'} - s_p f_{p'}),$$

where $W_{pp',kk'}$ is the probability rate for the scattering of a pair of electrons from the $k, k'$ states to the $p, p'$ states

$$W_{pp',kk'} = \frac{2\pi}{\hbar}\delta_{k+k',p+p'}\,\delta(E_k + E_{k'} - E_p - E_{p'})\,2\,V_{k-p}^2\,,$$

an additional factor of 2 takes into account the double degeneracy of the electronic states.

## 5. Electron–Electron Scattering Time Controlling the DPSR

In what follows we consider only a non-degenerate 2D electron gas in which case the zero-approximation spin-density matrix (8) can be written as

$$\hat{\rho}_k^0 = f_k^0(1 + 2\bar{s}\cdot\sigma)\,,\tag{30}$$

where $f_k^0$ is the Boltzmann distribution function, and $\bar{s}$ is the average spin per electron, $S/(n_+ + n_-)$. The nonequilibrium correction $\delta s_k$ satisfies the equation

$$\Omega_k \times (2f_k^0\bar{s}_0) + Q_k\{\delta s, f^0\} = 0\,.\tag{31}$$

It should be noted that in the 2D case the collision term (28) does not allow the quasi-elastic and relaxation time approximations and Eq. (31) must be solved directly.

The $\alpha$-component of the vector product $\Omega_k \times \bar{s}_0$ can be represented as $(\Omega_k \times \bar{s}_0)_\alpha = \Lambda_{\alpha\beta\gamma}k_\beta\bar{s}_{0\gamma}$, where the third-rank tensor $\Lambda$ in general case of both BIA and SIA linear-$k$ terms has four nonzero components

$$\Lambda_{xxz} = -\Lambda_{zxx} = \beta_+/\hbar,\quad \Lambda_{yyz} = -\Lambda_{zyy} = \beta_-/\hbar\,.$$

The function $(1/k_\beta)Q_k\{k_\beta F_k, f^0\}$ is independent of the azimuthal angle $\Phi_k$ (here $F_k$ is an arbitrary function of $k = |k|$) as the operator $Q_k\{\delta s_\alpha, f^0\}$ conserves the angular distribution in the $k$ space. In such case the solution may be written as follows

$$\delta s_\alpha(k) = -\Lambda_{\alpha\beta\gamma}\frac{k_\beta}{k}\bar{s}_{0\gamma}\,k_T\,\tau_{ee}^*\,e^{\mu/k_BT}\,v(K)\,.\tag{32}$$

Here we introduced the dimensionless wavevector $K = k/k_T$, $k_T = (2mk_BT/\hbar^2)^{1/2}$,

$$\tau_{ee}^* = \frac{\hbar k_B T\varkappa^2}{e^4 N}\,,\tag{33}$$

$N = n_+ + n_-$ and $v(K)$ satisfies the equation

$$Ke^{-K^2} = \int d^2K' \int d^2P\,\widetilde{W}_{PP',KK'}\left(v(K)e^{-K'^2} - \cos\Theta\,v(P)e^{-P'^2}\right),\tag{34}$$

where $\Theta$ is the angle between $\mathbf{K}$ and $\mathbf{P}$, $\mathbf{P}' = \mathbf{K} + \mathbf{K}' - \mathbf{P}$,

$$\widetilde{W}_{\mathbf{PP}',\mathbf{KK}'} = \frac{1}{|\mathbf{K} - \mathbf{P}|^2} \delta(K^2 + K'^2 - P^2 - P'^2).$$

Inserting Eq. (32) into Eq. (11) one obtains after summation over $k$ the principal values of the tensor of reciprocal spin relaxation times

$$\frac{1}{\tau_{x'x'}^s} = \left(\frac{\beta_+ k_T}{\hbar}\right)^2 \tau, \qquad (35)$$

$$\frac{1}{\tau_{y'y'}^s} = \left(\frac{\beta_- k_T}{\hbar}\right)^2 \tau,$$

$$\frac{1}{\tau_{zz}^s} = \frac{1}{\tau_{x'x'}^s} + \frac{1}{\tau_{y'y'}^s}.$$

The time $\tau$ which controls the spin relaxation is given by

$$\tau = \tau_{ee}^* I, \qquad I = \frac{1}{2} \int_0^\infty v(K) K^2 dK. \qquad (36)$$

The parameter $\tau_{ee}^*$ is present also in the $ee$-scattering time which determines the rate of energy exchange between $2D$ electrons [15]. The function $v(K)$ was expanded in series using a basis set $l_n(\varepsilon) = \sqrt{2} \exp(-\varepsilon) L_n(2\varepsilon)$, where $L_n(\varepsilon)$ are the Laguerre polynomials and $\varepsilon = K^2$. The expansion was substituted into the right-hand side of Eq. (34), the integration has been performed by Monte-Carlo method. The problem was reduced to a set of linear inhomogeneous equations for the expansion coefficients of $v(K)$. The resulting value of $I$ in Eq. (36) was found to be $\approx 0.027$. Allowance for the exchange interaction leads to a slight increase of this value to $\approx 0.028$.

*Comparison with ionized impurities scattering.* If the ionized impurities of the same concentration $N$ lie inside the 2D layer then the corresponding transport time is given by $\tau_{tr} = (2/\pi^2)\tau_{ee}^*$ (see [16]). The spin relaxation time, $\tau_s^{ii}$, controlled by scattering by the ionized impurities is given by Eq. (35) where $\tau$ is changed by $\tau_{tr}/2$, thus the ratio of the spin relaxation time, $\tau_s^{ee}$, governed by electron–electron collisions and the time $\tau_s^{ii}$ is $\approx 3.6$, i.e. the elastic scattering by impurities is less efficient. If the doped layer is separated from the quantum well by a spacer, the influence of the Coulomb potential of ionized impurities on $\tau_s$ is reduced and eventually can be neglected.

The above result was obtained for the 2D Coulomb potential $U^{2D}(\rho) = e^2/\varkappa\rho$, where $\rho$ is the distance between the electrons in the interface plane. It is this potential that leads to the Fourier transform given by Eq. (14). In order to analyze the role of the quasi-2D character of the electron wave function confined in a QW of the finite thickness $a$ one can replace $U^{2D}(\rho)$ by the effective potential obtained by averaging the three-dimensional Coulomb potential as follows

$$U(\rho) = \frac{e^2}{\varkappa} \iint \frac{\varphi_{e1}^2(z)\varphi_{e1}^2(z')}{\sqrt{\rho^2 + (z-z')^2}} \, dz dz',$$

where $\varphi_{e1}(z)$ is the electron envelope function at the lowest conduction subband $e1$. The straightforward calculation shows [18] that for the conduction band offset $V_c \gg \hbar^2/ma^2$ and $k_T a < 1$ a value of the time $\tau = \tau_{ee}^* I$ increases with widening the QW as the electron–electron interaction becomes weaker, but an order of magnitude of $I$ and $\tau$ remains the same as in the exact 2D case.

## Conclusion. Future Work

We have shown that electron–electron collisions control the D'yakonov-Perel' spin relaxation in the same way as any scattering processes do. The electron–electron collision integral has been derived for an arbitrary degeneracy and spin distribution of the 2D electron gas. The calculations has been performed for non-degenerate 2D and quasi-2D electrons confined in a quantum well. Calculations of the spin relaxation time controlled by electron–electron collisions in bulk semiconductors are in progress. An important next step is the extension of the calculations from non-degenerate to degenerate spin-polarized electron gas.

In agreement with our theory, the latest optical spin-dynamic measurements in an $n$-doped
GaAs/AlGaAs QW of high mobility give an experimental evidence of the electron–electron scattering to randomize the electron spin precession [19].

Note in conclusion that the time $\tau = \tau_{ee}^* I$ can be related physically with the momentum relaxation time of an electron by equilibrium holes of the density $N$ if the electron and hole effective masses are assumed to coincide. In the electron collisions with holes the directed electron momentum is transferred to the hole gas and decays within the time $\sim \tau$.

## Acknowledgements

The work was supported by RFBR, and by the programs of the Russian Ministry of Sciences and the Presidium of the Russian Academy of Sciences.

## References

1. Ivchenko, E.L. and Pikus, G.E. (1995) *Superlattices and Other Heterostructures: Symmetry and Optical Phenomena*, Springer-Verlag, second edition 1997.
2. Averkiev, N.S., Golub, L.E., and Willander, M. (2002) *J. Phys.: Condens. Matter* **14**, 271.
3. D'yakonov, M.I. and Perel', V.I. (1972) *Sov. Phys. Solid State* **13**, 3023.

4. D'yakonov, M.I. and Kachorovskii, V.Yu. (1986) *Sov. Phys. Semicond.* **20**, 110.
5. Ivchenko, E.L., Kop'ev, P.S., Kochereshko, V.P., Uraltsev, I.N., and Yakovlev, D.R. (1988) *JETP Lett.* **47**, 486.
6. Ohno, Y., Terauchi, R., Adachi, T., Matsukura, F., and Ohno, H. (1999) *Phys. Rev. Lett.* **83**, 4196.
7. Terauchi, R., Ohno, Y., Adachi, T., Sato, A., Matsukura, F., Tackeuchi, A., and Ohno, H. (1999) *Jpn. J. Appl. Phys.* **38**, 2549.
8. Britton, R.S., Grevatt, T., Malinowski, A., Harley, R.T., Perozzo, P., Cameron, A.R., and Miller, A. (1998) *Appl. Phys. Lett.* **73**, 2140.
9. Glazov, M.M. and Ivchenko, E.L. (2002) *JETP Lett.* **75**, 403.
10. Rashba, E.I. and Sheka, V.I. (1959) *Fiz. Tverd. Tela* Collected Papers, **2**, 162.
11. Rashba, E.I. (1960) *Sov. Phys. Solid State* **2**, 1109.
12. Averkiev, N.S., Golub, L.E., and Willander, M. (2002) *Sov. Phys. Semicond.* **36**, 91.
13. Averkiev, N.S. and Golub, L.E. (1999) *Phys. Rev. B* **60**, 15582.
14. Ivchenko, E.L. (1973) *Fiz. Tverd. Tela* **15**, 1566 [Sov. Phys. Solid State **15**, 1048 (1973)].
15. Esipov, S.E. and Levinson, I.B. (1985) *JETP Lett.* **42**, 239; (1986) *Sov. Phys. JETP* **63**, 191.
16. Lyo, S.K. (1986) *Phys. Rev. B* **34**, 10.
17. D'Amico, I. and Vignale, G. (2002) *Phys. Rev. B* **65**, 85109.
18. Glazov, M.M. (to be published).
19. Harley, R.T., Brand, M.A., Malinowski, A., Karimov, O.Z., Marsden, P.A., Shields, A.J., Sanvitto, D., Ritchie, D.A., and Simmons, M.Y. *Proc. ICSNN*, Toulouse, France, 2002.

# EVOLUTION OF THE 2DEG-FREE HOLE TO CHARGED EXCITON PHOTOLUMINESCENCE IN GaAs/AlGaAs QUANTUM WELLS

B.M. ASHKINADZE, V. VOZNYY, E. COHEN, ARZA RON
*Solid State Institute, Technion-Israel Institute of Technology,*
*Haifa 32000, Israel*

L.N. PFEIFFER
*Bell Laboratories, Lucent Technologies,*
*Murray Hill NJ 07974, USA*

Abstract. We report on a detailed study of the magneto-photoluminescence in high quality GaAs/AlGaAs quantum wells containing a two-dimensional electron gas. The 2DEG density was varied by optical depletion in the range of $n_{2D} = (0.01–2) \times 10^{11}$ cm$^{-2}$ with He-Ne laser illumination. The $n_{2D}$-dependent PL spectra were studied under an applied magnetic field ($B < 7$ T) at $T = 1.9$ K. In order to identify the PL origin, the PL changes induced by intense mw radiation were investigated. The evolution from a 2DEG — hole PL to neutral and charged exciton PL is attributed to the appearance of localized electrons with decreasing $n_{2D}$ or with increasing magnetic field.

Key words: two-dimensional electrons, charged excitons, magnetophotoluminescence, microwave-modulated photoluminescence

## 1. Introduction

A recombination of photogenerated holes with a two dimensional electron gas (2DEG) in heterostructures (single heterojunctions and quantum wells (QWs)) is widely used in the study of 2DEG properties at low temperature. These properties are strongly modified with varying the 2DEG density ($n_{2D}$) or under application of an external magnetic field ($B$) and this causes an appearance of various anomalies in the photoluminescence (PL) spectrum [1–4]. The patterns of the PL spectra, their evolution with $n_{2D}$ and with $B$ as well as their interpretation significantly differ in various experiments. Much attention has been paid to study the dilute 2DEG-hole system where at $B = 0$, the PL spectrum consists of the negatively charged exciton (trion, X$^-$) and neutral exciton (X$^0$) PL lines [5–7]. With increasing $B$, the triplet X$^-$ PL lines appear at higher energy in both circular polarizations [8, 9]. Recently, a low-intense PL band emerging between the singlet and triplet X$^-$ PL bands at temperatures below 2 K, was attributed to the dark triplet charged exciton [10].

The evolution from the 2DEG-free hole to the $X^-$ and $X^0$ PL with decreasing $n_{2D}$ was observed at $B = 0$ [6, 7] or when the magnetic field was applied perpendicularly to the plane of the 2DEG [11, 12]. However, a comprehensive understanding of this PL evolution with varying $n_{2D}$ and $B$ is still lacking. This motivated us to perform a detailed study of the PL spectral evolution in a high quality, single-sided modulation doped GaAs quantum well (MDQW) in which $n_{2D}$ was varied by optical depletion in the range of $(10^{10} - 2 \times 10^{11}$ cm$^{-2})$. In order to obtain lower $n_{2D}$ we also studied the PL and its excitation (PLE) spectral evolution at $B = 0$ in undoped mixed type I — type II QWs (MTQWs) where $n_{2D}$ was increased from $10^8$ to $10^{11}$ cm$^{-2}$ under He-Ne laser illumination [13].

## 2. Experimental Procedure

The GaAs/AlGaAs heterostructures were grown by molecular beam epitaxy on (001)-oriented, undoped GaAs substrates. Two samples containing a single 25 nm wide MDQW were investigated: one with $n_{2D}^0 = 0.7 \times 10^{11}$ cm$^{-2}$ and the other with $n_{2D}^0 = 2 \times 10^{11}$ cm$^{-2}$. Both have a dc-mobility of $\mu > 2 \times 10^6$ cm$^2$/V s at 4 K. The undoped MTQWs sample consists of 5 periods of narrow ($\sim 2.6$ nm) and wide ($\sim 20$ nm) GaAs QWs separated by barriers having two adjacent layers: AlAs ($\sim 10$ nm) and Al$_{0.3}$Ga$_{0.7}$As (40 nm).

Light beams from He-Ne and Ti-sapphire lasers simultaneously illuminate the sample. A defocused He-Ne laser light controls the 2DEG density; its intensity is varied in the range of $I_L = (0.05 - 10)$ mW/cm$^2$. The photon energy of He-Ne laser, $E_1 = 1.96$ eV, is greater than Al$_x$Ga$_{1-x}$As band gap (for MDQW samples) or than the bandgap of the narrow GaAs QWs (for MTQWs). In MDQW sample, the electron-hole (e–h) pairs photoexcited in the AlGaAs barrier are spatially separated by the built-in heterojunction electric field. The holes are collected in the GaAs QW, and this leads to a long-lived 2DEG density decrease (optical depletion): $n_{2D} = n_{2D}^0 - \Delta n$. In MTQWs, the electrons photoexcited in the narrow QW, transfer into the wide well. These electrons have a long lifetime due to a slow tunneling of the holes that are photoexcited in the narrow QW [13], and $n_{2D} = \Delta n$. A Ti-sapphire laser with intensity of $\sim 1$ mW/cm$^2$ at $E_2 = 1.6$ eV was focused (spot diameter $< 0.1$ mm) and excites e–h pairs in the wide QWs whose PL is studied. In most experiments the He-Ne laser light was used as a cw optical bias while the $E_2$-light was chopped, and the modulated PL spectra were detected by means of a photomultiplier and a lock-in amplifier.

In some experiments we used a pulsed-modulated $E_1$-light while the sample was illuminated by cw $E_2$-light. An acousto-optic modulator controls the He-Ne light pulse duration $\theta$ and the repetition frequency $f$ in the range of $(10^{-8} - 10^{-2})$ s and $(10^2 - 10^6)$ Hz, respectively. This technique allowed us to easily control $n_{2D}$ by varying $\theta$ or $f$. One can show that if the electron lifetime $\tau > (f)^{-1}$, the $n_{2D}$

change induced by He-Ne light is:

$$\Delta n = G\tau \frac{1 - \exp(-\theta/\tau)}{1 - \exp(-1/f\tau)},$$

where $G$ is the generation rate. The time-resolved PL experiments were performed by using a photon counter with a variable gate. The PL gate was separated from the $E_1$-photoexcitation pulse. This eliminates effects that are due to overheating of the photoexcited electrons by the light excitation since the PL is measured after the photoelectrons cooled down. In addition, this technique allowed to study the relaxation processes of the photoexcited system, in particular, to estimate the electron lifetime $\tau$ and the characteristic time of the in-plane electron diffusion.

The sample was placed in an 8-mm waveguide, and the $n_{2D}$ values (tuned by $E_1$-light intensity in MDQW), were obtained from the dimensional magneto-plasma resonance (DMPR) measurements [14]. The microwave (mw) power from a 36 GHz Gunn-diode was varied in the range of $P_{in} = (10^{-3}-50)$ mW, and it can be pulse-modulated by a p-i-n diode. The effect of mw irradiation on the PL spectrum was studied by increasing the mw power. The magnetic field $B$ was applied perpendicularly to the sample plane, while the mw electric field was in the plane.

## 3. Experimental Results

### 3.1. THE PL AND PLE EVOLUTION AT $B = 0$

Figure 1 shows the PL and PLE spectra for the MDQW at $B = 0$ and $T_L = 1.9$ K. Under $E_2$-light excitation, the PL spectrum (curve $a$) has a width of 2.4 meV, and this value is very close to the Fermi energy of the 2DEG with $n_{2D}^0 = 0.7 \times 10^{11}$ cm$^{-2}$. The corresponding PLE spectrum consists of two asymmetric bands associated with the heavy and light hole-electron (e–hh, e–lh) transitions that

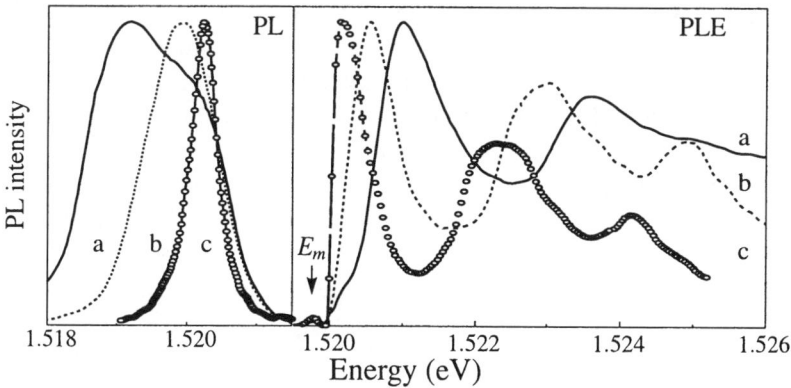

*Figure 1.* PL and corresponding PLE spectra for the MDQW under optical depletion (see text).

clearly display a Fermi-edge singularity [15]. The spectra for the lower $n_{2D}$-values (curves $b$ and $c$) were obtained under an additional $E_1$-light pulse photoexcitation by measuring the PL signal between pulses of $\theta = 2.10^{-7}$ and $1.10^{-6}$ s, respectively ($f = 500$ kHz). As $\theta$ increases, the PL and PLE bands become narrower, and the PL linewidth is reduced to 0.5 meV when $n_{2D}$ decreases down to $\sim 10^{10}$ cm$^{-2}$. One can see that the Fermi-edge singularity in the PLE spectra enhances as $n_{2D}$ decreases. We associate the narrowest PL and PLE bands with a dilute, low density 2DEG as it is concluded from our mw study (see below).

We were not able to decrease the 2DEG density below $10^{10}$ cm$^{-2}$ by optical depletion. In order to achieve lower $n_{2D}$ values, a sample (from the same wafer) was fabricated with a metal gate and electrical contacts, and then $n_{2D}$ could be reduced by applying an electric voltage between the gate and the 2DEG. With electric-induced strong depletion, we observed the appearance of the free exciton PL and PLE bands. However, both PL and PLE bands show a significant inhomogeneous broadening. It is plausible to assume that this method of tuning $n_{2D}$ by an electric bias in photoexcited heterostructures gives rise to a considerable in-plane 2DEG inhomogeneity because of the random distribution of the vertical (2DEG–spacer–doped layers) photoresistance and photovoltage. It results in the coexistence of areas with various $n_{2D}$ values, and the free exciton PL originates from regions with stronger depletion [7].

The low 2DEG density range was studied in the MTQWs sample. The PL and PLE spectra observed for increasing $E_1$-light pulse duration $\theta$ ($f = 30$ kHz) are shown in Fig. 2 (from bottom to the top, respectively). With only $E_2$-light excitation, the PL and PLE spectra are narrow, and this demonstrated the high MTQWs sample quality. The main PL line, $X^0$(linewidth of 0.5 meV) is due to free e–hh excitons. The low-intensity $X^-$ line shifted by $\sim 1.3$ meV to lower energy, is attributed to excitons bound to localized electrons (localized trions). Under $E_2$-light only, the density of localized electrons is low and the localized trion PL is weak. As the density of photoexcited electrons increases with adding $E_1$-light pulses, the $X^0$ intensity decreases while the $X^-$ PL increases (Fig. 2a). Simultaneously, the $X^-$ line emerges in the PLE spectra at the energy below the e–hh $X^0$ line (Fig. 2b). An energy shift and broadening of the e–lh exciton line is due to an appearance of the unresolved e–lh $X^-$ band. At longer $\theta$, the density of photoexcited electrons is so high that the $X^-$ PL and PLE lines transform into broad bands that originate in the 2DEG-hole recombination and photoexcitation, respectively.

The PL and PLE spectra measured at *various delays* after the $E_1$-light pulses terminated, were identical. This means that the density of the electrons injected by $E_1$-light pulses, $\Delta n$, does not change for a long time ($\sim 10^{-2}$ s). Thus, we conclude that the photoexcited electrons (and therefore the trions) are localized. If the photoexcited electrons were free, their density would have strongly decrease due to their in-plane diffusion from the illuminated region (for the time of $\sim 10^{-4}$ s). Additional evidence for the trion localization is obtained by studying the PL change

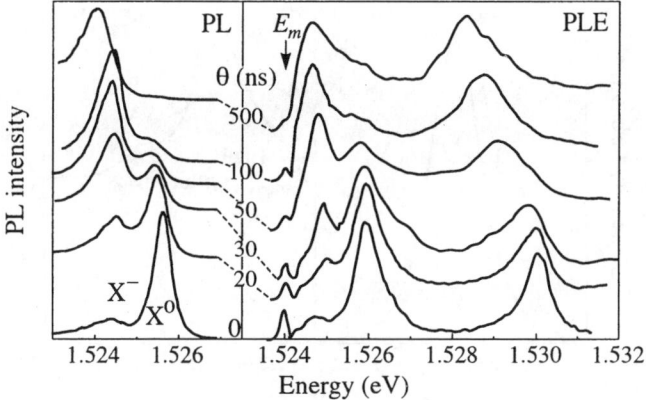

*Figure 2.* PL and PLE spectra measured in MTQWs sample at various $\theta$ (see text). Two strong PLE bands clearly seen at $\theta = 0$, correspond to the e–hh and e–lh exciton transitions.

induced by microwave irradiation (see below).

Recently, a shift of the excitonic PLE band to higher energy and an increased energy separation between the neutral and charged exciton absorption bands with increasing $n_{2D}$, was reported for GaAs MDQW [16]. We do not observe such changes in the PLE spectra neither in the electrically biased MDQW nor in the MTQWs samples (see Fig. 2b).

### 3.2. THE EFFECT OF MW IRRADIATION ON THE PHOTOLUMINESCENCE ($B = 0$)

The study of the PL changes induced by mw irradiation is very useful in identifying the different PL channels. In Figs. 3(a) and 3(b), the PL spectra obtained without and under mw irradiation of $\sim 0.5$ mW, are compared for the MDQW and MTQWs samples, respectively. One can see that the PL lineshape of the strong-depleted MDQW broadens while a PL intensity redistribution between the $X^0$ and $X^-$ lines is observed for the MTQWs sample.

The main effect of mw irradiation is free electron heating [14]. The PL lineshape of the 2DEG-hole radiative recombination is determined by the electron and hole distribution functions, and thus mw heating of the 2DEG causes PL spectral broadening. Indeed, such a mw heating effect was studied for a wide range of $n_{2D}$ in MDQW [14]. The PL spectral modification shown in Fig. 3a, evidences that the narrow (0.5 meV) PL line obtained under strong optical depletion in the MDQW, originates from the dilute 2DEG PL rather than from the charged excitons. In the MTQWs sample under mw irradiation, the intensity of the low-energy (trion) PL decreases, and the exciton PL increases (Fig. 3b). This is similar to the intensity redistribution between localized and free exciton PL bands that was studied for undoped QWs [17]. However, the mw-induced PL intensity redistribution observed in the MTQWs is much stronger.

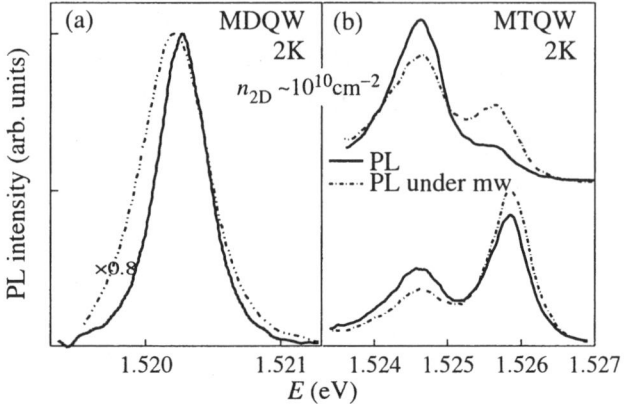

*Figure 3.* The effects of mw irradiation on the PL of the dilute 2DEG in the MDQW (a) and on the exciton-trion PL in the MTQWs (b). The mw effect in MTQW is shown for $\theta = 0.02$ (bottom) and 0.1 $\mu$s (top).

The mw heated electrons affect the exciton PL in the QWs by two mechanisms: (i) A direct interaction of the hot electrons with localized and free *excitons* leads to a PL intensity redistribution that is similar to that caused by thermal heating. (ii) Electron heating causes a decrease of the *localized electron* density. Then, the density of excitons bound to localized electrons decreases, and a PL intensity redistribution occurs. If an exciton would bind to a free electron and form a free trion, the mechanism of mw-induced PL redistribution would be direct heating. In order to distinguish between these two mechanisms, we applied short mw pulses and investigated the time-resolved mw-induced PL changes.

We found long-lived ($\sim 10^{-2}$ s) mw-induced PL changes for the MTQWs sample. The same PL spectral modification was observed during the mw pulse as well as after the pulse termination. This behavior cannot be explained by the first mechanism since the electrons rapidly lose their kinetic energy after the mw-pulse turns off. The long-lived mw-effect on the trion PL can thus be explained by the decrease of the localized electron density that leads to a decreased density of the localized trions. Namely, the electron heating causes a redistribution between free and localized electrons, and the steady-state value of the total electron density decreases under mw irradiation since the free electrons have a shorter lifetime than the localized ones. This experiment unambiguously shows that the excitons are captured by localized electrons and form localized charged exciton complexes. Thus, we conclude that at $B = 0$ *free* excitons do not coexist with *free* trions.

3.3. THE EVOLUTION OF THE 2DEG PL WITH INCREASING MAGNETIC FIELD

Figure 4 displays the evolution of the PL spectra with increasing $B$ at three $n_{2D}$ values for the same MDQW. The respective PL spectra for $B=0$ presented on the top of Fig. 4, can be used for $n_{2D}$-estimate. Figure 4(a) shows the circularly polarized

$\sigma^-$ PL spectra measured under $E_2$-laser excitation only. The disappearance of the second Landau level occurs at $B = 1.4$ T, and it identifies the position of a filling factor $\nu = 2$ that corresponds well to $n_{2D}^0 \sim 7 \times 10^{10}$ cm$^{-2}$. At $B \sim 3.0$ T ($\nu \sim 1$), a new PL band (H) emerges at higher energy side of the main S-PL line, and its intensity increases with $B$. The $\sigma^-$-polarized PL spectra under $E_1$-laser excitation of 0.025 mW and 0.1 mW are shown in Fig. 4(b) and 4(c), respectively. At lowest $n_{2D}$, H-band emerges at $\sim 1$ T and then, the other high energy bands ($X_t^-$ and $X^0$) appear with increasing $B$ (Fig. 4b, c). The H, $X_t^-$ and $X^0$ bands emerge at higher $B$ values as the $E_1$-laser intensity reduces (as $n_{2D}$ increases).

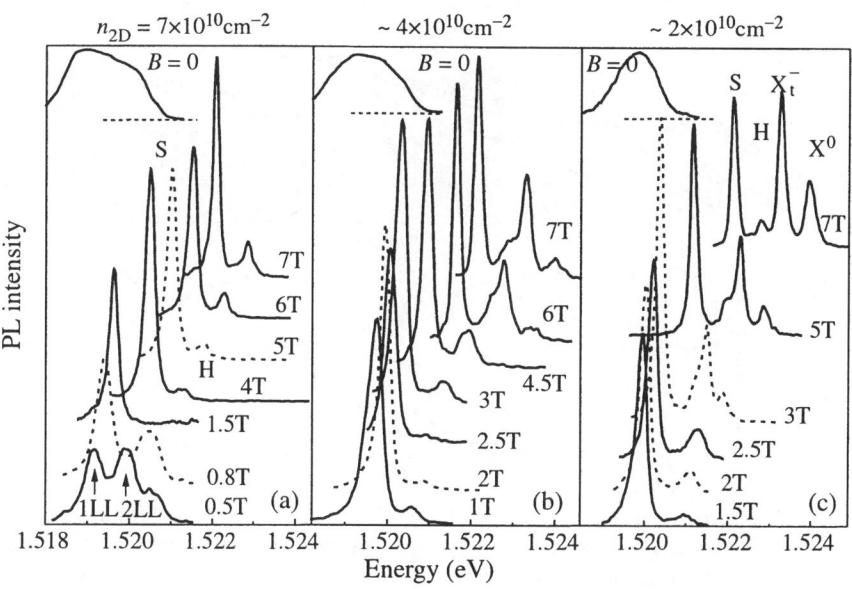

*Figure 4.* The evolution of the PL spectra with $B$ for the MDQW having $n_{2D}^0 = 7 \times 10^{10}$ cm$^{-2}$: (a) photoexcitation at $E_2 = 1.56$ eV, (b) and (c) photoexcitation at $E_1 = 1.96$ eV with laser power of 0.025 and 0.1 mW, respectively.

The $E(B)$ dependencies of all PL lines ($\sigma^-$ and $\sigma^+$ components) were measured for various $n_{2D}$. The S-line peak energy vs $B$ is presented in Fig. 5(a) for the highest and lowest $n_{2D}$ values. The $E(B)$ dependence at low $B$ ($< 2$ T) varies with $n_{2D}$ while it becomes independent of $n_{2D}$ above 4 T. At the highest $n_{2D}$, there is a change in the $E(B)$ S-line dependence from linear to approximately quadratic one [11, 12]. We associate this changeover with a transition to the magnetoexciton PL (at $\nu \sim 2$) and with the PL energy shift by the magnetoexciton energy proportional to $B^{1/2}$ [18]. At $B < 2$ T, the S-line shifts to higher energy with decreasing $n_{2D}$ as a result of a reduced 2DEG exchange energy. The $E(B)$ dependence of $\sigma^-$ H line is parallel to that of $\sigma^-$ S-line for $B > 4$ T and is separated by 0.7 meV. It should be stressed that the $E(B)$ dependencies of $X_t^-$ and $X^0$ coincide for all $E_1$-laser photoexcitation intensities (at various $n_{2D}$ values).

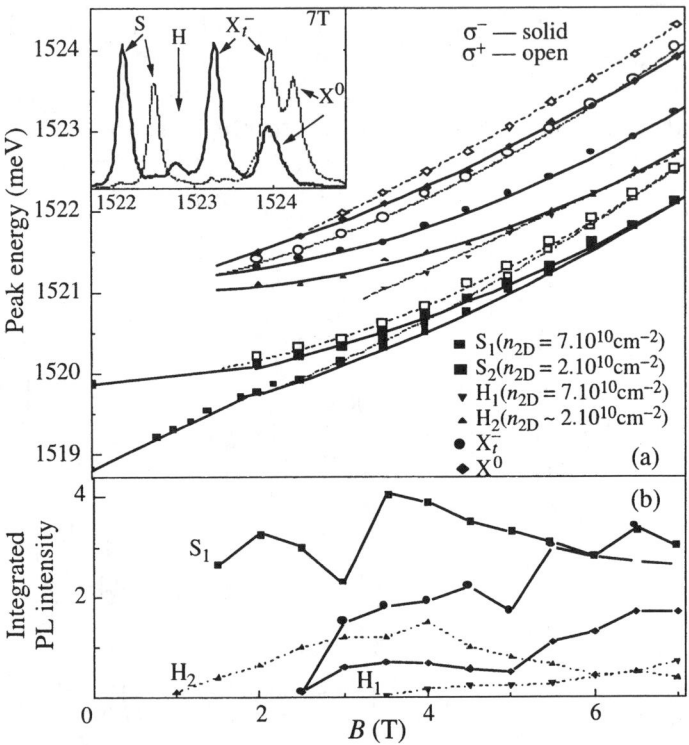

*Figure 5.* (a) Fan diagrams of the PL lines. The $E(B)$ dependencies of S and H lines are presented for two $n_{2D}$-values. $E(B)$ for $X_t^-$ and $X^0$ lines was obtained at $n_{2D} \simeq 3 \times 10^{10}$ cm$^{-2}$. Inset shows the PL spectra ($\sigma^-$ and $\sigma^+$) at 7 T for $n_{2D} \simeq 3 \times 10^{10}$ cm$^{-2}$. (b) The integrated PL intensity vs $B$. $T = 1.9$ K.

In Fig. 5(b) the integrated $\sigma^-$ PL intensity of the S, H, $X_t^-$ and $X^0$ bands versus magnetic field is shown. With increasing $B$, the H, $X_t^-$ and $X^0$ lines successively appear at higher $B$ and the S-line weakens. The H line intensity reveals a remarkable $B$-dependence presented in Fig. 5(b) for two 2DEG densities. For highest $n_{2D}$ ($E_2$-laser excitation only), the H-line intensity increases while $X_t^-$ and $X^0$ lines are not observed up to 7 T. At low $n_{2D}$ (0.1 mW He-Ne laser power), as $B$ is increased, the H-line intensity increases and then, decreases when $X_t^-$ and $X^0$ lines appear.

The integrated PL intensity as a function of $B$ (Fig. 5b) and as a function of photoexcitation intensity ($n_{2D}$) were studied for both circularly polarized components, and this allowed us to correctly assign the Zeeman-split components of the neutral exciton ($X^0(\sigma^-)$ and $X^0(\sigma^+)$ lines) and of the triplet-exciton ($X_t^-(\sigma^-)$ and $X_t^-(\sigma^+)$ lines), as shown in Fig. 5(a). The assignment of the $X^0(\sigma^+)$ and $X_t^-(\sigma^+)$ lines differs from previously reported ones [8, 9]. The $\sigma^-$ and $\sigma^+$ S-PL lines for $B > 4$ T are associated with the charged singlet-exciton $X_s^-(\sigma^-)$ and $X_s^-(\sigma^+)$ PL

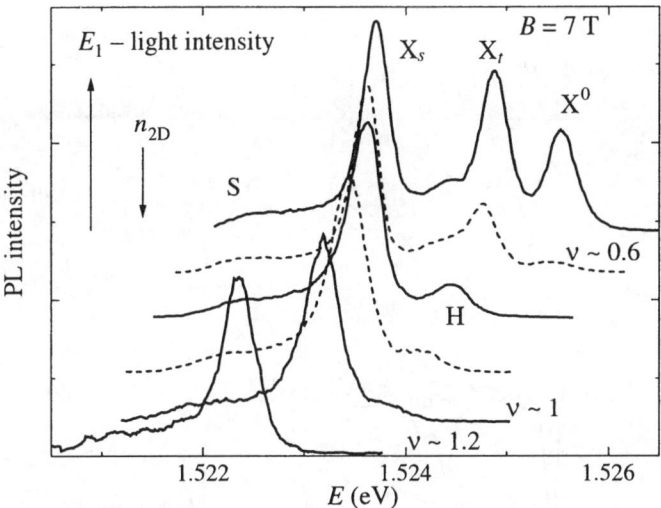

*Figure 6.* The evolution of the PL spectra with $n_{2D}$ at $B = 7$ T for the MDQW having $n_{2D}^0 = 2 \times 10^{11}$ cm$^{-2}$.

lines. At 7 T the Zeeman splitting of $X_t^-$ is larger than that of $X^0$ and $X_s^-$. The nature of the S-PL line at $B < 4$ T and the H line, will be discussed below. We would like to noted that the $\sigma^-$ and $\sigma^+$ components have about the same PL intensities, thus, the PL occurs in the non-thermalized 2DEG-hole system.

Well resolved PL spectra of the studied MDQW, allow us to clearly observe the magneto-PL evolution with decreasing $n_{2D}$. Figure 6 shows the PL evolution with $n_{2D}$ at 7 T for the MDQW having $n_{2D}^0 = 2 \times 10^{11}$ cm$^{-2}$. The PL spectra are measured at decreasing $n_{2D}$ (from the bottom to the top). The H-line first appears as a high energy shoulder of S-line. As $n_{2D}$ decrease further, the H-line increases, and then, $X_t^-$ and $X^0$ lines emerge.

### 3.4. THE EFFECT OF MW RADIATION ON THE MAGNETOPHOTOLUMINESCENCE

The nature of the PL bands that appear with increasing $B$ or decreasing $n_{2D}$ was investigated by the PL response to mw irradiation.

A variety of the mw-induced PL changes are observed in the MDQW magneto-photoluminescence. As was discussed above, the electron temperature $T_e$ increases due to mw absorption by free electrons. The absorbed mw power decreases at $B > \sim 1$ T (above any microwave electron resonance absorption bands [14]), because the electron mobility drops as $B^{-2}$. Thus, $T_e$ decreases and the mw-induced PL modulation (MPL) becomes weaker as $B$ increases. The MPL spectra were measured using a lock-in amplifier and a modulated mw irradiation (modulation frequency varies in the range of 9–5000 Hz). At $B = 7$ T, the largest MPL amplitude

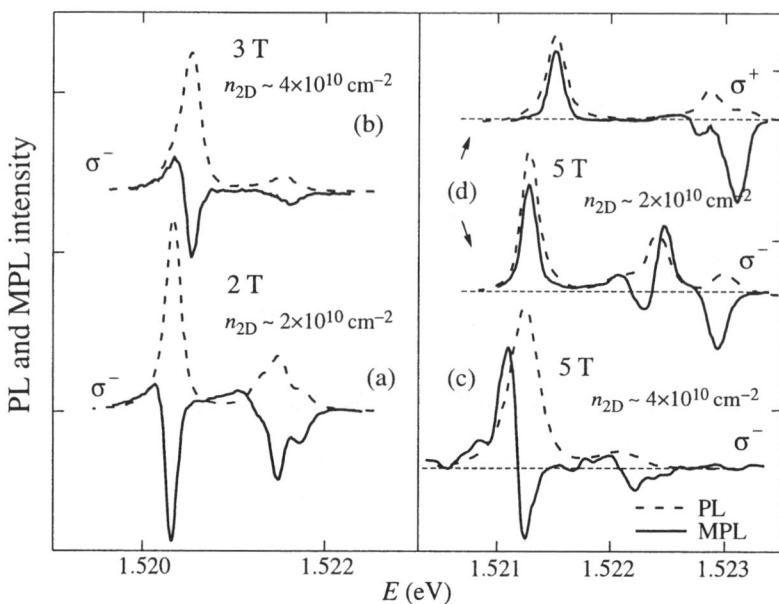

*Figure 7.* The effect of mw irradiation on the magnetophotoluminescence of the MDQW. PL and its mw-modulated (M PL) spectra are shown for two 2DEG densities and three $B$-values. The mw power is $\sim 5$ mW.

was less than $10^{-3}$ at an incident mw power of 10 mW. Figure 7 shows the PL and typical MPL spectra observed at three $B$-values.

The MPL spectra drastically change with varying $B$ and $n_{2D}$. The observed MPL spectra are associated with a broadening of lines S and H (Fig. 7a, b), with a shift of line S to lower energy (Fig. 7c) and with an intensity redistribution between the $X^0$, $X_t^-$ and $X_s^-$ PL bands (Fig. 7d).

At low $B$ and high $n_{2D}$, the S and H MPL spectra reflect a broadening due to increased 2DEG temperature (Figs. 7a and 7b). At $B > 5$ T, all MPL spectra obtained at various $n_{2D}$ can be explained by invoking a mw-induced increase of $n_{2D}$. Indeed, in the absence of mw irradiation, as $n_{2D}$ increases by reducing He-Ne laser intensity, we observe a low-energy shift of the lines S and H and the PL redistribution (Figs. 4 and 6). The MPL spectra shown in Fig. 7(d) and 7(c), reveal similar PL changes under mw radiation. The physical mechanism responsible for the mw-induced increase of $n_{2D}$ is not clear. It is observed at high incident mw-power (above 5 mw) and at high $B$-values when the mw absorption by the *high mobility* 2DEG is insignificant. The increase of $n_{2D}$ might be caused by a non-equilibrium phonon flux generated by mw-heated, *low-mobility* photoelectrons in the undoped layers of the heterostructure. Such a phonon flux may cause a delocalization of electrons in the QW.

An important finding of the MPL study is the different response of the bands S, H and the lines $X^0$, $X_t^-$ to the mw irradiation. The $X^0$ and $X_t^-$ MPL show a intensity redistribution between the $X^0$ and $X_t^-$ lines without their spectral modification at all $B$-values. The mw radiation causes an S-band broadening at low $B$ (Fig. 7a). At $B > 4$ T and at high $n_{2D}$, S-band exhibits a low-energy shift (Fig. 7c) while for lower $n_{2D}$, the intensity modulation only occurs (Fig. 7d). The H-MPL shows mainly PL line broadening.

Thus, the S-PL band is attributed to a radiative 2DEG-hole recombination in the range of $B$ and $n_{2D}$ where its MPL spectrum broadens. It is the 2DEG PL spectrum whose linewidth is very sensitive to the electron temperature. The S-band transforms into $X_s^-$ line with increasing $B$ or decreasing $n_{2D}$, and then, the mw irradiation affects only the intensity of the $X_s^-$ PL (see above). At a given $B$-value, this transformation can be detected as a saturation of the S line energy shift with decreasing density (see Fig. 6).

H PL band is due to a radiative recombination hole with the 2DEG at ($\nu < 1$). Its intensity reduces as S-band transforms into the $X_s^-$. It should be stressed that H-band emerges with varying $B$ and $n_{2D}$, *before* the charged triplet exciton line appears (see Fig. 6) and that only $\sigma^-$) H-component is observed, in the contrast to the $\sigma^-$) and $\sigma^+$) components of $X_t^-$ line. Recently, the PL band similar to the discussed H-band was attributed to the dark $X_t^-$ state [10]. The studied evolution of the H-band and its MPL spectrum is likely to rule out this assignment. The additional peak at the higher energy similar to the H-band, was previously observed and associated with "a free-hole state" [3, 18]. More experiments are needed to fully understand the nature of H PL band.

## 4. Discussion

With increasing magnetic field, the PL and MPL spectra of the nonequilibrium system consisted of 2DEG and hole, evolve into those that are due to the exciton complexes $X_s^-$, $X_t^-$ and $X^0$. The higher is the 2DEG density, the higher magnetic field is needed for such a transition, and the transition occurs at the filling factor $\nu \sim 0.5$. This phenomenon resembles a metal-insulator transition. An approximate criterion for an electron–hole binding into an exciton (like the Mott transition criterion) is $a_B < (n_{2D})^{-0.5}$ ($a_B$ is the Bohr radius). As $B$ increases, the effective Bohr radius decreases, and thus, the transition occurs at higher $n_{2D}$ [11]. We believe that an appearance of a charged triplet exciton and a neutral exciton PL lines takes place near the metal-insulator transition (Figs. 2 and 4). Thus, we propose that the 2DEG-hole PL evolution to the $X_s^-$, $X_t^-$ and $X^0$ exciton PL is associated with electron (or hole) localization.

It should be noted that electron localization is an essential condition for the appearance of a strong charged exciton PL line near the metal-insulator transition (2DEG–hole–exciton). Indeed, as $n_{2D}$ decreases, the free exciton has to appear first, and it cannot form a free trion since mobile 2DEG screens the trions having a larger Bohr radius. Moreover, the experiments have shown that the neutral and charged triplet exciton PL lines emerge approximately at the same $B$-values (see

Fig. 5b). This can be understood if the 2DEG localization leads to exciton and trion formation. The study of the trion-PL in the MTQWs presented above, clearly demonstrates that the electron localization induces the trion formation. On the other hand, the observation of the higher energy H, $X_t^-$ and $X^0$ PL lines at low temperatures shows that the system is in a nonequilibrium state. The latter can result from the short lifetime of the holes, excitons and trions as well as from an effective localization of the photoexcited particles by the inherent disorder.

## 5. Conclusions

We studied the evolution of the photoluminescence and mw-modulated PL spectra with varying $n_{2D}$ and $B$ for the 2DEG-hole system in high quality MDQWs. As $n_{2D}$ decreases down to $10^{10}$ cm$^{-2}$, the PL, PLE and MPL spectra at $B = 0$ are attributed to a low density 2DEG plasma. The trions appear at even lower 2DEG density as electron localization occurs. We propose that the evolution of the 2DEG-free hole PL into the charged and neutral exciton PL with increasing magnetic field is also accompanied by a localization of the electrons.

## Acknowledgements

The research at the Technion was done in the Barbara and Norman Seiden Center for Advanced Optoelectronics and was supported by the Israel Science Foundation (founded by the Israel Academy of Science and Humanities). B.M.A. acknowledges support by a grant under framework of the KAMEA Program.

## References

1. Heitman, D. *et al.* (1994) *Physica B* **201**, 315.
2. Goldberg, B. *et al.* (1991) *Phys. Rev.* **B44**, 4006.
3. Plentz, F. *et al.* (1998) *Physica B* **249**, 558.
4. Munteanu, F. M. *et al.* (2000) *Phys.* **B61**, 4731.
5. Kheng, K. *et al.* (1993) *Phys. Rev. Lett.* **71**, 1752.
6. Shield, A. J. *et al.* (1995) *Phys. Rev.* **B51**, 18049.
7. Finkelstein, G. (1995) *Phys. Rev. Lett.* **74**, 976.
8. Shield, A. J. (1995) *Phys. Rev.* **B52**, 7841.
9. Hayne, M. *et al.* (2001) *Phys. Rev.* **B63**, 125331
10. Yusa, G. *et al.* (2001) *Phys. Rev. Lett.* **87**, 216402
11. Yoon, H. W. *et al.* (1997) *Solid State Comm.* **104**, 287.
12. Gekhtman, D. *et al.* (1996) *Phys. Rev.* **B54**, 10320.
13. Manassen, A. *et al.* (1996) *Phys. Rev.* **B54**, 10609.
14. Ashkinadze, B.M. *et al.* (2000) *Phys. Rev.* **B62**, 10310; (2001) **B64**, 161306.
15. Brum, J and Hawrylak, P. (1997) *Comments Cond. Mat. Phys.* **19**, 135.
16. Yusa, G. *et al.* (2000) *Phys. Rev.* **B62**, 15390.
17. Ashkinadze, B. M. *et al.* (1993) *Phys. Rev.* **B47**, 10613.
18. Cooper, N. R. and Chklovskii, D. B. (1997) *Phys. Rev.* **B55**, 2436.

# RADIATIVE DECAY, DIFFUSION, LOCALIZATION AND DEPHASING OF TRIONS AND EXCITONS IN CdTe QUANTUM WELLS

M.T. PORTELLA-OBERLI, V. CIULIN, J.-D. GANIÈRE,
B. DEVEAUD
*Institute of Quantum Electronics and Photonics,*
*Swiss Federal Institute of Technology Lausanne,*
*CH1015 Lausanne-EPFL, Switzerland*

P. KOSSACKI
*Institute of Experimental Physics, Warsaw University,*
*Hoza 69, 00-681 Warsaw, Poland*

M. KUTROWSKI, T. WOJTOWICZ
*Institute of Physics, Polish Academy of Science,*
*aleja Lotnikow 32/46, 02-668, Warsaw, Poland*

Abstract. Excitons and trions in CdTe modulation-doped quantum well are investigated, as a function of temperature and in the low density regime, by picosecond photoluminescence as well as three-pulse four-wave mixing experiments. Luminescence decay can be interpreted in the framework of delocalised trions, except at low temperatures where deviations from the behaviour expected for free particles is evidenced. Four-wave mixing experiments allow to show that trions are mobile for high enough temperatures. The diffusivity of trions and excitons increases with temperature, and is shown to be mainly limited by acoustic phonon scattering. The mobility of free trions therefore decreases with temperature above 10 K. Consistently, studying the decay of the interband coherence reveals an increase of the homogeneous broadening as the temperature is raised. At low temperatures, excitons and trions are localized by interface roughness scattering, the later also being localized by the electrostatic potential due to remote ionized donors.

Key words: quantum well, exciton, trion, diffusion, four wave mixing, luminescence, mobility

## 1. Introduction

The optical response of undoped semiconductors is dominated, at low temperatures, by excitonic resonances, due to the binding of one electron and one hole via Coulomb interaction. In quantum wells, the increase of the exciton binding energy as well as the reduction of their interaction with optical phonons, allows, even in III–V materials, the observation of the excitonic resonance up to room temperature.

This opens the way to a number of possible uses of these excitonic resonances [1], and has boosted the research in the field over the last twenty years. In lightly doped semiconductors, excitons might bind to extra charges. The corresponding charged quasiparticle, usually called trion, is not observed in bulk systems, but 2D confinement allows the binding of trions, which has therefore been observed in II–VI quantum wells first [2] and soon after in GaAs quantum wells [3].

Trions in semiconductors [4] constitute interesting objects for a number of reasons. One being that, like excitons they show optical resonances which can be easily accessed. Trions however possess the advantage of being a charged quasiparticle, possibly sensitive to applied electric fields. Another interest of the trion system is that it constitutes the first step in the transition of the optical susceptibility of semiconductor quantum structures from the undoped case to the case of an arbitrary doping level. The third being that, in the same way excitonic resonances evidence very interesting non-linear effects of different kind [5, 6], trions might also show interesting non-linearities. Trion resonances indeed show a number of effects similar to excitons, such as strong coupling in microcavities [7] (and absorption bleaching [8]).

One of the many questions to be answered, for our better understanding of the properties of trions in real samples, is similar to the question still under debate in the case of excitons in quantum wells. Do they behave like ideal quasi particles? Or on the contrary do they show evidence that the defects and unavoidable disorder of real samples change their optical and electronic properties. As an example for the case of excitons in quantum wells, the same samples showed an ubiquitous behavior by having radiative properties in very good agreement with what is expected from free excitons [9] and also evidencing the fact that, in the near field, the quantum wells do behave more or less as a collection of quantum boxes [10].

As a matter of fact, the reports on the possible free character of trions are rather conflicting. Some studies find that, at least in the low density limit, trions are confined in the potential fluctuations induced by the remote donors (or acceptors), consistently with the idea that trions are being formed in a regime where the 2D gas conductivity is reduced [11, 12]. Other reports claim that the observed behavior may be traced back to "free trions", moving freely in the plane of the quantum well [13]. Time resolved measurements of the decay of the trion luminescence also reveals an ubiquitous behavior where the trion appears to be free at high temperatures and localized at low temperatures [14].

Recently, trions have been observed to drift upon applying a voltage between the source and drain of a high electron-mobility transistor [15]. However, the observation of a movement of the trions under application of an electric field is not a demonstration as simple as it may seem in a system where three kinds of quasi-particle cohabit: excitons, electrons and trions. Current drag between electrons and holes has been reported earlier [16], showing that one quasi particle might be able to drag another. The observed drift effect may then not be a direct measure of the true trion properties. Diffusion experiments on trions in luminescence have also been interpreted as mediated by excitonic diffusion through a process of dissociation/formation of the trions [17].

## 2. Discussion and Results

Here, we present the results of a careful measurement of the properties of trions in a carefully chosen CdTe quantum well, aiming at the determination of the possible mobility of trionic quasiparticles. Radiative decay is studied through time resolved luminescence whereas diffusion and dephasing are measured using three-pulse four-wave mixing techniques. We show that trions behave as free particles as soon as the lattice temperature is increased up to 10 K, and that, at low temperature, they are more localized than excitons. The trion radiative decay as well as their mobility is compared to standard models showing reasonable agreement. Acoustic phonon scattering via deformation potential is shown to provide a very good account of the observed mobilities. This is further confirmed by a direct measure of the dephasing times of excitons and trions. We show that for trions and excitons, the homogeneous broadening becomes important with increasing temperature as expected for free complexes. The combination of the results from the different measurements mentioned above provides a detailed picture of the trion and exciton dynamics, related to both dephasing and diffusion processes.

The sample studied here has been fully characterized in previous studies [18, 19]. It is a one sided modulation-doped CdTe/CdMgTe heterostructure containing one single quantum well of 8 nm. Two monolayers of iodine were incorporated in the barrier separated from the quantum well by a spacer of 10 nm, resulting in an electron concentration in the well of about $2 \times 10^{10}$ cm$^{-2}$. The reflectivity and luminescence spectra evidence two spectral lines, corresponding respectively to the exciton at 1.625 eV and to the trion at 1.622 eV. The lifetimes of excitons and trions have been investigated using a streak camera under strictly resonant excitation [14, 19].

We have also used time-resolved three-pulse four-wave mixing (3P-FWM) technique in order to investigate both diffusion and dephasing processes in the same experimental arrangement [20]. This technique combines time and space resolution and allows a selection of a pure (population) or mixed (coherence) quantum state for the system under investigation. The excitation is provided by a laser pulse of 2 picoseconds delivered by a Ti: Sapphire laser, matching the absorption linewidths. The laser beams are linearly polarised and the signal is detected in the reflection configuration. This study has been carried out for densities of excitons and trions of about $4 \times 10^9$ cm$^{-2}$.

As reported in Fig. 1, the radiative recombination time of trions is basically linearly varying with temperature. As for excitons, this corresponds to the fact that trions are only able to recombine over a narrow energy range, even if the remaining

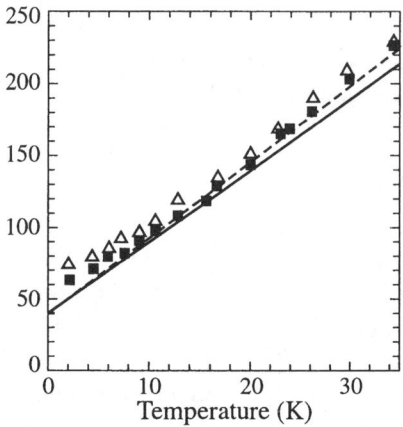

*Figure 1.* Radiative lifetime of trions in CdTe quantum wells, as a function of temperature. A linear increase is observed as the temperature is raised, as expected from theory, except for the lowest temperatures. Triangles and squares correspond to two different electron densities. The continuous line correspond to the computed theoretical lifetime, the dashed lines corresponds to the same calculation including a possible electric field of 15 kV/cm in the sample.

electron is able to take away the excess momentum. As the temperature is raised, a thermal population of trions has less and less proportion of particles in the radiative zone.

The deviation observed at low temperatures might be due to localization effects, preventing trions to retain their full coherence volume at low temperatures, and thus rendering the lifetime longer than expected from theory [21].

The lifetime of excitons has comparatively been studied with the same technique with however the difficulty that at the same time excitons recombine, they are able to form trions. The dynamics are therefore not straightforward to analyze and need some modeling. We have used the simplest possible model, sketched on the inset of Fig. 2. Basically, we model the excitons as a collection of two sub systems: the radiative excitons around $k=0$, and the excitons not coupled to light. Doing this, we forget about spin (a reasonable assumption because we use linearly polarized excitation and detection). The model gives the radiative lifetime of the excitons and the formation time of trions both reported in Fig. 2. The trion formation is observed to increase with temperature, which is expected in the framework of the multiphonon capture model proposed by Lax [22]. The radiative lifetime of excitons is observed not to depend on temperature, which is expected as we only consider radiative states here, however the deduced times are much longer than expected (80 ps instead of 8 ps for the case of CdTe, the same trend being observed in the case of GaAs quantum wells). This might partly be due to the simplicity of the model used to retrieve the radiative time, but the large discrepancy seems to indicate at least some localization.

Diffusion of excitons and trions has then been studied via transient grating

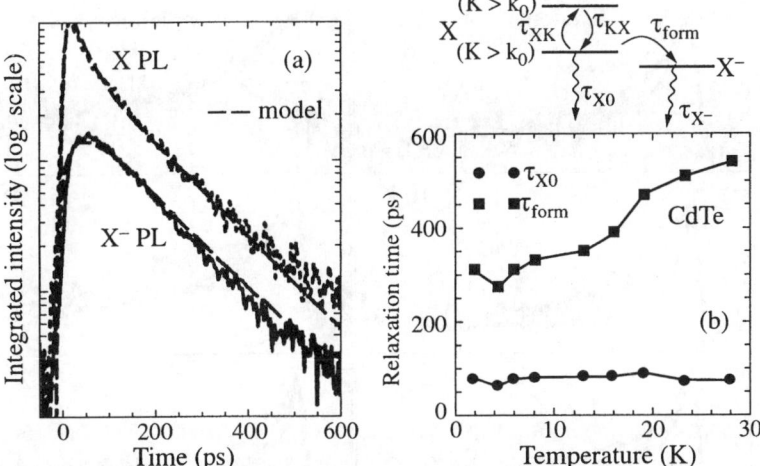

*Figure 2.* (a) Temporal behavior of exciton and trion luminescence under excitation resonant with the exciton. (b) The radiative lifetime of excitons and formation time of trions are extracted using the model shown in the inset.

experiments [23]. The intensity of the signal diffracted on the population grating depends on the relaxation and diffusion processes. In these experiments the angle between the directions $k_1$ and $k_2$ of the two first beams is varied so as to change the grating fringe spacing, any change between two grating periods being the signature that some species move. The experiments are performed under resonant excitation with either the trion or the exciton absorption lines, the width of the pulses being adjusted to the absorption linewidth. This allows to study selectively the dynamics of trions or of excitons.

In Figure 3, we show the 3P-FWM signals from the diffusivity experiments performed on exciton (Fig. 3a) and trion (Fig. 3b) resonances at 2 K, 10 K and 20 K and for grating periods of 5 $\mu$m and 15 $\mu$m. While a change in the grating period does not affect the signal decay for trions at 2 K, the decay is accelerated for both trions and excitons at higher temperatures when a smaller $\Lambda$ is chosen. This behavior is the signature that the excitonic complexes diffuse at $T \geq 10$ K.

By fitting the experimental curves we can directly obtain the diffusion coefficient taking for the lifetime the results obtain in the luminescence measurements, carried out under the same conditions. The fast part of the signal at short times corresponds to some coherence effect which is fitted with an independent parameter that does not affect the results.

The diffusion coefficients of trions and excitons as a function of temperature are plotted in Fig. 4a. Within experimental precision, trions are fully localized at 2 K, whereas excitons are slightly mobile. A clear increase of the diffusion coefficient is observed, from 2 to 10 K, both for excitons and trions. It is interpreted as the release of both quasi-particles from their localization sites by the thermal energy. When

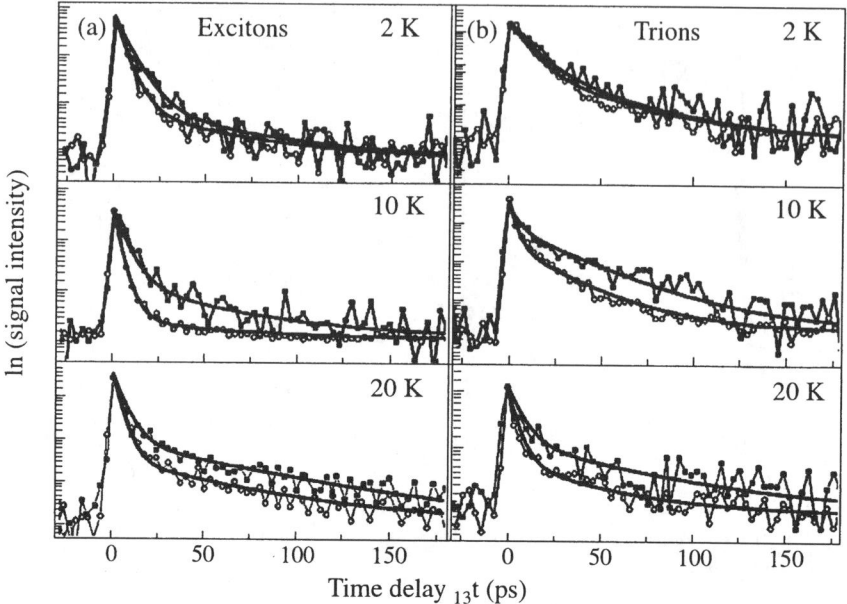

*Figure 3.* Diffracted signal intensity as a function of delay for (a) excitation resonant with excitons and (b) excitation resonant with trions. Temperatures range from 2 K to 20 K. Clear evidence for diffusion is seen in all cases, except for trions at low temperatures.

the temperature is further increased to 20 K the diffusion constants do not increase significantly. The diffusion coefficient of trions can be related to the mobility $\mu$ by the Einstein relation, $\mu = eD/kT$ leading to the mobilities reported in Fig. 4b. Using the same relation we have also calculated an "effective mobility" for excitons which has been plotted in fig. 4c. The trion mobility increases from 2 to 10 K and by increasing the temperature further to 20 K we observe that the trion mobility decreases slightly. For excitons the mobility decreases constantly over the whole temperature range from 2 to 20 K.

Amongst the possible scattering mechanisms only the scattering with acoustic phonons is expected to give rise to a decrease of mobility in such a temperature range. Impurity and interface scatterings are increasingly important at low temperature and LO phonons are not activated at such low temperature [24]. Interactions such as trion-electron and exciton-electron may also limit the mobility. The experiments have been performed in low-density excitations and low 2DEG concentration to avoid these interactions. At low temperature, the mobility of trions, as charged particles, may also be partly determined by scattering with the potential fluctuation from the remote donors in the barrier. The donors being randomly distributed give rise to a random electrostatic potential in the plane with an amplitude of the order of 1 meV, which will contribute less as the temperature is increased.

The expected mobility of excitons and trions considering only scattering by

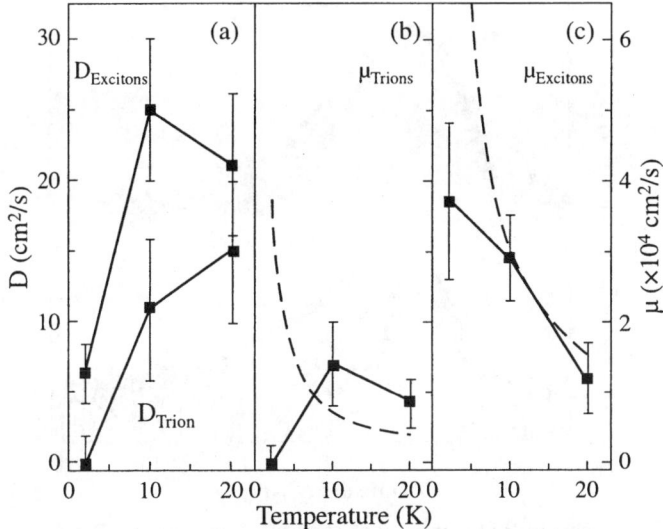

*Figure 4.* (a) Diffusion coefficients for trions and excitons. Deduced mobilities for trions (b) and excitons (c) compared with the computed mobilities limited by acoustic phonon scattering.

acoustic phonons via deformation potential has been calculated using the usual equation [25], only changing the mass of the trions and the interaction potential to take into account the second electron [26]. Note that the mobility of trions in such a case is expected to be lower than that of excitons because of the larger mass of the quasi-particle, and of the larger interaction potential. The calculated mobilities of trions and excitons are reported in Fig. 5b and 5c, respectively. They are in quite good agreement with experiments above 10 K for both excitons and trions. This result strongly supports our above expectation that, in this temperature range, excitons and trions mainly scatter with acoustic phonons via deformation potential. This is not the case at 2 K: at low temperatures excitons and trions are more sensitive to fluctuations in the static potential of two-dimensional confinement. Interface roughness scattering is known to play an important role in reducing the exciton motion, thereby determining their mobility at 2 K. One may wonder whether trions may also be localized by the same potential fluctuations. The above values of mobility show that trions are subject to stronger potential fluctuations than excitons.

When interface disorder is weak, it only affects the center-of-mass motion of excitons or trions because their relative electron-hole motion can be assumed to be unaffected [27]. In this regime, the effective disorder potential results from an averaging of the true microscopic potential by the exciton (trion) wave function [28]. As a result, the spatial fluctuations over lengths smaller than the Bohr radius ($r_B$) of excitons (trions) are smoothed out. Therefore, the effective disorder potential seen by trions is smoother out than for excitons due to their different Bohr radius

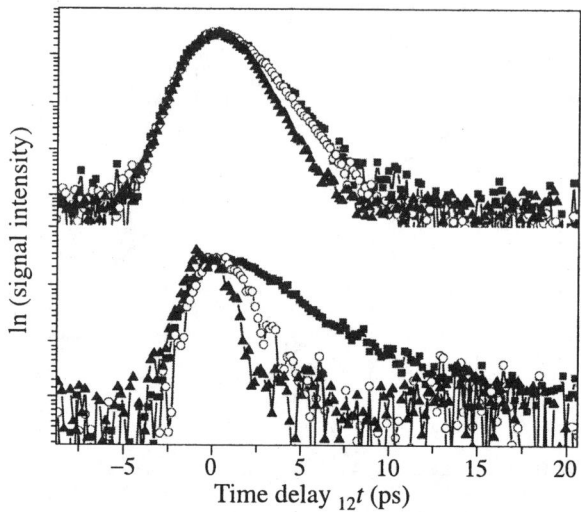

*Figure 5.* (a) Three beam FWM for excitons at 3 different temperatures (■ 2 K, ○ 10 K, ▲ 20 K); (b) Same for the case of trions. A clear asymmetry of the signal is observed, specially in the case of trions, indicating an inhomogeneous broadening.

(200 Å for trions and 65 Å for excitons). For this reason, interface roughness, not sufficient to localize excitons in our sample, cannot localize trions at 2 K. We attribute the trion localization to the additional potential fluctuations due to the electrostatic influence of remote ionized donors to which excitons are not sensitive because they are neutral quasi-particles.

Such localization and scattering processes should have a sizeable influence on the exciton and trion absorption lines. The lines should be mainly inhomogeneously broadened at low temperatures. When diffusion takes place, excitons and trions interact more strongly with their surrounding, which should progressively bring the system to a more homogeneous character. To assess such properties, we have studied the trion and exciton dephasing dynamics in a 3P-FWM configuration, the third pulse being kept at the same delay as the second one and the diffracted signal being recorded as a function of the delay $\tau_{12}$ between the first pulse and the two other pulses. In such a configuration, the shape of FWM signal gives direct information about the nature of the spectral broadening of the absorption lines [29, 30]. An asymmetric shape of the signal with respect to the zero time delay is the signature of an inhomogeneous broadening, on the contrary, a symmetric temporal shape indicates the dominance of homogeneous dephasing.

3P-FWM signals from coherence experiments are represented in Fig. 5. We observe an evolution from an asymmetric towards a symmetric temporal shape of the signal about zero delay time as the temperature is increased. These data bring direct evidence that the system is inhomogeneously broadened at 2 K and that homogeneous broadening is increasingly important as temperature is increased up to 20 K. The decay time constant of the signal is equal to $T_2/2$ in the limit of pure homogeneous broadening and to $T_2/4$ for inhomogeneous broadening ($T_2$ being the dephasing time of exciton or trion). The dephasing times have then been calculated considering, as appropriate, the system as either purely homogeneously or purely inhomogeneously broadened depending on the temperature. At 2 K, the dephasing time of trions is much longer than that of excitons. This result is in agreement with the observed localization of trions at this temperature leading to the fact that their interaction with the surrounding is less effective than that of excitons. As the temperature is increased, the dephasing gets faster for both excitons and trions. This dephasing behavior is expected for free particles interacting with acoustic phonons. Furthermore, at 10 K and 20 K, the trion dephasing times are shorter than those of exciton, demonstrating directly that the interaction of acoustic phonon with trion is stronger than with exciton, as already found in the diffusion data. Let us finally mention that our values are in agreement with the measured dephasing of positively charged excitons reported in Ref. [11]. The measured mobilities can be transformed into dephasing time using a very simple Drude model. As shown in Ref. [20], the extracted dephasing times are in very good agreement with the dephasing times directly measured using four wave mixing. This results is in favor of our interpretation of the fact that diffusion is indeed limited by acoustic phonon scattering.

This is a result which may seem amazing in different aspects. We will discuss a few in the following of this contribution. At 20 K a sizeable portion of trions may dissociate into excitons and a thermal equilibrium may be reached on the same time scale as diffusion takes place [31]. The following question may be raised: Are trions dragged by excitons? This is certainly not the case because at 10 K, when only a negligible amount of trions dissociates, we found that trions have already an increased mobility. We therefore conclude that the trion mobility increase and the concomitant diffusion process is not related to the presence of excitons. Also the mobility reduction at higher temperatures can be well explained by the trion-acoustic phonon interaction alone.

The absence of clear effects of electrons is also amazing. One would expect that trions, as negatively charged particles, would be very sensitive to the presence of the other electrons, and therefore would be very quickly dephased. One order of magnitude might be obtained from the work of Honold *et al* [32], which would give dephasing times of less that 1 ps for electron densities of a few $10^{10}$ cm$^{-2}$. However, our results together with those of Ref. [11] show that the dephasing of trions is not as fast as this reference would predict. We also show that the dephasing of excitons also in the presence of electrons is not as fast as expected. So that clearly the effect of an electron Fermi see on the dephasing of trions and excitons is not as strong as what was previously estimated.

## 3. Summary

In summary, the present investigation combining both diffusion and coherence decay of excitons and trions shows directly that trions are mobile when $K_B T$ is on the order of the localization potential, but are localized when the temperature is decreased to 2 K. We expect this behaviour to be generic for trions in two-dimensional semiconductors. It is important to realize that the localization properties inferred from the present picosecond FWM experiments are in full agreement with those obtained from trion luminescence lifetimes on the same sample [14]. Excitons are not fully localized at low temperature, unlike trions, and are also free at higher temperatures. Localization of trions is caused by interface roughness scattering and remote ionized donor potential. As a consequence of the disorder effects, at low temperature, the exciton and trion absorption lines are predominantly inhomogeneously broadened. At higher temperatures, homogeneous broadening is largely increased and competes with the inhomogeneous broadening for both types of quasi-particles as a result of acoustic phonon scattering. The same scattering mechanism limits the mobility values for both excitonic complexes, and to has an expectedly stronger effect on the trions.

## Acknowledgements

We would like to acknowledge numerous discussions with S. Haacke, A. Esser and M.A. Dupertuis. This work was supported in part by Swiss National Science Foundation and Office Federal de L'Education et de la Science.

## References

1. Chemla, D.S. and Miller, D.A.B. (1985) Room-Temperature Excitonic Nonlinear-Optical Effects in Semiconductor Quantum-Well Structures, *J. Opt. Soc. Am. B-Opt. Phys.* **2**, 1155–1173.
2. Kheng, K., Cox, R.T., Merle d'Aubignè, Y., Bassani, F., Saminadayar K., and Tatarenko S. (1993) Observation of Negatively Charged Excitons X- in Semiconductor Quantum-Wells, *Phys. Rev. Lett.* **71**, 1752–1755.
3. Finkelstein, G., Shtrikman, H., and Bar Joseph, I. (1995) Optical Spectroscopy of a 2-Dimensional Electron-Gas near the Metal-Insulator-Transition, *Phys. Rev. Lett.* **74**, 976–979.
4. Lampert, M. A. (1958) Mobile and immobile effective mass particle complexes in non metallic solids, *Phys. Rev. Lett.* **1**, 450–453.
5. Schmittrink, S., Chemla, D.S., and Miller D.A.B., (1989) Linear and Nonlinear Optical-Properties of Semiconductor Quantum Wells, *Adv. in Physics* **38**, 89–188.
6. Deveaud, B., Quochi, F., Saba, M., Ciuti, C., and Staehli, J.L. (2001) Femtosecond dynamics and non-linearities of exciton-photon coupling in semiconductor microstructures *CRAS, series IV* **2**, 1439–1451.
7. Rapaport, R., Qarry, A., Cohen, E., Ron, A., and Pfeiffer, L.N. (2001) Charged excitons and cavity polaritons, *Phys. Stat. Solidi* **227**, 419–427.
8. Portella-Oberli, M.T., Ciulin, V., and Deveaud, B., unpublished

9. Deveaud, B., Clerot, F., Roy, N., Satzke, K., Sermage, B., and Katzer, D.S. (1991) Enhanced Radiative Recombination of Free-Excitons in Gaas Quantum-Wells, *Phys. Rev. Lett.* **67**, 2355–2358.
10. Gammon, D., Snow, E.S., Shanabrook, B.V., Katzer, D.S., and Park,D. (1996) Fine structure splitting in the optical spectra of single GaAs quantum dots, *Phys. Rev. Lett.* **76**, 3005–3008.
11. Eytan, G., Yayon, Y., Rappaport, M., Shtrikman, H., and Bar Joseph, I. (1998) Near-field spectroscopy of a gated electron gas: a direct evidence for electron localization *Phys. Rev. Lett.* **81**, 1666–9.
12. Brinkmann, D., Kudrna, J., Gilliot, P., Hönerlage, B., Arnoult, A., Cibert, J., and Tatarenko, S. (1999) Trion and exciton dephasing measurements in modulation-doped quantum wells: A probe for trion and carrier localization, *Phys. Rev.* **B 60**, 4474–4477.
13. Ron, A., Yoon, W., Sturge, M.D., Manassen, A., Cohen, E., and Pfeiffer, L. N. (1996) Diffusion of free trions in mixed type GaAs/AlAs quantum wells, *Solid State Comm.* **97**, 741–747.
14. Ciulin, V., Kossacki, P., Haacke, S., Ganiere, J.D., Deveaud, B., Esser, A., Kutrowski, M., and Wojtowicz, T. (2001) Radiative behavior of negatively charged excitons in CdTe-based quantum wells: A spectral and temporal analysis, *Phys. Rev.* **B62**, R16310–R16313.
15. Sanvitto, D., Pulizzi, F., Shields, A.J., Christianen, P.C.M., Holmes, S.N., Simmons, M.Y., Ritchie, D.A., Maan, J.C., and Pepper, M. (2001) Observation of charge transport by negatively charged excitons, *Science* **294**, 837.
16. Hopfel, R.A., Shah, J., Wolff, P.A., and Gossard, A.C. (1986) Negative absolute mobility of holes in n-doped GaAs quantum wells, *Appl. Phys. Lett.* **49**, 572–574.
17. Pulizzi, F., Thijssen, W.H.A., Christianen, P.C.M., Maan, J.C., Yakovlev, D.R., Ossau, W., Wojtowicz, T., Karczewski, G., and Kossut, J. (2001) Motion of neutral and negatively charged excitons in high magnetic fields, *Physica* **B 298**, 397–401.
18. Wojtowicz, T., Kutrowski, M., Kaczewski, G., and Kossut, J. (1998) Modulation-doped $Cd_{1-x}Mn_xTe/Cd_{1-x}Mg_yTe$ quantum well structures with spatial in-plane profiling of the well width and the doping intensity, *Appl. Phys. Lett.* **73**, 1379–1381.
19. Ciulin, V., Kossacki, P., Haacke, S., Ganière, J.-D., Deveaud, B., Esser, A., Kutrowski, M., and Wojtowicz, T. (2001) Radiative behavior of negatively charged excitons in CdTe-based quantum wells: A spectral and temporal analysis, *Phys. Stat. Solidi.* (b) **227**, 307.
20. Portella-Oberli, M-T., Ciulin, V., Kossaki, P., Haacke, S., Kutrowski, M., Wojtowicz, T., Ganière, J-D., and Deveaud, B. (2002) Dynamics of excitons and trions in CdTe quantum wells: direct observation of diffusion and localization, *Phys. Stat. Sol.* (a) **190**, 787–791.
21. Esser, A., Runge, E., Zimmermann, R., and Langbein, W. (2000) Photoluminescence and radiative lifetime of trions in GaAs quantum wells, *Phys. Rev.* **B62**, 8232–8235.
22. Lax, M. (1960) Cascade capture of electrons in solids, *Phys. Rev.*, **119**, 450–453.
23. Salcedo, J.R., Siegmann, A.E., Dlott, D.D., and Fayer, M.D. (1978) Dynamics of energy-transport in molecular crystals – picosecond transient grating method, *Phys. Rev. Lett.* **41**, 131–134.
24. Basu, P.K. and Ray, P. (1991) Calculation of the mobility of 2-dimensional excitons in a GaAs/AlGaAs quantum well, *Phys. Rev.* **B 44**, 1844–1849.
25. Hillmer, H., Forchel, A., Hansmann, S., Morohashi, M., Lopes, E., Meier, H.P., and Ploog, K. (1989) Optical investigations on the mobility of two-dimensional excitons in GaAs/GaAlAs quantum wells, *Phys. Rev.* **B 39**, 10901–10912.
26. Portella-Oberli, M. *et al.* to be published
27. Zimmermann, R., Grosse, F., and Runge, E. (1997) Excitons in semiconductor nanostructures with disorder, *Pure Appl. Chem.* **69**, 1179–1186.
28. Intonti, F., Emiliani, V., Lienau, C., Elsaesser, T., Savona, V., Runge, E., Zimmermann, R., Nötzel, R., and Ploog, K.H. (2001) Quantum mechanical repulsion of exciton levels in a disordered quantum well, *Phys. Rev. Lett.* **87**, 1–4.

29. Mukamel, S. (1995) *Principles of Nonlinear Optical Spectroscopy*, Oxford University Press, New York.
30. Weiner, A.M., De Silvestri, S., and Ippen, E.P. (1985) 3-pulse scattering for femtosecond dephasing studies – Theory and experiment, *J. Opt. Soc. Am.* **B 2**, 654–662.
31. Siviniant, J., Scalbert, D., Kavokin, A.V., Coquillat, D., and Lascaray, J.-P. (1999) Chemical equilibrium between excitons, electrons, and negatively charged excitons in semiconductor quantum wells, *Phys. Rev.* **B 59**, 1602–1605.
32. Honold, A., Schultheis, L., Kuhl, J., and Tu, C.W. (1989) Collision broadening of two-dimensional excitons in a GaAs single quantum well, *Phys. Rev.* **B 40**, 6442–6445.

# OPTICAL STUDIES OF SPIN POLARIZED 2DEG IN MODULATION-DOPED (Zn,Mn)Se/(Zn,Be)Se QUANTUM WELLS IN HIGH MAGNETIC FIELDS

D. KELLER, G.V. ASTAKHOV[†], D.R. YAKOVLEV[†‡], L. HANSEN, W. OSSAU
*Physikalisches Institut der Universität Würzburg,*
*D-97074 Würzburg, Germany*
[†] *Ioffe Physico-Technical Institute, Russian Academy of Sciences, 194021, St Petersburg, Russia*
[‡] *Since April 2002 affiliation at Experimentelle Physik 2, Universität Dortmund, D-44221 Dortmund, Germany*

S.A. CROOKER
*National High Magnetic Field Laboratory, Los Alamos, 87545 New Mexico, USA*

Abstract. Optical properties of (Zn,Mn)Se/(Zn,Be)Se quantum wells containing a two-dimensional electron gas with densities varying from $10^9$ to $5.5 \times 10^{11}$ cm$^{-2}$ have been studied in magnetic fields up to 45 T. A strong $s-d$ exchange interaction of free electrons with magnetic Mn ions results in a giant Zeeman splitting of conduction band states, that has been exploited for manipulation with a spin polarization of the electron gas. Critical behavior of the photoluminescence and reflectivity linewidths and of the energy shift of the lines in magnetic fields in vicinity of integer filling factors 2 and 1 has been found and analyzed.

Key words: quantum well, diluted magnetic semiconductor, excitons, trions

## 1. Introduction

Diluted magnetic semiconductors (DMS) based on II–VI materials, like (Cd,Mn)Te or (Zn,Mn)Se, are known for their giant magneto-optical effects. The effects are due to the strong exchange interaction of electrons localized on the $d$-shells of magnetic Mn-ions with free carriers: either electrons in the conduction band ($s-d$ exchange interaction) or holes in the valence band ($p-d$ exchange interaction). At present magneto-optical properties of undoped heterostructures (e.g. quantum well (QW) structures) based on wide-gap DMS are known in great detail. Although few investigations have been performed for the modulation-doped DMS quantum wells [1–4] the complete picture of the associated effects and phenomena is still missing. One of the challenges to study DMS QWs with free carriers lies in the possibility to achieve a fully spin-polarized carrier gas for high densities of free carriers at low magnetic fields [5].

Due to the stronger Coulombic interaction in ZnSe (exciton binding energy in bulk is 20 meV compared to 10 meV in CdTe), ZnSe-based structures are particularly suitable for studies of Coulombic correlation effects in a 2DEG. However, the available data for modulation-doped DMS QWs on (Zn,Mn)Se basis are limited to (Zn,Cd,Mn)Se/ZnSe structures with a quaternary alloy in the well layer [3, 4]. In these structures the band edge states are strongly inhomogeneously broadened due to alloy fluctuations of Cd, which restricts significantly the information received by optical methods. To avoid these limitations, we study (Zn,Mn)Se-based QWs with a low Mn content and small inhomogeneous linewidth.

## 2. Experimentals

We studied a nominally undoped $Zn_{0.995}Mn_{0.005}Se/Zn_{0.96}Be_{0.04}Se$ QW (sample #1) and two modulation-doped QWs with different electron densities of $3.2\times10^{11}$ cm$^{-2}$ (sample #2) and $5.5\times10^{11}$ cm$^{-2}$ (sample #3). The value of electron concentration was evaluated by utilizing the characteristic changes of optical spectra at critical filling factors [6]. The samples have been grown by molecular beam epitaxy on (100)-oriented GaAs substrates. All samples have an identical design. They consists of a 100-Å-thick single quantum well (SQW) confined by 1000-Å-thick $Zn_{0.96}Be_{0.04}Se$ barrier layers. The barrier band gap exceeds the well band gap by 65 meV. The depth of the QW for the electrons can be estimated to 50 meV, using an conduction band offset of 78% of the total band gap discontinuity [7]. In the modulation-doped samples each barrier contains a 20-Å-thick Iodine doped layer at a distance of 100 Å to the QW. This structure is embedded between two $Zn_{0.92}Be_{0.08}Se$ layers with a large band gap to prevent loss of carriers to the substrate and due to surface recombination.

The samples were studied by polarized photoluminescence (PL) and reflectivity at a temperature of 1.6 K. Magnetic fields up to 45 T were generated by a capacitor-driven mid pulsed magnet (400 ms decay) of the National High Magnetic Field Laboratory (Los Alamos, USA). Fields were applied along the structure growth axis in the Faraday geometry. Fiber optics were used and the signal was analyzed by spectrometer and associated charge-coupled device (CCD). Circular polarization degree of emitted/reflected light was analyzed. A complete set of field-dependent PL and reflectivity spectra was collected during each magnetic field pulse (for details see Ref. [8]).

## 3. Properties of the Undoped Structure

In this part we concentrate on the optical properties of the undoped DMS sample to determine important sample parameters, necessary for analyzing experimental results for a 2DEG. Parameters determined here are mainly related to exchange interaction of band states with localized Mn-spins. For the "nonmagnetic" properties of excitons and charged excitons (binding energy, oscillator strength, energy shift in external magnetic fields, values of $g$ factors, effective mass, etc.) in nonmagnetic ZnSe-based QWs, we refer to our previous studies [9, 10].

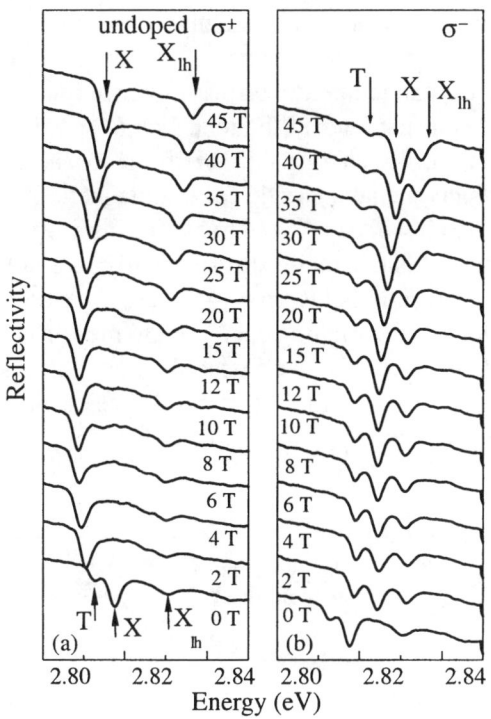

*Figure 1.* Reflectivity spectra of an undoped 100-Å-thick $Zn_{0.995}Mn_{0.005}Se/Zn_{0.96}Be_{0.04}Se$ QW at different magnetic fields detected in $\sigma^+$ and $\sigma^-$ polarization. Resonances of the charged exciton (T), heavy-hole exciton (X) and light-hole exciton ($X_{lh}$) are marked by arrows. $T = 1.6$ K.

Figure 1 shows reflectivity spectra of the undoped reference sample for $\sigma^+$ and $\sigma^-$ polarizations. At zero magnetic field the strong resonance at 2.8176 eV corresponds to the heavy-hole exciton (X). The light-hole exciton ($X_{lh}$) is shifted by 13 meV to higher energies due to strain and quantum confinement effects. The resonance shifted by 5.2 meV to lower energies with respect to the heavy-hole exciton corresponds to the negatively charged exciton or trion (T). The negatively charged exciton is a complex of two electrons bound to one hole [9]. The small full width at half maximum (FWHM) of the neutral and charged exciton lines of

2.1 meV in luminescence spectra [see Fig. 4(a)] indicates high structural quality of the sample. The formation of a trion requires additional background electrons, which can be provided by background impurities and/or generated by an above-barrier illumination in the undoped sample. In external magnetic fields the trion resonance is strongly polarized due to the spin polarization of the background electrons [11]. Whereas the trion line has completely vanished at $B = 0.5$ T in $\sigma^+$ polarization, it is present in the whole range of magnetic fields in $\sigma^-$ polarization, with a weak decrease of oscillator strength for high fields.

The energy position of the exciton and trion resonances in dependence of magnetic field is shown in Fig. 2. For small fields ($B \leq 6$ T) the energy shift of the lines is dominated by the $sp - d$ exchange interaction between the carriers and the localized electrons of the magnetic Mn ions. The part of the Zeeman splitting induced by exchange interaction with the Mn ions will be referred to as "exchange" Zeeman splitting later in the text. For $B \geq 6$ T, where the exchange splitting is in saturation, the energy shift of the resonances is dominated by a Zeeman splitting, which is characteristic for nonmagnetic ZnSe-based structures [9]. In the following text it will be referred to as "intrinsic" Zeeman splitting. Apart of splitting a pronounced diamagnetic shift is clearly seen for all transitions. We stress here that high magnetic fields up to 45 T are essential for investigation of strongly bound excitons with a binding energy of 30 meV in (Zn,Mn)Se QWs. The diamagnetic shift of the exciton is well described assuming an exciton in-plane effective mass of $\mu = 0.15 m_0$, which is in good agreement with values, derived for nonmagnetic ZnSe-based structures with similar QW parameters [9].

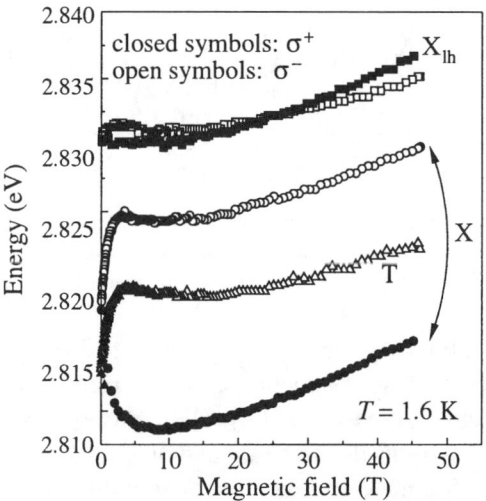

*Figure 2.* Energy of the trion (T), heavy-hole exciton (X) and light-hole exciton ($X_{lh}$) of the undoped QW vs. magnetic field. Data determined from reflectivity spectra detected under $\sigma^+$ ($\sigma^-$) polarization are presented by closed (open) symbols.

In the mean field approximation the Zeeman splitting of the heavy-hole exciton due to exchange splitting is described by

$$E_{ZM}^{hh}(B) = (\alpha - \beta) N_0 x \langle S_z \rangle, \qquad (1)$$

where $x$ is the Mn concentration and $N_0\alpha = 0.26$ eV and $N_0\beta = -1.31$ eV are the exchange constants for the conduction and valence band of $Zn_{1-x}Mn_xSe$, respectively [12]. For low magnetic fields the thermal average value of the Mn spin $\langle S_z \rangle$ can be described by a modified Brillouin function

$$\langle S_z \rangle = -S_{eff}(x) \, B_{5/2}\left[\frac{5 g_{Mn} \mu_B B}{2 k_B (T_{Mn} + T_0(x))}\right]. \qquad (2)$$

$S_{eff}(x)$ and $T_0(x)$ are parameters for the effective spin and effective temperature of the Mn ions that phenomenologically describe the effect of the antiferromagnetic coupling of Mn ions caused by the $d - d$ exchange interaction [13]. In DMS structures with high Mn concentration the mean spin of Mn ions can no longer be described with the modified Brillouin function for high fields. The magnetization of samples with high Mn concentration shows an additional step like behavior, because the antiferromagnetic coupling between pairs of Mn spins is broken at certain magnetic fields [14]. The breaking of antiferromagnetic coupled pairs of Mn spins gives no significant contribution to the Zeeman splitting in our samples, because of the low Mn concentration of $x = 0.005$.

Closed circles in Fig. 3 show experimental values of the Zeeman splitting of the heavy-hole exciton. Solid lines represent a fit of experimental data. A superposition of the exchange splitting described by Eqs. (1), (2) and the intrinsic Zeeman splitting with a g factor of $g_{Xhh} = +0.4$ determined for nonmagnetic ZnSe/(Zn,Be,Mg)Se QWs of the same thickness [9] have been used for the fit. For comparison a simulation of the pure exchange splitting is shown by a dotted line. For high fields the exchange Zeeman splitting is reduced by the intrinsic splitting, because of the opposite sign of the $g$ factors of holes for intrinsic and exchange splitting. With the value of the Mn concentration, obtained by a fitting of the Zeeman splitting of the exciton, and the ratio of the exchange constants of conduction and valence band states 1:5, we simulate the splitting of the conduction band electrons. The result is plotted in the inset of Fig. 3. For electrons both exchange and intrinsic splittings have the same sign. Dashed and dotted lines represent the pure intrinsic splitting and the pure exchange splitting of electrons, respectively. A dot-dashed line shows the superposition of both effects. The value of $g_e = +1.15$ for the intrinsic g factor in ZnSe QWs was taken from literature [9].

Open circles in Fig. 3 show the splitting of the light-hole exciton. In this case exchange splittings in conduction- and valence bands tend to compensate each other. As a result the exchange part of the Zeeman splitting of the light-hole exciton is strongly reduced with respect to the heavy-hole exciton with a maximum value of $E_{ZM}^{lh} = (\alpha - \beta/3) N_0 x$. At $B = 24$ T the intrinsic splitting

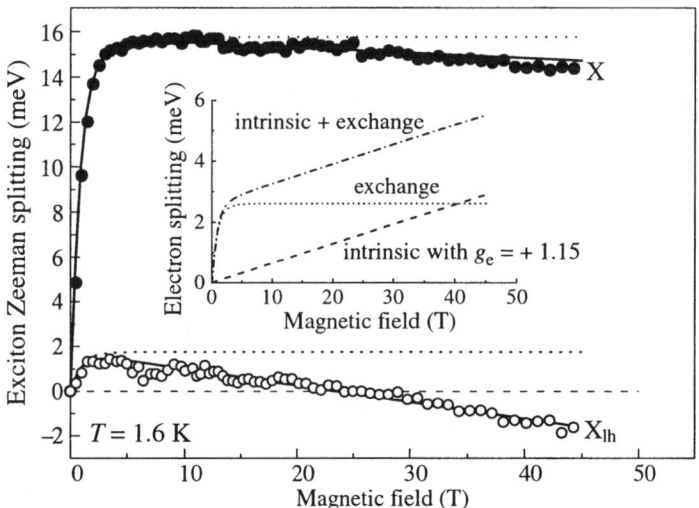

*Figure 3.* Experimental values of the heavy-hole (closed circles) and light-hole (open circles) exciton Zeeman splitting vs. magnetic field. Solid lines represent fittings of the splitting using a superposition of the "exchange" Zeeman splitting (due to $sp - d$ exchange interaction with the Mn ions) and the "intrinsic" Zeeman splitting (characteristic for nonmagnetic ZnSe-based QWs). Dotted lines represent simulations of the pure exchange splitting used in the fit. The inset shows the "intrinsic" (dashed line) and "exchange" (dotted line) Zeeman splitting and the superposition of both effects (dot-dashed line) for electrons.

overcomes the exchange splitting and the total Zeeman splitting changes its sign. Using the parameters for exchange splitting determined for the heavy-hole exciton we evaluate the intrinsic Zeeman splitting of the light-hole exciton to be described by a g factor $g_{Xlh} = +1.2$.

Keeping in mind the value for the electron g factor $g_e = +1.15$ we come to the g factors for heavy-hole $g_{hh} = g_e + g_{Xhh} = +1.6$ and light-hole $g_{lh} = g_e + g_{Xlh} = +2.4$. Note that the bulk relation for the degenerated valence band $g_{hh} = 3g_{lh}$ is not valid any more for the QW structures. Reason for that is an admixture of the higher heavy-hole states to the light-hole energy state for the QW structures [15].

## 4. Properties of Modulation Doped Samples

Let us now turn to the modification of the optical spectra by the presence of a 2DEG. Figure 4 shows PL and reflectivity spectra of the three investigated samples at zero magnetic field. The PL spectrum of sample #1 shows two lines due to the recombination of the exciton (X) and the negatively charged exciton (T). Both resonances are very pronounced in reflectivity spectra as already shown above. Similar to the case of ZnSe-based QWs [6], in the PL spectra with increasing 2DEG density the exciton line vanishes and the charged exciton line evolves into a broad luminescence band. A linewidth of this band (namely a full width at half

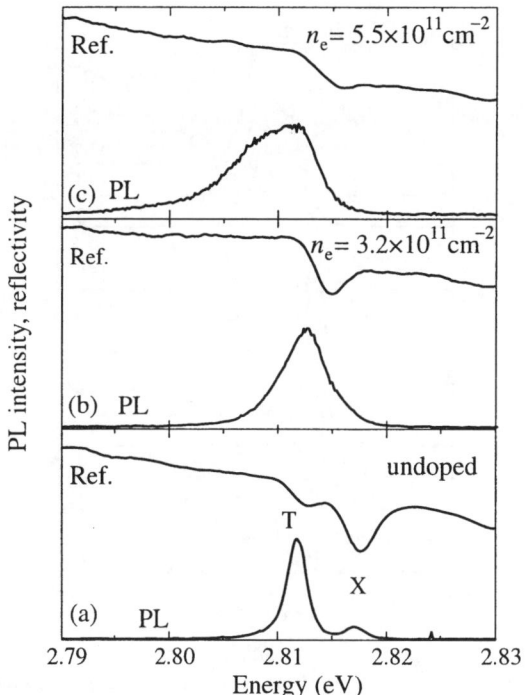

*Figure 4.* Reflectivity and luminescence spectra of three $Zn_{0.995}Mn_{0.005}Se/Zn_{0.96}Be_{0.04}Se$ QWs with different doping levels at zero magnetic field.

maximum) is 4.7 meV in the sample #2 and 8.8 meV in #3. These values coincide with high accuracy with the Fermi energies of 4.9 meV and 8.4 meV evaluated for the 2DEG with densities of $3.2 \times 10^{11}$ cm$^{-2}$ (sample #2) and $5.5 \times 10^{11}$ cm$^{-2}$ (sample #3), respectively. An electron effective mass of $m_e = 0.15 m_0$ was used in this evaluation. We note here that the linewidth of PL bands in modulation-doped structures exceeds the inhomogeneous broadening of exciton line of about 2 meV caused by QW width fluctuations. Therefore, the linewidth contains information on the 2DEG itself. In more detail we will discuss this point in respect to the data given in Fig. 8.

Figure 5 shows typical PL spectra for the sample #2 with $n_e = 3.2 \times 10^{11}$ cm$^{-2}$ for different magnetic fields. For small fields ($B \leq 6$ T) the luminescence lines in different polarizations shift according to the exchange Zeeman shift. Between 7 and 14 T the linewidth of the $\sigma^+$ polarized line decreases while keeping constant the integral PL intensity. For $B \geq 14$ T ($\nu \leq 1$) a new line (labeled as X) appears in the $\sigma^+$ polarization at the high-energy wing of the main line. This line gains intensity with increasing magnetic field at the expense of the main line (labeled as T-line in high fields). Energy positions of the two lines in high fields change similar to the exciton and trion lines in the undoped reference sample, which gives us reasons for their labeling. The effect of ionization of the bound exciton complexes like trion

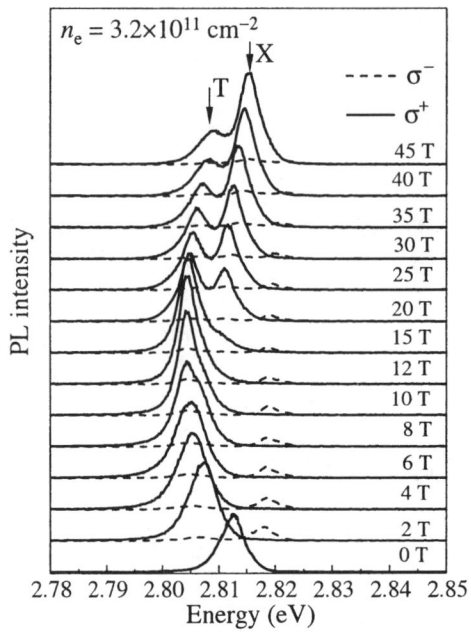

*Figure 5.* Photoluminescence spectra of a modulation-doped $Zn_{0.995}Mn_{0.005}Se/Zn_{0.96}Be_{0.04}Se$ QW with electron density of $n_e = 3.2 \times 10^{11}$ cm$^{-2}$ at different magnetic fields.

or exciton bound to neutral donor ($D^0$,X) is known for DMS bulk materials and heterostructures [16, 17]. These complexes become unbound when the Zeeman splitting in conduction band (contributed by both exchange and intrinsic parts) exceeds the binding energy. In the case of the sample #2 with a trion binding energy of 6 meV the ionization condition will be achieved in magnetic fields above 50 T (compare with inset of Fig. 3). For the lower fields trion is still bound, however it looses its intensity with growing field in favor of the exciton line.

In $\sigma^-$ polarization the shift of the PL line in Fig. 5 corresponds to the behavior of the trion resonance in the undoped sample #1. Although the PL linewidth decreases strongly already at low magnetic fields, where a giant Zeeman splitting controls the shift of the line. The linewidth variations for different polarizations in modulation-doped samples will be presented in more details in Fig. 8. Polarization properties of PL lines also differ in the doped and undoped samples. For the undoped case the trions intensity vanishes in $\sigma^-$ polarization at $B > 6$ T. For the doped sample #2 the intensity ratio of the trion components in two polarizations $I^+/I^-$ measured at filling factors $\nu < 1$, i.e. for a fully polarized electron gas, is four times smaller than in the undoped sample. Trions that contribute to the $\sigma^-$ polarized emission are formed from background electrons with spin $-1/2$ and photo-generated exciton with spin $-1$. The exciton with spin $-1$ is energetically unfavorable as it occupies the upper Zeeman sublevel. Usually it is not observed in PL spectra of DMS structures at higher magnetic fields due to the fast spin relaxation of the holes. The

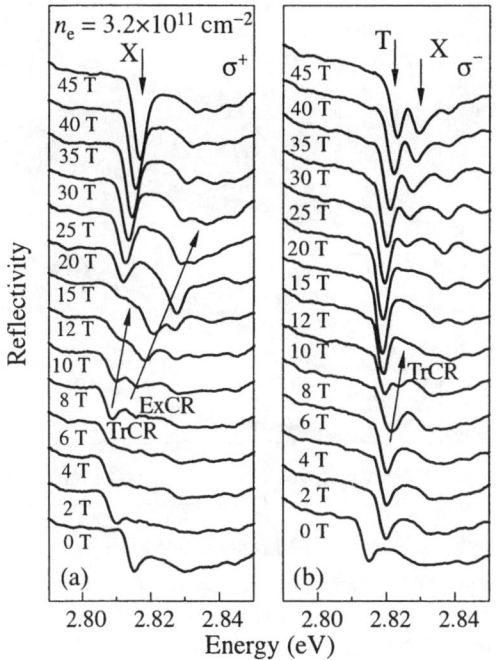

*Figure 6.* Reflectivity spectra of a $Zn_{0.995}Mn_{0.005}Se/Zn_{0.96}Be_{0.04}Se$ QW with $n_e = 3.2 \times 10^{11}$ cm$^{-2}$ at different magnetic fields detected in $\sigma^+$ and $\sigma^-$ polarizations.

decrease of the intensity ratio $I^+/I^-$ of the trion components with electron density indicates a nonequilibrium occupation of Zeeman sublevels by trions. This allows us to suggest a slowing of spin relaxation of heavy-holes in the doped structures. Mechanism responsible for that still requires clarification.

Let us now turn to the reflectivity spectra of the sample #2. Their evolutions in magnetic fields are given in Fig. 6 for two circular polarizations. Comparing these data with the spectra for the undoped sample from Fig. 1 one can see that they are rather similar in the high-field limit and differs significantly at small fields. At $B = 0$ T only one resonance is presented in spectra of doped sample. Futhermore, at intermediate fields from 6 to 25 T two new lines can be traced in the doped sample (they are labeled as ExCR and TrCR), which do not show up in the sample #1. Specifics of these lines is in their linear shift with magnetic fields, which can be clearly seen in Fig. 7, where the energy shift of observed transitions in reflectivity (squares) and luminescence (triangles) spectra are plotted versus magnetic fields strength. It has been shown in our previous studies for nonmagnetic QWs with a 2DEG, that these lines are due to the interaction of excitons and trions with free electrons on Landau levels. E.g. when the exciton is photogenerated in the vicinity of a background electron this electron can be excited from the lowest to higher Landau levels. The energy necessary for of such an excitation is equal to the cyclotron energy of electrons. Simultaneous generation of exciton and

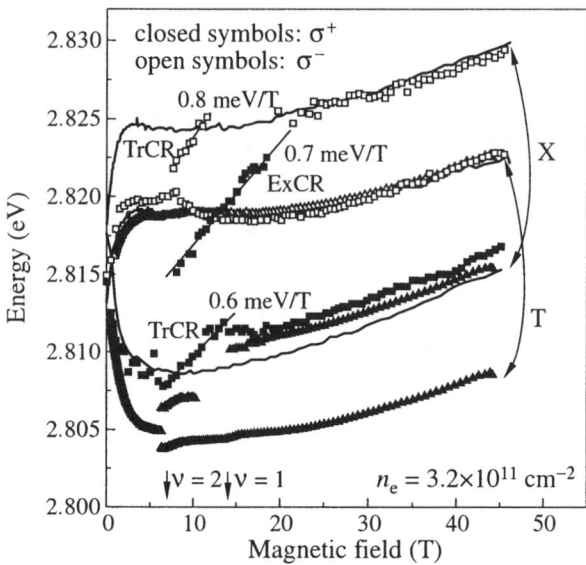

*Figure 7.* Energy position of the optical transitions in reflectivity (squares) and luminescence (triangles) spectra of a $Zn_{0.995}Mn_{0.005}Se/Zn_{0.96}Be_{0.04}Se$ QW with $n_e = 3.2 \times 10^{11}$ cm$^{-2}$. The exciton and trion energies for the undoped reference sample are shown by lines for comparison.

excitation of a background electron provide a new state, which will give resonance in the reflectivity spectra. This process is known as combined exciton-cyclotron resonance (ExCR) [18]. Similar process involving trion states was identified as combined trion-cyclotron resonance (TrCR) [19]. Intensities of these resonances depend on the properties of the 2DEG, e.g. if the upper Landau level is occupied the ExCR and TrCR will be suppressed. Rather specific criteria should be met for observation of TrCR. On one hand the upper Landau levels should have unoccupied states, on the other hand on the site of exciton generation two additional electrons should be located. As a result TrCR-line could be observed in the magnetic field range where the filling factor is $1 < \nu < 2$. This property give us another optical method for the evaluation the electron densities in the studied structures. Arrows at $\nu = 1$ and $\nu = 2$ in Fig. 7, evaluated from the linewidth of PL line for the sample #2, are in very good agreement with the existence range of TrCR line.

Solid lines in Fig. 7 trace the energy shift of exciton and trion lines in undoped sample #1. They are given for comparison with the behavior of the doped sample. One can see that in high magnetic fields for $\nu < 1$ both structures demonstrate pretty similar shifts, as already reported for nonmagnetic ZnSe-based QWs [6]. However, differences are obvious for the lower magnetic fields ($\nu > 1$), where the combined transitions and cusps at integer filling factors of the trion line are present for the doped structure.

Figure 8 shows the FWHM of the luminescence line of the two modulation-doped samples. For the sample #2 with $n_e = 3.2 \times 10^{11}$ cm$^{-2}$ the FWHM increases

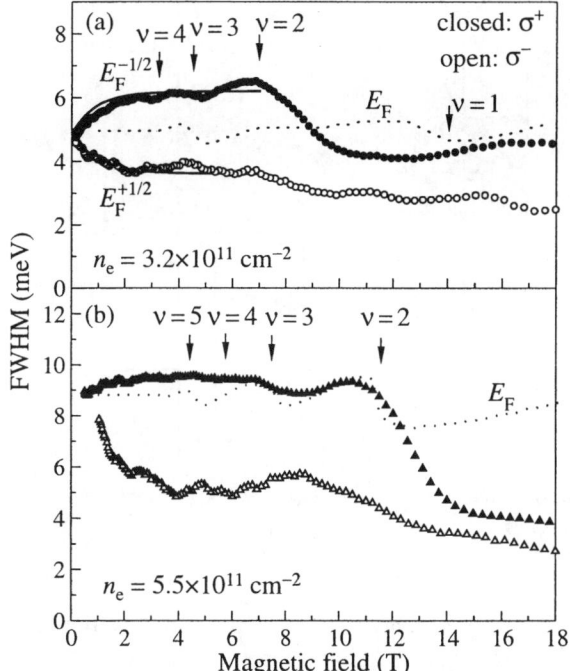

*Figure 8.* Full-width-at-half-maximum of the PL line of samples #2 and #3 with different electron densities as a function of magnetic field. Dotted lines represent simulations of the Fermi energy in the samples, calculated for a Landau quantized electron gas with inhomogeneously broadened Landau levels with a width of 2 meV. Solid lines in panel (a) correspond to simulations of Fermi levels of electrons with spin $+1/2$ and $-1/2$ calculated for spin splitted electron levels without Landau quantization.

for the $\sigma^+$ polarization, while it decreases symmetrically in the $\sigma^-$ polarization for small magnetic fields. The different width of the luminescence is caused by the spin polarization of the 2DEG, which is illustrated in Fig. 9. The giant Zeeman splitting of the conduction band results in a different occupation of the two spin levels and, therefore, in different Fermi energies $E_F^{+1/2}$ and $E_F^{-1/2}$ for electrons with spin $+1/2$ and $-1/2$.

To model the linewidth behavior at low magnetic fields we have simulated the Fermi levels for two spin components at a constant density of states. Required parameters of the electron Zeeman splitting were taken as determined for the undoped sample #1. The result of this simulation, which is shown by solid lines in Fig. 8(a), is in good agreement with the experimental values of the FWHM of the luminescence for the sample with $n_e = 3.2 \times 10^{11}$ cm$^{-2}$. The sample #3 with an electron density of $n_e = 5.5 \times 10^{11}$ cm$^{-2}$ shows different width of the luminescence in both polarizations as well. However, the increase and decrease is not symmetrical for the two polarizations as in sample #2 with lower doping. The oscillations of the linewidth at higher magnetic fields are due to the Landau quantization of the

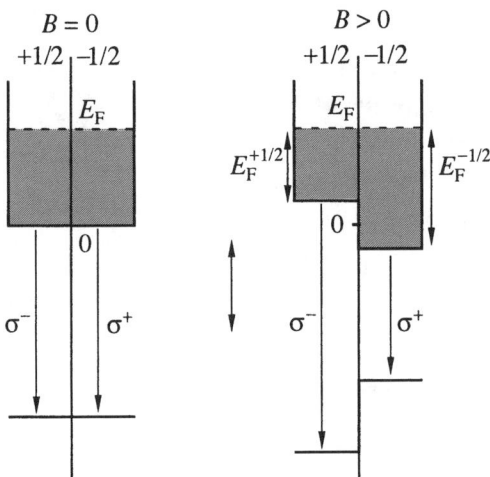

*Figure 9.* Spin-polarization of a 2DEG in magnetic fields, caused by the giant Zeeman splitting of the conduction band states.

2DEG. The dotted lines in Fig. 8(a) and (b) represent the calculated Fermi energies $E_F$ with respect to the conduction band minimum at $B = 0$ T (see Fig. 9). It was calculated for a 2DEG quantized on Landau levels in magnetic fields for $T = 1.6$ K. Inhomogeneous broadening of 2 meV has been introduced for the Landau levels in the calculations to be closer to the realistic conditions of the studied samples. It can be seen that the oscillations in the linewidth are related to the oscillations of the Fermi energy at integer filling factors, when the Fermi energy crosses the Landau levels. Both samples show a strong decrease of the FWHM in two polarizations at magnetic fields where one Landau level of the corresponding spin is occupied only, i.e. $\nu \leq 3$ for spin $+1/2$ and $\nu \leq 2$ for spin $-1/2$. For high magnetic fields (e.g. $B > 9$ T for sample #2 and $B > 14$ T for #3) the luminescence width is dominated by the inhomogeneous broadening and tends to approach the linewidth of undoped structure.

In order to have a closer look into the polarization properties of a 2DEG in the studied DMS quantum wells we model the occupation of electronic spin levels for different magnetic fields. Figure 10 displays the modeling of the Landau level fan chart. Solid and dashed lines give electron states with different spin orientations. Thick solid lines represent the calculated Fermi energies for samples #2 and #3 at $T = 0$ K. For both samples at magnetic fields above filling factor 2 the two lowest occupied Landau levels are levels with $n = 0$ and opposite spin. That means that the full spin polarization of the 2DEG, i.e. 100% of electrons have the same spin orientation will be achieved for $\nu < 1$ only. Qualitatively this situation is similar to the properties on the 2DEG in nonmagnetic structures.

Specifics of the DMS structures shows up for the case (i.e. for the range of magnetic fields) where the giant Zeeman splitting of Landau levels exceeds the

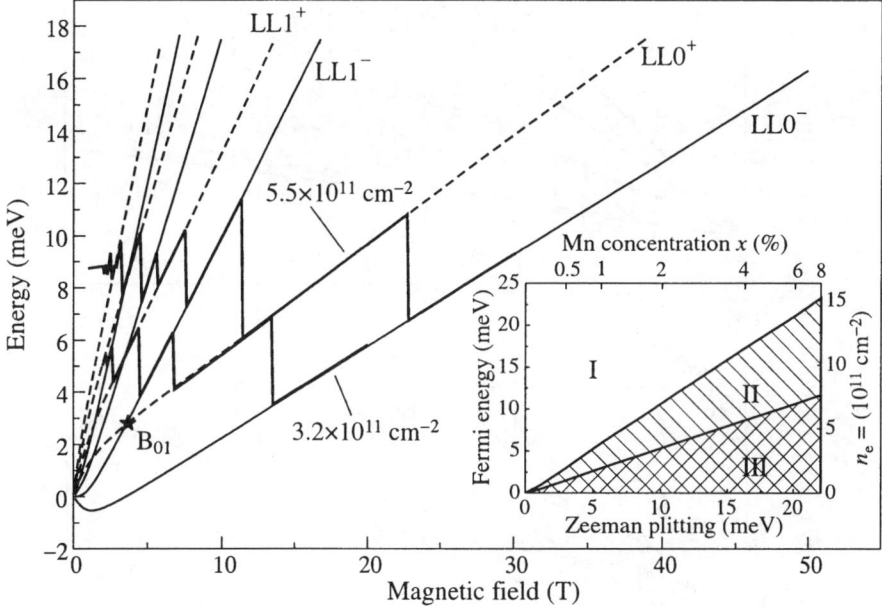

*Figure 10.* Landau level fan chart for electrons in conduction band of the studied structures. Solid and dashed lines represent Landau levels for spin down and spin up electrons, respectively. The highest occupied Landau level for sample #2 and #3 at $T = 0$ K is shown by a thick line. The inset gives an overview of the polarization properties of $Zn_{1-x}Mn_xSe$ QWs with different Mn content $x$ and electron densities $n_e$ as explained in the text.

cyclotron energy ($\hbar\omega_c$). Under this condition two (or even more) spin levels can be occupied ($\nu = 2$) but the 2DEG is fully spin-polarized. The particular behavior depends on the ratio of the Fermi energy $E_F$ and the Zeeman splitting $\Delta E_{ZM}$.

Let us now consider the sample parameters for which the specifics of DMS QWs can be realized. Figure 11 shows the distribution of the 2DEG over Landau levels and its degree of spin polarization $P = (N^+ - N^-)/(N^+ + N^-)$ for nonmagnetic and $Zn_{0.99}Mn_{0.01}Se$ DMS QWs. $N^{\pm}$ denotes the total number of electrons with spin $\pm 1/2$. For the nonmagnetic case $\hbar\omega_c \gg \Delta E_{ZM}$ for any magnetic field. This means that the spin splitting of the Landau levels with the same quantum number $n$ never exceeds the splitting between two consecutive Landau levels with different $n$ and same spin. Thus, the polarization of the Landau quantized 2DEG will be always equal to $P = 0$ at even filling factors. For the semimagnetic case the situation can be different. Even for filling factors $\nu > 1$ the electrons can be fully spin-polarized ($P = \pm 1$). Figure 11(b) shows the Landau levels fan chart when an additional Zeeman splitting, induced by Mn-ions is present. The filling factor dependence is the same as for the nonmagnetic case of panel (a) as both panels are modeled with the same electron density of $1.7 \times 10^{11}$ cm$^{-2}$. However, the energy position

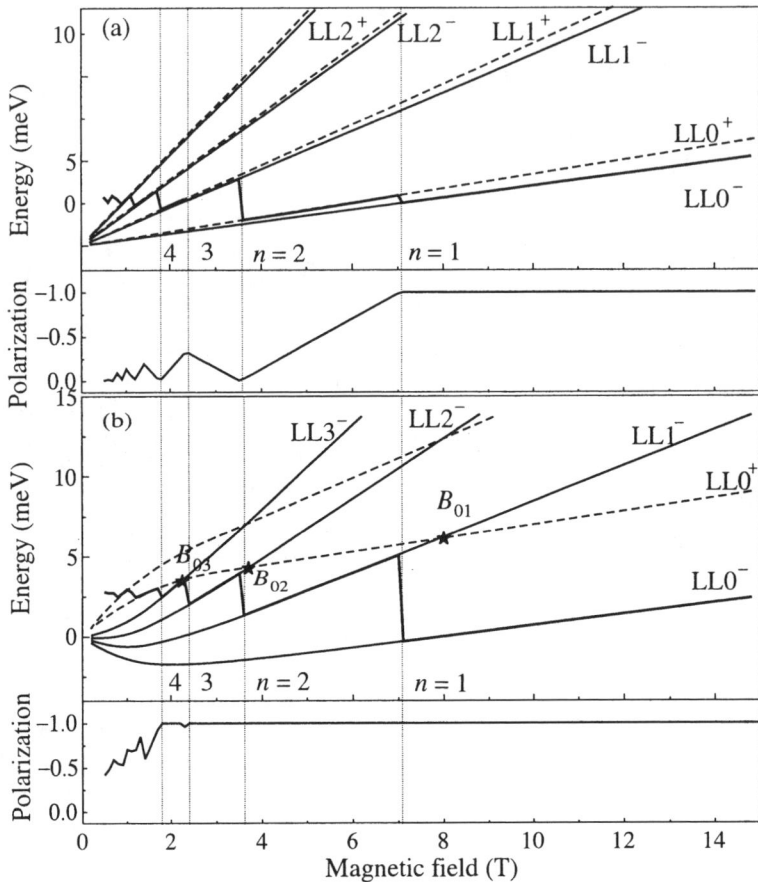

*Figure 11.* (a) Landau level fan chart for a nonmagnetic ZnSe-based QW, with the filling factor dependence for an electron gas with $n_e = 1.7 \times 10^{11}$ cm$^{-2}$. (b) Landau level fan chart for a $Zn_{0.99}Mn_{0.01}Se$ QW with same electron concentration. The degree of spin polarization of the 2DEGs at $T = 0$ K is shown below each panel.

of the Landau levels has changed. The factor which determines the polarization at integer filling factors is the value of magnetic field $B_{01}$ ($B_{02}$), for which the second spin down Landau level LL1$^-$ with $n = 1$ (third with $n = 2, \ldots$) crosses the lowest spin up Landau level LL0$^+$ with $n = 0$. If $\nu \leq 1$ for $B_{01}$, then for the range of $1 < \nu \leq 2$ two Landau levels with spin down are occupied and no spin up Landau level. If $\nu \leq 2$ for $B_{02}$, three Landau levels with spin down are occupied and spin up Landau level between $2 < \nu \leq 3$ are empty, etc. If $1 < \nu \leq 2$ for $B_{01}$, the two lowest occupied Landau levels for $\nu = 2$ have the same spin ($P = -1$), but with increasing magnetic field the situation changes and returns to a situation where the two lowest occupied Landau levels have opposite spins. That means that the situation changes from a situation specific for semimagnetic structures around $\nu = 2$ to a situation typical for nonmagnetic structures around $\nu = 1$.

The behavior that can be expected for (Zn,Mn)Se QWs with different electron concentration is summarized in the inset of Fig. 10. As already mentioned above this behavior depends on the ratio between Fermi energy and Zeeman splitting and can be categorized into three classes. For sample parameters in region I the structures behave similar to nonmagnetic samples, i.e. in high magnetic field the two lowest Landau levels have opposite spins. For sample parameters in region III structures show a specific behavior for DMS, i.e. a fully polarized 2DEG at filling factors $\nu > 1$. For sample parameters of section II, the behavior changes from semimagnetic around $\nu = 2$ to nonmagnetic around $\nu = 1$ for magnetic fields between $1 \leq \nu \leq 2$. One should mention that the suggested classification is only valid strictly for a Landau quantized electron gas at $T = 0$ K, with discrete Landau levels. At finite temperatures and for inhomogeneously broadened Landau levels the polarization properties will be modified.

## 5. Conclusions

(Zn,Mn)Se/(Zn,Be)Se DMS QWs with low Mn content are shown to be a very reliable model system for the examination of the strongly polarized 2DEG of moderate and high densities. High magnetic fields up to 45 T are essential for these studies as they provide a possibility to tune in wide range the cyclotron energy of electrons. This is important for attending the regime where the cyclotron energy exceeds the giant Zeeman splitting of the conduction band.

## Acknowledgements

This work was supported in part by the Deutsche Forschungsgemeinschaft (SFB 410).

## References

1. Lemaître, A., Testelin, C., Rigaux, C., Wojtowicz, T., and Karczewski, G. (2000) Magneto-optical evidence of many-body effects in a spin-polarized two-dimensional electron gas, *Phys. Rev. B* **62**, 5059–5065.
2. Wojtowicz, T., Kutowski, M., Karczewski, G. Kossut, J., Teran, F.J., Potemski, M. (1999) g-factor dependence of the evolution of magneto-optical spectra with the density of quasi-two-dimensional electrons in $Cd_{1-x}Mn_xTe/Cd_{1-y}Mg_yTe$ heterostructures, *Phys. Rev. B* **59**, R10437–R10440.
3. Salib, M.S., Kioseoglou, G., Chang, H.C., Luo, H., Petrou, A., Dobrowolska, M., Furdyna, J.K., and Twardowski, A. (1998) Magnetoluminescence study of a two-dimensional electron gas confined in diluted-magnetic-semiconductor quantum wells, *Phys. Rev. B* **57**, 6278–6281.
4. Crooker, S.A., Johnston-Halperin, E., Awschalom, D.D., Knobel, R., and Samarth, N. (2000) Stability of trions in strongly spin-polarized two-dimensional electron gases, *Phys. Rev. B* **61**, R16307–R16310.
5. Kulakovskii, V.D., Tyazhlov, M.G., Dite, A.F., Filin, A.I., Forchel, A., Yakovlev, D.R., Waag, A., and Landwehr, G. (1996) Interparticle interaction in spin-aligned and spin-degenerate exciton systems and magnetoplasmas in II-VI quantum wells, *Phys. Rev. B* 54, 4981–4987.

6. Ossau, W., Yakovlev, D.R., Astakhov, G.V., Waag, A., Meinig, C.J., Nickel H.A., McCombe, B.D., and Crooker, S.A. (2002) High magnetic field studies of 2DEG in modulation-doped ZnSe quantum wells, *Physica E* **12**, 512–514.
7. König, B., Zehnder, U., Yakovlev, D.R., Ossau, W., Gerhard, T., Keim, M., Waag, A., and Landwehr, G. (1999) Magneto-optical properties of $Zn_{0.95}Mn_{0.05}Se/Zn_{0.76}Be_{0.08}Mg_{0.16}Se$ quantum wells and $Zn_{0.91}Mn_{0.09}Se/Zn_{0.972}Be_{0.28}Se$ spin superlattices, *Phys. Rev. B* **62**, 2653–2660.
8. Crooker, S.A., Rickel, D.G., Lyo, S.K., Samarth, N., Awschalom, D.D. (1999) Magnetic semiconductor quantum wells in high fields to 60 Tesla: Photoluminescence linewidth annealing at magnetization steps, *Phys. Rev. B* **60**, R2173.
9. Astakhov, G.V., Yakovlev, D.R., Kochereshko, V.P., Ossau, W., Faschinger, W., Puls, J., Henneberger, F., Crooker, S.A., McCulloch, Q., Wolverson, D., Gippius, N.A., and Waag, A. (2002) Binding energy of charged excitons in ZnSe-based quantum wells, *Phys. Rev. B* **65**, 165335–165352.
10. Astakhov, G.V., Kochereshko, V.P., Yakovlev, D.R., Ossau, W., Nürnberger, J., Faschinger, W., and Landwehr, G. (2000) Oscillator strength of trion states in ZnSe-based quantum wells, *Phys. Rev. B* **62**, 10345–10352.
11. Astakhov, G.V., Kochereshko, V.P., Yakovlev, D.R., Ossau, W., Nürnberger, J., Faschinger, W., Landwehr, G., Wojtowicz, T., Karczewski, and G., Kossut, J. (2002) Optical method for the determination of carrier density in modulation-doped quantum wells, *Phys. Rev. B* **65**, 115310 (1-9).
12. Twardowski, A., von Ortenberg, M., Demianiuk, M., and Pauthenet, R. (1984) Magnetization and exchange constants in ZnMnSe, *Solid State Commun.* **51**, 849–852.
13. Keller, D., Yakovlev, D.R., König, B., Ossau, W., Gruber, Th., Waag, A., and Molenkamp, L.W. (2002) Heating of magnetic ion system in (Zn,Mn)Se/(Zn,Be)Se semimagnetic quantum wells by means of photoexcitation, *Phys. Rev. B* **64**, 035313–035320.
14. Shapira, Y., Foner, S., Ridgley, D.H., Dwight, K., and Wold, A. (1984) Technical saturation and magnetization steps in diluted magnetic semiconductors: Predictions and observations, *Phys. Rev. B* **30**, 4021–4023.
15. Kiselev, A.A., Ivchenko, E.L., Sirenko, A.A., Ruf, T., Cardona, M., Yakovlev, D.R., Ossau, W., Waag, A., and Landwehr, G. (1998) Electron and hole *g* factor anisotropy in CdTe/CdMgTe quantum wells, *J. Crystal Growth* **184-185**, 831–834.
16. Heiman, D., Becla, P., Kershaw, R., Ridgley, D., Dwight, K., Wold, A., and Galazka, R.R. (1986) Field-induced exchange effects in (Cd,Mn)Te and (Cd,Mn)Se from photoluminescence measurements, *Phys. Rev. B* **34**, 3961–3969.
17. Kulakovskii, V.D., Tyazhlov, M.G., Gubarev, S.I., Yakovlev, D.R., Waag, A., and Landwehr, G. (1995) Magnetic field induced dissociation of bound excitons in semimagnetic semiconductor quantum wells, *Il Nuovo Cimento* **17D**, 1549–1553.
18. Yakovlev, D.R., Kochereshko, V.P., Suris, R.A., Schenk, H., Ossau, W., Waag, A., Landwehr, G., Christianen, P.C.M., and Maan, J.C. (1997) Combined Exciton-Cyclotron resonance in Quantum Well Structures, *Phys. Rev. Lett.* **20**, 3974–3977.
19. Kochereshko, V.P., Astakhov, G.V., Yakovlev, D.R., Ossau, W., Faschinger, W., and Landwehr, G. (2002) Combined exciton-electron and trion-electron excitations in ZnSe/ZnMgSSe modulation doped quantum wells, *Phys. Stat. Sol. (b)* **229**, 543–547.

# FINE STRUCTURE OF EXCITONS IN QUANTUM WIRES

R.I. DZHIOEV, V.L. KORENEV, M.V. LAZAREV, V.F. SAPEGA
*Ioffe Physico-Technical Institute RAS,*
*Politechnicheskaya 26, St Petersburg, Russia*

R. NÖTZEL, K. PLOOG
*Paul-Drude-Institut für Festkörperelectronik,*
*Hausvogteiplatz 5-7, 10117 Berlin, Germany*

Abstract. The fine structure of the localized exciton ground state is studied in GaAs/AlAs quantum wires. Polarization of photoluminescence under the polarized excitation and spin-flip Raman scattering versus external magnetic field were measured. Exciton longitudinal $g$-factor and values of the level splitting due to anisotropic exchange interaction were determined. The exchange splitting values $\delta_2$ of optically active states and the distribution of split dipole in their directions can be probed optically.

Key words: localized excitons, quantum wires, spin-flip Raman scattering

The exciton ground state e1–hh1(1s) in quantum wells is fourfold degenerated taking into account spins of electron ($s = \pm 1/2$) and hole ($j = \pm 3/2$) and is characterized by angular momentum projections $m = s + j = \pm 1, \pm 2$. An electron–hole exchange interaction splits the e1–hh1(1s) state into two optically inactive singlets and optically active doublet. The latter also splits by anisotropic electron-hole exchange interaction into two states dipole-active in orthogonal directions when the exciton is localized in anisotropic island. The splitting energy $\delta_2$ [1] as well as the directions of the split dipoles depend on localizing island size and shape.

A several tens of papers were devoted to the fine structure of excitons localized in quantum sized structures. The splitting energy was measured by near field spectroscopy method in GaAs/AlGaAs (001) QW [2] and in type-II GaAs/AlGaAs (001) QW by optically detected magnetic resonance [3]. Exciton fine structure and cascade process of exciton localization were investigated in GaAs/AlAs (001) type-II superlattice by polarization spectroscopy [4]. Recently, the fine structure of exciton localized on single GaAs/AlGaAs V-shaped quantum wire was also detected by polarization-resolved micro-photoluminescence experiments [5].

Here we present the results of the polarization spectroscopy experiments carried out on structure with GaAs quantum wires. The longitudinal $g$-factor of the

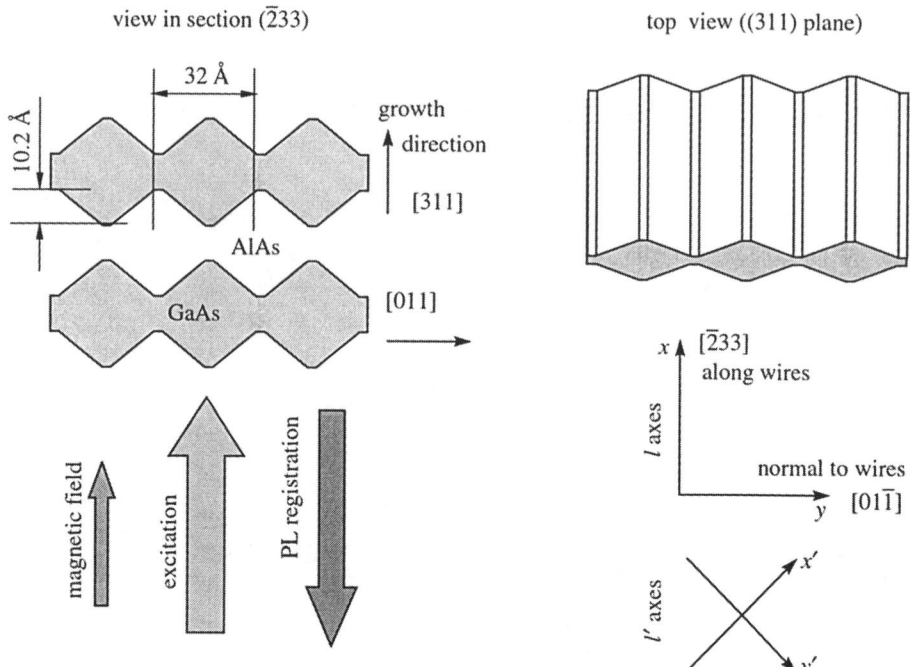

*Figure 1.* Views of the sample under study in two sections. The directions of the excitation, the registration and of the magnetic field are shown, also.

optically active exciton was found from exciton spin-flip Raman scattering and the values of $\delta_2$ were determined with the use of polarization spectroscopy. The sample under study (Fig. 1) was grown by MBE on (331) surface of GaAs substrate. The GaAs wires were streaked along the $[\bar{2}33]$ axes with lateral periodicity of 32 Å in $[01\bar{1}]$ direction (for more information see [6]).

We used two experimental methods based on investigation of: 1) Raman light scattering and 2) PL polarization under polarized excitation, both under applied magnetic field. The longitudinal $g$-factor was determined by analyzing of Raman scattering behavior in magnetic field and PL polarization dependences were used to determine the values of $\delta_2$. This allowed us to determine the values of energy splitting without a spectral resolution of lines of the fine structure. All experiments were carried out under steady-state excitation using polarized beam of Ti:Sapphire tunable laser. The exciting light was incident along normal to surface of sample. The radiation was detected in backscattered geometry. Magnetic field was applied along the normal to surface (Faraday geometry). The sample was cooled up to 2 K. The PL spectrum of the sample exhibits a broad line at 1.627 eV with width of $\approx$ 12 meV (Fig. 1b).

Under the resonant (1.627 eV) excitation in applied magnetic field the Raman spectra (see insert in Fig. 2) exists of highly polarized line (X) with width of

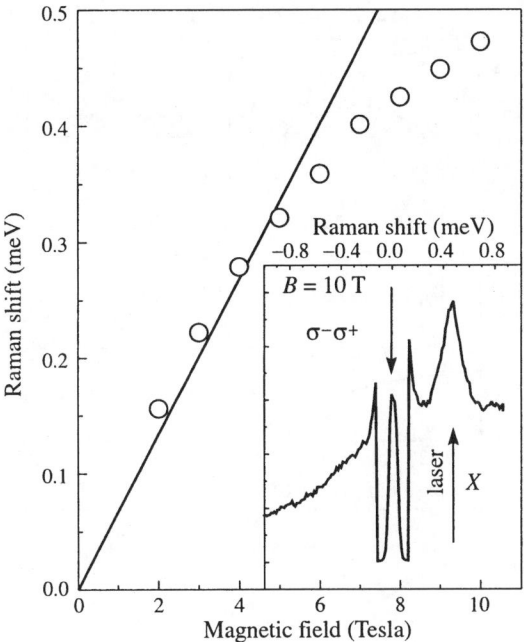

*Figure 2.* Raman scattering. The insert shows the Raman scattering spectra. The dependence of the shift of X line is shown by circles.

$\approx 0.12$ meV. We consider this line as a light scattering related to flipping of angular momentum of exciton upon interaction with acoustical phonon [7]. In this process, a photon with angular momentum projection $\pm 1$ is absorbed and an exciton is born in the state $\pm 1$; then, the exciton scatters in the state $\mp 1$ by acoustical phonons; and then, it annihilates emitting a photon with the momentum projection $\mp 1$. The experimental dependence of shift of this line on the magnetic field is shown in Fig. 2(a) by circles. The curve begins from 2 Tesla because of the line X was not resolved at lower fields. The value of this shift is the energy of splitting of optical active exciton and it is governed by Zeeman splitting at the fields more then 2 T. So, one can define the longitudinal g-factor of the exciton using the low field ($< 5$ T) part of the dependence: $g_\parallel = 1.16$.

The polarization spectroscopy has been used to determine the fine structure of optically active exciton. In these experiments we used the modulation technique where the analyzer was in a fixed position and the sample was pumped by the incident light changing its polarization from circular or linear to orthogonal at frequency 26.61 kHz. The sample was excited quasi-resonantly in band of exciton PL. The PL of the same line was detected a little below of the excitation (on $\approx 10$ meV). There were two experiments carried out in configurations (energy of excitation (eV)/energy of detection (eV): 1.631/1.621 (case a) and 1.621/1.616 (case b).

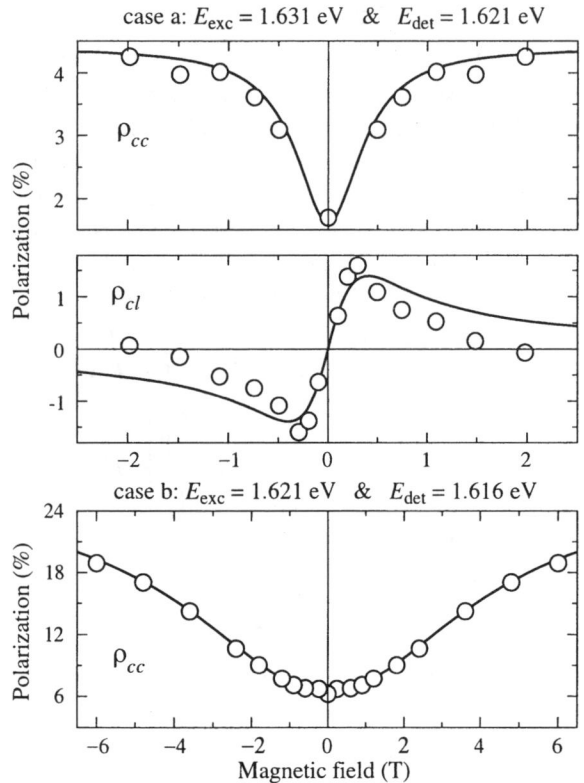

*Figure 3.* The experimental dependences on magnetic field of degrees of PL polarizations are shown by circles. Two cases (a) and (b) differ in energies of excitation ($E_{exc}$) as well as of PL detection ($E_{det}$). Here $\rho_{cc}$ and $\rho_{cl}$ indicates, respectively, circular and linear polarizations of PL under the circular polarized excitation.

The measured values are then the effective polarization degrees:

$$\rho_{c\alpha} = \frac{I_\alpha^{\sigma_+} - I_\alpha^{\sigma_-}}{I_\alpha^{\sigma_+} + I_\alpha^{\sigma_-}}, \quad \rho_{l\alpha} = \frac{I_\alpha^{x} - I_\alpha^{y}}{I_\alpha^{y} + I_\alpha^{y}}, \quad \rho_{l'\alpha} = \frac{I_\alpha^{x'} - I_\alpha^{y'}}{I_\alpha^{x'} + I_\alpha^{y'}},$$

where $I_\alpha^\beta$ is the intensity of $\alpha$-polarized component of PL under the $\beta$-polarized excitation. The symbols $\alpha$ and $\beta$ stand for light polarization: circular $\sigma_+$, $\sigma_-$, or linear along axes $x$, $y$, $x'$, $y'$. Here the axes $x$ and $y$ ($l$ axes) are perpendicular one to another and directed, respectively, along wires (direction [$\bar{2}33$]) and normal to wires (direction [$01\bar{1}$]); $x'$ and $y'$ ($l'$ axes) rotated relatively the latter on 45°. The experimental dependences on longitudinal magnetic field of the degrees of PL polarizations are shown by circles in Fig. 3 for both cases a) and b).

Note, that the measured degrees are equal to respective degrees of exciton polarization. So, the circular polarization indicates the exciton spin orientation, and the linear polarization reflects the exciton dipole alignment.

The dependences of PL circular polarization under the circular polarized excitation were detected in both cases (a) and (b) (Fig. 3, case (a) ($\rho_{cc}$) and (b) ($\rho_{cc}$), respectively), i.e. in both cases the restoration of exciton spin orientation goes on. Here we explain it by the presence of the exchange splitting of optical active exciton due to localization on anisotropic islands: optical orientation vanishes at zero magnetic field due to splitting of exciton into two linear polarized dipoles; when the longitudinal magnetic field is applied, the Zeeman splitting makes the dipoles circular and, so, prevents the depolarization of orientation, and the degree of the letter increases coming up to saturation at large magnetic fields. When the dipoles have preferential directions the magnetic field induces the conversion from exciton spin orientation to dipole alignment. In this case the linear polarization of PL appears under the circular polarized excitation [4]. Indeed, that conversion (to dipole alignment in $l$ axes) was detected in configuration (a) (Fig. 3, case (a) ($\rho_{cl}$)). It points to that the optically active excitons detected in that configuration really split by exchange interaction into linear dipoles directed along $l$ axes predominantly (one along and another normal to wire axis), that reflects the quantum wire potential. So, those excitons are localized on anisotropic islands. We evaluate a value of the exchange splitting using the $g$-factor found previously by Raman scattering: $\delta_2 = 30$ $\mu$eV.

The excitons detected in configuration (b) (energy of excitation is 1.631 eV and of registration is 1.621 eV) smaller then in (a) (1.621 and 1.616 eV, respectively) must be localized stronger then in case (a). So, the energy of their exchange splitting must be more then 30 $\mu$eV. We obtain from Fig. 3 (case (b)) the value 300 $\mu$eV. However, no linear polarization (neither in $l$ as well as in $l'$ axes) appears under the circular polarized excitation (the orientation-to-alignment conversions). It is naturally to explain that fact by isotropic distribution of dipoles in their directions. Indeed, the strong localized excitons mainly are under the potential of zero-sized island then under the ordered potential of wires.

In conclusion, the anisotropic exchange splitting of optically active exciton was measured with the use of the Raman scattering and PL polarization under the polarized excitation in the magnetic field.

It is shown, the splitting value $\delta_2$ and character of distribution of split dipoles in directions depend on the exciton localization strength on anisotropic island. Weekly localized excitons have a smaller $\delta_2$ and their dipoles directed predominantly along the principal axes of the wire as a result of the action of potential of quantum wire. On the contrary, the excitons that strongly localized have an enhanced splitting $\delta_2$ and their dipoles are distributed isotropically because of they undergo mainly a zero-sized potential changing randomly from island to island.

## Acknowledgements

Authors acknowledge gratefully partial support of the Russian Ministry of Science (contract N 40.072.11.1177) and also of RFBR 02-02-17636.

## References

1. Ivchenko, E.L. (1995) *Pure Appl. Chem.* **67**, 463.
2. Gammon, D. et al. (1996) *Phys. Rev. Lett.* **76**, 3005.
3. van Kesteren, H.W. et al. (1990) *Phys. Rev. B* **41**, 5283.
4. Dzhioev, R.I., et al. (1997) *Phys. Rev. B* **56**, 13405.
5. Guillet, T., et al. (2001) *Physica E* **9**, 686.
6. Nützel, R., et al. (1991) *Phys. Rev. Lett.* **67**, 3812.
7. Sapega, V.F., et al. (1992) *Phys. Rev. B* **45**, 4320.

# ELECTROMAGNETIC FIELD RESONANCE IN SEMICONDUCTOR NANOCRYSTALS

P. LAVALLARD
Groupe de Physique des Solides, CNRS,
UMR 7588,
Université Denis Diderot and Université Pierre et Marie Curie,
2 place Jussieu, 75251 Paris Cedex 05, France

G. LAMOUCHE
Institut des Matériaux Industriels, CNRC,
75 rue de Mortagne, Boucherville, Quebec, Canada J4B-6Y4

S.V. GOUPALOV
Georgia Institute of Technology,
Atlanta, GA 30332-0250 USA
Ioffe Physico-Technical Institute,
194021 St. Petersburg, Russia

Abstract. We model the inhomogeneous dielectric permittivity of a semiconductor nanocrystal by considering a dielectric sphere with a core and a concentric spherical shell of different permittivities. We show that the penetration of the electric field is resonantly enhanced when the laser frequency approaches an electromagnetic surface mode of the sphere. The absorption lines are shifted to higher energy and the transition oscillator strengths are changed by the resonance. The line shift is the same as the line shift that originates from the long-range electron-hole exchange interaction and was calculated by Goupalov and Ivchenko [1]. Taking into account both the short-range exchange interaction and the resonance shift, we find an excellent agreement with experimental data for CdSe nanocrystals. The line splitting in CdTe nanocrystals is calculated.

Key words: semiconductor, nanocrystal, exchange interaction

## 1. Introduction

Optical properties of semiconductor nanocrystals (NC) embedded in glass or in polymer films have been intensively studied [2]. The large scale structure of the absorption spectrum in CdSe NCs was well explained in the framework of the effective mass approximation [3–5]. Nevertheless the nature of the emitting state has remained controversial for a long time. The luminescence line is redshifted with respect to the resonant excitation energy of the lowest electron-hole pair state and has an unusual long lifetime. It was first attributed to surface-localized carriers [6]. But the well-defined polarization of the line and the increase of the Stokes shift with decreasing NC size have led to the conclusion that it has an intrinsic origin. The line was attributed to an optically forbidden state split from the first allowed state by the electron-hole exchange interaction which is strongly enhanced by quantum confinement [7]. Further studies in magnetic field have confirmed the nature of the luminescent state [8]. The whole fine structure of the exciton state was explained by Efros *et al.* [9] within a model which takes into account the crystal field, the short-range electron-hole exchange interaction and the anisotropic NC shape. Assuming a size-dependent NC ellipticity, they obtained a reasonable fit of the experiments. The disagreement for small size NCs was attributed to an additional contribution to the Stokes shift by phonons, which is not accounted for in the model. Recently Goupalov and Ivchenko [1] reconsidered the calculation of the electron-hole exchange interaction. They showed that the contribution of the long-range exchange interaction to the splitting between the optically allowed and optically forbidden exciton states can even exceed the short-range exchange contribution.

It is the purpose of this paper to give an intuitive and simple understanding of the long-range exchange interaction as a resonance of the local electromagnetic field in the NC and to show that, by taking into account both short-range and long-range contributions in a spherical NC, one gets a very good agreement with the experimental data.

## 2. Model

As shown by Cho [10], the interaction of the longitudinal component of the electromagnetic field with matter polarization is a part of Coulomb interaction among the charged particles. Two schemes can be used to describe the interaction of matter with light. In the first scheme, the interaction is taken between Coulomb excitons, in which direct Coulomb, short range and long range interactions are included, with the transverse part of the electric field. In the second scheme, the interaction is taken between mechanical excitons, in which direct Coulomb and short range interactions are included, with total electric field. This second scheme is equivalent to considering the interaction between matter polarization and its depolarization field.

To calculate the amplitude of the electromagnetic field in a NC, we use the method of electrostatics which is equivalent to assuming a Coulomb interaction between charges [11]. As the NC size is much smaller than the wavelength of light, retardation effects can be neglected.

The dielectric function is not homogeneous in the whole NC. To simplify its description, we consider the spherical NC as a dielectric sphere with a concentric spherical shell of different dielectric permittivity. The core dielectric function takes into account the exciton transition contribution. The shell dielectric function is the background dielectric permittivity.

The polarizability of the coated sphere is [12]:

$$\alpha = 3v_{NC}\varepsilon_h \frac{3\varepsilon_b(\varepsilon - \varepsilon_h)v_{core} + (\varepsilon + 2\varepsilon_b)(\varepsilon_b - \varepsilon_h)(v_{NC} - v_{core})}{3\varepsilon_b(\varepsilon + 2\varepsilon_h)v_{core} + (\varepsilon + 2\varepsilon_b)(\varepsilon_b + 2\varepsilon_h)(v_{NC} - v_{core})} \quad (1)$$

where $\varepsilon_h$ and $\varepsilon_b$ are the permittivities of the host material and shell volume respectively. $\varepsilon$ is the frequency-dependent dielectric function of the NC core and $v_{core}$ and $v_{NC}$ are the core volume and the NC volume, respectively.

The absorption coefficient of light passing through a host medium containing an assembly of such coated spheres is:

$$\gamma = 4\pi \frac{Nk}{\sqrt{\varepsilon_h}} \text{Im}(\alpha) \quad (2)$$

where $N$ is the NC concentration and $k$ the wave vector of light in vacuum. The symbol $\text{Im}(\alpha)$ means imaginary part of $\alpha$.

The polarizability of the coated sphere has resonant modes which correspond to negative values of the real part of $\varepsilon$. The frequencies of these so-called surface modes are larger than the transition frequencies $\omega_i$ in the medium. At resonant frequency, the imaginary part of $\alpha$ is maximum. The resonance frequency is then the frequency of the peak of $\gamma/\omega$. The frequency shift of the transition has the same origin as the shift of the optical phonon frequency in a ionic NC or the shift of the plasmon resonance in a metallic sphere [13].

We describe specifically cubic-structure NCs with simple and complex valence bands and wurtzite-structure NCs.

## 2.1. CUBIC-STRUCTURE NANOCRYSTALS WITH SIMPLE VALENCE BAND

Let us, first, consider a cubic semiconductor (CuCl type), with a simple conduction band $\Gamma_6$ and a simple valence band $\Gamma_7$. The optically active exciton state has the symmetry of a vector, $\Gamma_5$. It is separated from the optically inactive state $\Gamma_2$ by the short-range exchange interaction. Only the transition to the $\Gamma_5$ level makes a contribution to the NC core dielectric permittivity:

$$\varepsilon = \varepsilon_b \left(1 + \frac{\hbar\omega_{NC}}{\hbar\omega_1 - \hbar\omega + i\Gamma}\right) \quad (3)$$

$\Gamma$ is the homogeneous line width of the transition; $\hbar\omega_1$ is the energy of the $\Gamma_5$ level; $\hbar\omega_{NC}$ characterizes the oscillator strength. It is proportional to the longitudinal-transverse energy splitting, $\hbar\omega_{LT}$, in bulk material and inversely proportional to the core volume [14]:

$$\hbar\omega_{NC} = \hbar\omega_{LT} \frac{\pi a_B^3}{v_{core}}. \quad (4)$$

In the strong confinement approximation, the exciton wavefunction is the product of the electron and hole wavefunctions. For an infinitely high barrier, the envelope of the ground-state wavefunction of electron or hole is:

$$\psi(r) = \frac{1}{\sqrt{2\pi R}} \frac{\sin(\pi r/R)}{r} \quad (5)$$

where $R$ is the NC radius.

The core volume is calculated by keeping constant the total probability to find the electron and the hole at the same position:

$$v_{core}^{-1} = \frac{1}{(3\pi R)^2} \int_0^R \frac{\sin^4 \pi r/R}{r^2} 4\pi \, dr = \frac{0.67}{R^3}; \quad \frac{v_{core}}{v} = 0.36. \quad (6)$$

Inserting $\varepsilon$ as given by Eq. (3) into the expression of $\alpha$, we calculate the absorption coefficient $\gamma$ (Eq. (2)) as a function of $\omega$. We find that the absorption peak is shifted from to higher energy by:

$$\begin{aligned}\Delta_1 &= \hbar\omega_{NC} \left(\frac{1}{3} + \frac{2}{3}\frac{v_{core}}{v_{NC}}\frac{\varepsilon_b - \varepsilon_h}{\varepsilon_b + 2\varepsilon_h}\right) \\ &= \left(\frac{1}{3}0.67\pi\hbar\omega_{LT} + \frac{1}{2}\hbar\omega_{LT}\frac{\varepsilon_b - \varepsilon_h}{\varepsilon_b + 2\varepsilon_h}\right)\left(\frac{a_B}{R}\right)^3.\end{aligned} \quad (7)$$

The expression (7) is identical to the expression (37) given in [1] for the long-range exchange interaction.

## 2.2. CUBIC-STRUCTURE NANOCRYSTALS WITH COMPLEX VALENCE BAND

In a spherical NC of cubic structure, the first electron level is an isotropic state, doubly degenerate with respect to its spin direction and the first hole level is fourfold degenerate with respect to the projection of its total angular momentum. In the strong confinement approximation, the electron-hole pair or exciton envelope function is the product of the electron and hole envelope wavefunctions. When taking into account the direct Coulomb interaction, the exciton level is eightfold degenerate. The short-range electron-hole exchange interaction breaks this degeneracy into a fivefold and a threefold degenerate states. The two manyfolds can be labeled by the magnitude of the exciton total angular momentum, $J = 2$ and $J = 1$. The lowest energy state, $J = 2$, is separated from the highest energy state, $J = 1$, by the energy difference [9]:

$$4\eta = 4\left(\frac{a_{ex}}{R}\right)^3 \hbar\omega_{ST}\chi(\beta) \quad (8)$$

where $a_{ex}$ is the exciton Bohr radius, $\hbar\omega_{ST}$, the energy difference between the corresponding levels in bulk material and $\chi(\beta)$, a function of the heavy and light holes effective mass ratio, $\beta$.

Only the transition to the $J = 1$ level is optically allowed and makes a contribution to the NC core dielectric permittivity:

$$\varepsilon = \varepsilon_b \left( 1 + \frac{4/3 \hbar \omega_{NC}}{\hbar \omega_1 - \hbar \omega + i\Gamma} \right). \tag{9}$$

To take into account the weakening of the oscillator strength due to the incomplete overlap of the electron and hole wavefunctions, we write:

$$\hbar \omega_{NC} = \hbar \omega_{LT} \frac{\pi a_B^3}{v_{core}} K(\beta) \tag{10}$$

where the factor $K(\beta)$ is equal to the overlap integral squared [15],

$$K(\beta) = 2 \left[ \int_0^1 R_0(r) \sin(\pi r) r \, dr \right]^2 \tag{11}$$

and $R_0(r)$ is the isotropical part of the hole wave function.

The dipole field is not averaged to zero in the NC core because of the anisotropy of the hole wavefunction. Since the relation between the induction and the macroscopic electromagnetic field is not local, the probability of transition is not proportional to the product of the probabilities of presence of electron and hole. Nevertheless it is possible to define a core volume by taking into account explicitly the dipole-dipole interaction in the calculation of the transition oscillator strength. In order to make a comparison with the long-range exchange interaction calculated by Goupalov and Ivchenko [1], we define the core volume by the relation:

$$K(\beta) \frac{v_{NC}}{v_{core}} = \frac{4\pi}{3} Z(\beta) \tag{12}$$

where $Z(\beta)$ is a dimensionless function which was introduced in [1].

The calculation of the absorption coefficient is done along the same lines as in section 2.1. We find:

$$\Delta_1 = 4 \left( \frac{a_B}{R} \right)^3 \hbar \omega_{LT} \left[ \frac{\pi}{9} Z(\beta) + \frac{1}{3} \frac{\varepsilon_b - \varepsilon_h}{\varepsilon_b + 2\varepsilon_h} \frac{K(\beta)}{2} \right]. \tag{13}$$

The same result was obtained in [1].

It is interesting to note that the absorption peak value does not depend on the core volume:

$$\gamma_{max} = \frac{3}{4} \frac{2\pi}{\lambda} \frac{1}{\varepsilon_h^{1/2}} \left( \frac{3\varepsilon_h}{\varepsilon_b + 2\varepsilon_h} \right)^2 \varepsilon_b \frac{\hbar \omega_{LT}}{\Gamma} \pi a_B^3 K(\beta). \tag{14}$$

The numerical values for CdTe NCs are obtained from literature. The exciton energy splitting varies as $R^{-3}$ and reaches 140 meV for $R = 1$ nm. Assuming

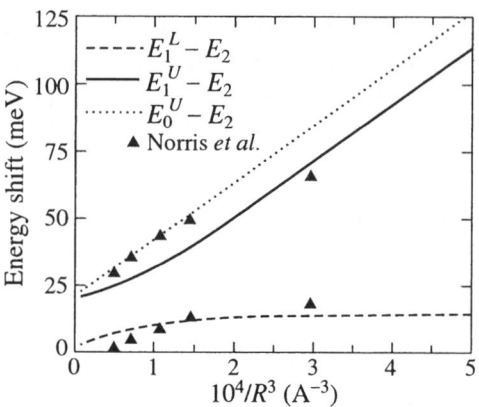

*Figure 1.* Size dependence of the exciton transition lines in CdSe NCs. The reference energy is $\hbar\omega_2$, the energy of the lowest energy line which is optically forbidden: dashed line, $\hbar\omega_1^L - \hbar\omega_2$; dotted line, $\hbar\omega_1^U - \hbar\omega_2$; solid line, $\hbar\omega_0^U - \hbar\omega_2$.

an homogeneous line width of 0.003 meV at low temperature [16], the absorption cross-section of a nanocrystal is evaluated to be 1000 nm². The radiative lifetime [15] is calculated to be 7 ns.

### 2.3. WURTZITE-STRUCTURE NANOCRYSTALS

In a wurtzite-structure spherical NC, the crystal field and the short-range electron-hole exchange interaction split the eightfold degenerate exciton, into five levels which can be labeled by the magnitude of the exciton total angular momentum projection on the c-axis: one level with $J = \pm 2$, two levels with $J = \pm 1$: $\pm 1^U$ and $\pm 1^L$ and two levels with $J = 0$: $0^U$ and $0^L$. The transition to the $0^U$ level is optically allowed with a polarization of light parallel to the c-axis. The transitions to the $\pm 2$ and $0^L$ levels are optically forbidden [9].

The NC core permittivities for parallel and perpendicular polarizations are:

$$\varepsilon_\parallel = \varepsilon_b \left(1 + \frac{4/3\hbar\omega_{\rm NC}}{\hbar\omega_0 - \hbar\omega + i\Gamma}\right) \quad (15)$$

$$\varepsilon_\perp = \varepsilon_b \left(1 + \frac{q^L \hbar\omega_{\rm NC}}{\hbar\omega_1^L - \hbar\omega + i\Gamma} + \frac{q^U \hbar\omega_{\rm NC}}{\hbar\omega_1^U - \hbar\omega + i\Gamma}\right) \quad (16)$$

where $\hbar\omega_0$, $\hbar\omega_1^L$ and $\hbar\omega_1^U$ are the transition energies to the $0^U$, $\pm 1^L$ and $\pm 1^U$ states, respectively. The size dependence of the energies, $\hbar\omega_0$, $\hbar\omega_1^L$, $\hbar\omega_1^U$ and the coefficients $q^L$, $q^U$ are given in [9] and will not be reproduced here.

By inserting $\varepsilon_\parallel (\varepsilon_\perp)$ in (1), we calculate the polarizability, $\alpha_\parallel (\alpha_\perp)$ for a parallel (perpendicular) polarization of light. The absorption coefficient of a medium containing a random assembly of such NCs is obtained by replacing $\alpha$ by $\frac{2\alpha_\perp + \alpha_\parallel}{3}$ in the expression of $\gamma$ (Eq. (3)). The calculation of the poles of $\alpha_\parallel (\alpha_\perp)$ and show that

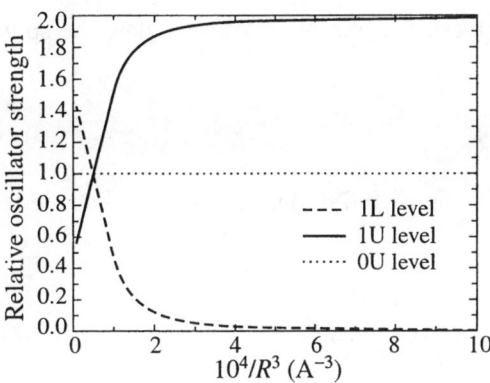

*Figure 2.* Size dependence of the oscillator strengths of the $\pm 1^U$ and $\pm 1^L$ states, relative to that of the $0^U$ line, for CdSe NCs.

the shift of the lines is exactly the same as the shift which was calculated in [1] for wurtzite structure.

We find that the transition line which is allowed for parallel polarization of light is shifted by the same amount of energy as the transition line for an assembly of cubic-structure spherical NCs: $\Delta = \Delta_1$ (Eq. (12)). In perpendicular polarization, the two allowed transitions interact. There is an enhancement in both the strength and the shift of the upper line and a reduction of the strength and the shift of the lower line [17].

A numerical estimation is done for CdSe NCs. The resulting size dependence of the exciton transition lines is shown in Fig. 1. On the same figure we have plotted the data obtained by Norris *et al.* [18] from photoluminescence excitation spectra on NCs whose size was accurately measured by electron microscopy. The agreement between experiments and theory is very good. Even for the lowest size NCs ($R = 15$ Å), the determination of the line splitting is well reproduced by our model. Figure 2 shows the size dependence of the transition oscillator strengths. As expected, there is a redistribution of the oscillator strength among the two lines which are allowed in the same polarization, normal to the NC c-axis.

## 3. Conclusion

We have modeled semiconductor NCs by assuming that the exciton contribution to the dielectric permittivity is concentrated in the core of the NC. We have shown that the electromagnetic field inside the NC is resonantly enhanced when the laser frequency approaches an electromagnetic surface mode of the coated sphere. The resonance shifts the transition lines by the same amount as it was found by Goupalov and Ivchenko [1], when considering the long-range exchange interaction between electron and hole. Taking into account both the short-range interaction and the resonance shift of the lines we reproduce very well the experimental data obtained for spherical CdSe NCs. We calculate the splitting of the line for CdTe NCs.

**Acknowledgements**

Support of INTAS (Code 99-0-0858) is gratefully acknowledged.

**References**

1. Goupalov, S.V. and Ivchenko, E.L. (2000) The Fine Structure of Excitonic Levels in CdSe Nanocrystals, *Phys. Solid State* **42**, 2030–2038.
2. Woggon, U. (1996) *Optical Properties of Semiconductor Quantum Dots*, Springer Tract in Modern Physics **136**, Springer, Berlin.
3. Grigoryan, G.B., Kazaryan, E.M., Efros, Al.L., and Yazeva, T.V. (1990) Quantized holes and the absorption edge in spherical semiconductor microcrystals with a complex valence band structure, *Sov. Phys. Solid State* **32**, 1031–1035.
4. Ekimov, A.I., Hache, F., Schanne-Klein, M.C., Ricard, D., Flytzanis, C., Kudryavtsev, I.A., Yazeva, T.V., Rodina, A.V., and Efros, Al.L. (1993) Absorption and intensity-dependent photoluminescence measurements on CdSe quantum dots-assigment of the 1st electronic-transitions, *J. Opt. Soc. Am. B* **10**, 100–107.
5. Norris, D.J., and Bawendi, M.G. (1996) Measurement and assignment of the size-dependent optical spectrum in CdSe quantum dots, *Phys. Rev. B* **53**, 16338–16346.
6. Bawendi, M.G., Wilson, W.L., Rothberg, L., Carroll, P.J. Jedju, T.M., Stegerwald, M.L., and Brus. L.E. (1990) Electronic structure and photoexcited-carrier dynamics in nanometer-size CdSe clusters, *Phys. Rev. Lett.* **65**, 1623–1626.
7. Chamarro, M., Gourdon, C., Lavallard, P., Lublinskaya, O., and Ekimov, A.I. (1996) Enhancement of electron-hole exchange interaction in CdSe nanocrystals: A quantum confinement effect, *Phys. Rev. B* **53**, 1336–1342.
8. Nirmal, M., Norris, D.J., Kuno, M., Bawendi, M.G., Efros, Al.L., and Rosen, M. (1995) Observation of the "Dark Exciton" in CdSe Quantum Dots, *Phys. Rev. Lett.* **75**, 3728–3731.
9. Efros, Al. L., Rosen, M., Kuno, M., Nirmal, M., Norris, D.J., and Bawendi, M. (1996) Band-edge exciton in quantum dots of semiconductors with a degenerate valence band: Dark and bright exciton states, *Phys. Rev. B* **54**, 4843–4856.
10. Cho, K. K. (1999) Mechanisms for LT splitting of polarization waves: a link between electron-hole exchange interaction and depolarization shift, *J. Phys. Soc. Japan* **68**, 683–691.
11. Born, M. and Huang, K. (1954) *Dynamical theory of crystal lattices*, Clarendon Press, Oxford.
12. Van de Hulst, H.C. (1981) *Light scattering by small particles*, Dover Publications, New York.
13. Raether, H. (1988) *Surface Plasmons*, Springer Tracts in Modern Physics **111**, Springer-Verlag, Berlin, Heidelberg.
14. Henry, C.H. and Nassau, K. (1970) Lifetimes of Bound Excitons in CdS, *Phys. Rev. B* **1**, 1628–1634.
15. Efros, Al.L. (1992) Luminescence polarization of CdSe microcrystals, *Phys. Rev. B* **46**, 7448–7458.
16. Goupalov, S.V., Suris, R.A., Lavallard, P., and Citrin, D.S. (in press) On the zero phonon line homogeneous broadening in two-level systems.
17. Ramaniah L.M., Nair S.V., Rustagi K.C.(1989) Linear and nonlinear optical response of spherical anisotropic semiconductor microcrystallites, *Phys. Rev. B* **40**, 12423–12432.
18. Norris, D.J., Efros, Al., Rosen, M., and Bawendi, M.G. (1996) Size dependence of exciton fine structure in CdSe quantum dots, *Phys. Rev. B* **53**, 16347–16354.

# CO-MANIFESTATION OF INTERFACES ASYMMETRY AND MAGNETIC FIELD INFLUENCE ON LUMINESCENCE POLARIZATION ANISOTROPY OF [100]-ORIENTED QW WITH SEMIMAGNETIC BARRIERS

S.M. RYABCHENKO, O.V. TERLETSKIJ
*Institute of Physics NAS of Ukraine,*
*Kiev, Ukraine*

YU.G. SEMENOV
*Institute of Semiconductor Physics NAS of Ukraine,*
*Kiev, Ukraine*

F.V. KIRICHENKO
*Institute of Physics of Polish Academy of Sciences,*
*Warsaw, Poland*

Abstract. The polarization anisotropy (PA) of the 1e–1hh optical transitions in [100]-quantum wells (QWs) $A^2B^6$ with semimagnetic barriers $A^2_{1-x}Mn_xB^6$ in a magnetic field applied in the QW plane is studied. This anisotropy stems from the mixing of light- and heavy-hole (HH) states by interactions, which reduces the in-plane symmetry of structure. Zeeman energy of carriers in a magnetic field **G**, which is exchange-enhanced due to presence of the semimagnetic barrier, as well as in-plane strains and effects of an interface asymmetry are taken into account. Comparison of $\propto (J_x J_y + J_y J_x)$ and $\propto (J_x^3 G_y + J_y^3 G_x)$ $C_{2v}$-invariants manifestations in the luminescence PA in the presence of in-plane magnetic field is reported (here **J** is a hole angular moment).

We predict and analyze a new effect of PA singularity caused by interference of HH splitting due to Zeeman energy in third order $\propto G^3$ and linear on $G$ contribution mediated by $C_{2v}$ interaction. The quantitative evaluation of 1e–1hh luminescence PA is carried out in terms of electron and hole spin temperatures. It is shown that fluctuations of potential due to QW interface roughness can result in redistribution of $\pi$-periodic and $\pi/2$-periodic contributions to the PA. Theoretical analysis compares with our experiments carried out on QW $Cd_{1-x}Mn_xTe/CdTe/Cd_{1-x}Mn_xTe$ as well as with the data of other papers.

Key words: semiconductors, magnetic field, semimagnetic, luminescence polarization, polarization anisotropy, quantum well, low symmetry effect

## 1. Introduction

Optical properties of semiconductors with a zinc blend structure are isotropic in most cases. Magnetic field applied perpendicular the light propagation has to lead to some optical anisotropy connected with this field. In semimagnetic compounds $Cd_{1-x}Mn_xTe$, luminescence polarization is shown to depend on a field direction in a cleaving plane of crystal [011] in the case of high enough Mn content $x \geq 0.15$ (see Ref. [1] and references therein). There was assumed that anisotropy of Mn–Mn exchange interaction results in optical anisotropy observed in Ref. [1].

A strong enough polarization anisotropy (PA) of luminescence from [001] oriented quantum wells (QW) $Cd_{1-x}Mn_xTe/CdTe/Cd_{1-x}Mn_xTe$ was observed in Refs. [2, 3]. The magnitude of this PA increases strongly with increasing of in-plane magnetic field while its direction does not influence on PA principal axes, which coincide with axes [110] and [1$\bar{1}$0]. The $C_{2v}$ symmetry hole interaction $V_{q2} = q_2(J_x^3 B_x^* - J_y^3 B_y^*)$ with effective magnetic field $B^*$ enhanced due to sp-d interaction in semimagnetic barrier assumes to be responsible for these properties ($x$- and $y$-axes direct along [110] and [1$\bar{1}$0] axes in Refs. [1, 2], $J$ is a hole angular moment). The origin of interaction $V_{q2}$ as a possible manifestation of in plane strains and/or low symmetry effects of interfaces was discussed qualitatively while the microscopic model for $V_{q2}$ interaction was not analyzed in Refs. [2, 3].

The lowering of interface potential symmetry from $D_{2d}$ up to $C_{2v}$ in the [001] oriented heterostructures of zinc-blend type was found and justified in Refs. [4–6]. The principal axes of the $C_{2v}$ interface potential constituent correspond to [110] and [1$\bar{1}$0] directions, which lie in the plane of heterostructure. This interaction mixes the heavy hole (HH) with $J_z = \pm 3/2$ and light hole (LH) with $J_z = \pm 1/2$ states. If the materials QW and barriers have common atom (anion or cation), the signs of the $C_{2v}$ contributions of the interface potential are opposite for "right" and "left" interfaces. However, annihilation of these two contributions does not take place if the HH states are not symmetrical with respect to QW centre. Last situation always arises for non bilaterally symmetrical interfaces.

In the case of QW in-plane strain, the principal axes of hole deformation potential of the $C_{2v}$ symmetry correspond to the strain directions, which do not obligatory coincide with [110] and [1$\bar{1}$0]. Such $C_{2v}$ components of a potential mix the HH and LH states with non-orthogonal envelope functions. The optical transitions between the electron and HH confined states reveal in plane PA defined by deformation or/and interface potential.

One should be also mentioned that $C_{2v}$ components in QW potential has influence on the heavy hole exciton splitting that related with electron - hole exchange interaction (Refs. [7, 8]).

The authors of Refs. [2, 3, 7, 8] consider the magnetic field applied perpendicular to direction of QW growth. In such a case Hamiltonian takes into account the $D_{2d}$ and $C_{2v}$ invariant in the form of $q_1(J_x^3 B_x^* + J_y^3 B_y^*) + q_2(J_x^3 B_y^* + J_y^3 B_x^*)$

without Zeeman term $J_x B_x^* + J_y B_y^*$. At the same time, the only Zeeman term with a transversal magnetic field leads to the HH doublet splitting in the third order of the perturbations theory. So, a few different kinds of contributions to the HH splitting can be revealed in PA. In this situation we can expect manifestation of particularities caused by competition between magnetic field induced low symmetrical and Zeeman contributions to PA. Moreover, a combined account of Zeeman term and non-magnetic low-symmetry interactions can result in nontrivial singularities in polarization of the 1e–1hh exciton transition components with growth of a transversal magnetic field. As to PA of photo-luminescence, it is pertinent to note that the PA reveals anisotropy as a sum of few quantum transitions between electron and HH spin sublevels. Therefore, distribution of populations between spin sublevels associated with hole $T_h$ and electron $T_e$ spin temperatures has to be manifested in an experiment.

The theoretical problems mentioned in preceding paragraph are undecided till now and constitute the matter of present paper. In next section we introduce the Hamiltonian with low symmetry terms, which are responsible for PA. Then, we find HH splitting and eigenstates modified by these interactions and an effective field $\boldsymbol{B}^* = \boldsymbol{B} + \boldsymbol{B}_{ex}$. In terms of eigenvectors, we calculate the matrix element for optical transitions with two perpendicular polarizations planes and accounts their contribution to optical polarization according to populations of spin levels. Experimental results and discussion are presented at last section of the paper.

## 2. The Theoretical Consideration

Let we consider a case of large enough spin-orbital splitting $\Delta_{SO}$ that allows to describe HH–LH splitting $\Delta \ll \Delta_{SO}$ in a QW in terms of Luttinger Hamiltonian at zero in-plane wave vector. Solution of this problem assumes to be known. Simplicity sake, we consider only eigen-functions of two-fold degenerated HH and LH confined ground (1HH and 1LH) states that can be symbolical written as a set $\{\psi_i\} = \{|HH\rangle, \hat{K}|HH\rangle, |LH\rangle, \hat{K}|LH\rangle\}$, where $\hat{K}$ is a time inversion operator. Assuming Zeeman energy as well as other low-symmetry interactions is small perturbation, the 4×4 perturbation matrix can be represented in terms of hole angular moment

$$\hat{V} = G_{hx} J_x + G_{hy} J_y + T\{J_{x'} J_{y'}\} + q_1 \left[ G_{hx} J_x^3 + G_{hy} J_y^3 \right] \\ + q_2 \left[ G_{hx'} J_{y'}^3 + G_{hy'} J_{x'}^3 \right] \quad (1)$$

where $\{J_x J_y\} \equiv (1/2)(J_x J_y + J_y J_x)$, $T\{J_{x'} J_{y'}\} = \Delta t_{l,r} \{J_x J_y\} + d\varepsilon_{x_d, y_d} \{J_{x_d} J_{y_d}\}$; $T$ is crystal field $C_{2v}$ invariant [4]; $d$—deformation potential; $\varepsilon_{x_d, y_d}$—the strain with $x_d$ and $y_d$, principal axes, which turn on some angle $\phi$ in respect to $x$ and $y$ ones. $\Delta t_{l,r} = t_l - t_r$, $t_l$ and $t_r$ are the left and right interface $C_{2v}$-potential parameters; the sum of interface and deformation contributions can be represented in the similar form $T\{J_{x'}J_{y'}\}$ with $x'$ and $y'$ axes turned on some other angle $\varphi \neq \phi$. The term, which is proportional to $q_1 \equiv q_{D2d}$ stems from Luttinger Hamiltonian, while the terms with parameter $q_2 \equiv q_{C2v}$ is $C_{2v}$ invariant being linear on the magnetic field. According to the model we develop, the admixture of the states that are not included into the basis of Hamiltonian (1) by deformation and interface potentials can be responsible for this term appearance. We assume that principal axes of this $C_{2v}$ invariant and $T$-invariant coincide. The exchange enhanced magnetic field $G_h$ (in the energy units) is applied in the plane of QW, so $G_h = G_h\{\cos\theta, \sin\theta, 0\}$. One should be mentioned that the matrix elements of Zeeman terms in Hamiltonian (1) depend on overlap of 1HH- and 1LH-enveloping $f_{1hh}(z)$ and $f_{1lh}(z)$ of wave functions. We neglect their difference that allows writing the $G$-depended parameters in Eq. (1) in the form

$$G_h = \kappa H + \frac{1}{3}N_0\beta \int_{-\infty}^{\infty} x(z)\langle S_H(z)\rangle |f_{1h}(z)|^2 dz, \quad (2)$$

where $\kappa$ is a Luttinger parameter; $\beta$ is a hole-Mn exchange interaction parameter, $N_0$ the cation concentration; $x(z)$ and $\langle S_H(z)\rangle$ the content of Mn and average Mn spin that depend on $z$ (see, for instant, Refs. [9, 10]).

Spin-Hamiltonian for the confinement conductivity electron has the usual form

$$H_e = \frac{1}{2}\left(G_{ex}\sigma_x + G_{ey}\sigma_y\right), \quad (3)$$

where $\sigma_x$ and $\sigma_y$ are the Pauli matrixes; the components of exchange enhanced in-plane magnetic field acting on the electron spin $G_{ex}$ and $G_{ey}$ are defined similar to Eq. (2) with changing of $\kappa$ on $g_e\mu_B$, $\beta$ on $\alpha$ and $f_{1h}(z)$ on $f_e(z)$.

The eigen-energies of electron spin are

$$E_{e1,2} = E_{1e} \pm \frac{1}{2}h_e\Delta, \quad (4)$$

where $E_{1e}$ is $1e$ state energy at zero magnetic field.

The in-plane magnetic field splits the 1HH state due to $q_1$- and $q_2$-terms in first order over perturbation (1), due to interference of $T$-term and Zeeman term in second order, and due to only Zeeman term in the 3-th orders. As a result, the HH splitting is:

$$E_{hh1,2} = E_{hh0} \pm G_{hh}/2 \quad (5)$$

where $G_{hh} = (3/2)hR\Delta$, $E_{hh0}$ is "center of gravity" of 1HH states energy, which shifted from its zero magnetic field value on the $-3/4h^2\Delta$ due to interaction with 1LH. The $R$ is defined by expression

$$\begin{aligned}R^2 &= h^4 + q_1^2 + (q_2 - 2t)^2 + t^4 - 2(q_2 - 2t)\{(h^2 + t^2)\sin[2(\theta - \varphi)] \\ &\quad - q_1\sin[2(\theta - \varphi)]\} + 2h^2 q_1\cos[4\theta] \\ &\quad - 2t^2 q_1\cos[4\theta] - 2h^2 t^2\cos[4(\theta - \varphi)].\end{aligned} \quad (6)$$

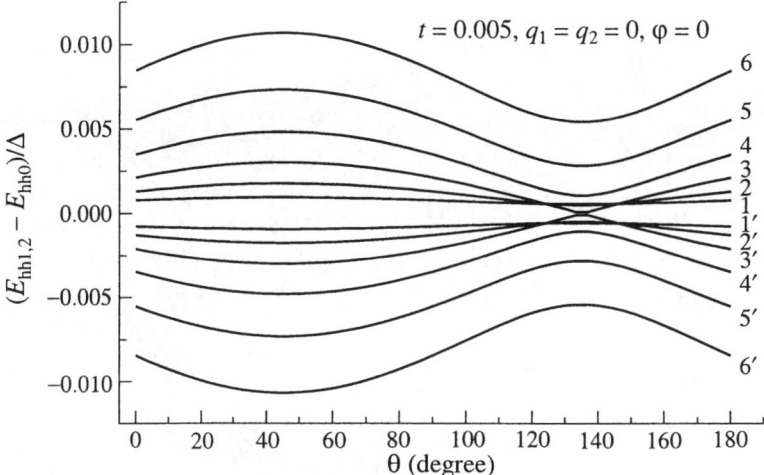

*Figure 1.* The position of 1HH energy doublet in a magnetic field vs. its orientation. The $h$ values are: 0.05; 0.075; 0.1; 0.125; 0.15 and 0.175 for pair of curves 1, 1′; 2, 2′; 3, 3′; 4, 4′; 5, 5′ and 6, 6′ respectively.

Here and in the expressions for $G_{hh}$ and $E_{hh0}$ we used dimensionless values $h = G_h/\Delta$ and $t = T/\Delta$. We can see that in the case of relatively weak field, $2h \ll \sqrt{2|q_2 - 2t|}$, the HH splitting $G_{hh} \approx (3/2)h(q_2 - 2t)\Delta$ is induced by interference of $C_{2v}$ invariants and Zeeman term. In the limit $2h \gg \sqrt{2|q_2 - 2t|}$, the HH splitting, $G_{hh} \approx (3/2)h^3\Delta$, is not sensitive to $C_{2v}$ invariants. In the intermediate field region the splitting values depend from the $\theta$ and $\varphi$ angles. At the $2h = \sqrt{2|q_2 - 2t| - (q_2 - 2t)^2}$ (with $q_1 = 0$) the $G_{hh}$ tends to zero at $\theta = \pi/4$ for $(q_2 - 2t) > 0$ or $3\pi/4$ for $(q_2 - 2t) < 0$. The parameter $q_1$ changes slightly these "critical points" of $\theta$. The dependencies of $E_{hh1,2} - E_{hh0}$ on $\theta$ are shown in Fig. 1 for the series of $h$.

All four possible transitions between electron and HH states with energies (4) and (5) are allowed. Their probabilities for different light polarizations can be calculated by usual way (see for inst. [11]). The light polarizations in two pair of mutually perpendicular planes, for instance $\rho_0$ and $\rho_{45}$, determine the PA in a unique fashion. If we introduce the intensities $I_\parallel$, $I_\perp$, $I_{+45}$ and $I_{-45}$ of optical transitions in polarization planes being normal to QW and formed the angles of 0, 90, +45 and −45 degrees with direction of $G$, abovementioned polarizations read

$$\rho_0 = \frac{I_\parallel - I_\perp}{I_\parallel + I_\perp} \quad \text{and} \quad \rho_{45} = \frac{I_{+45} - I_{-45}}{I_{+45} + I_{-45}}. \tag{7}$$

Table 1 displays polarizations $\rho_{0ij}$ and $\rho_{45ij}$ calculated with Eqs. (7) as a series of small parameters $t$ and $h$ for each pair of indexes $i$ and $j$. The indexes $i$ and $j$ refer to the states 1 and 2 of doublets (4) and (5) respectively. Notifications used

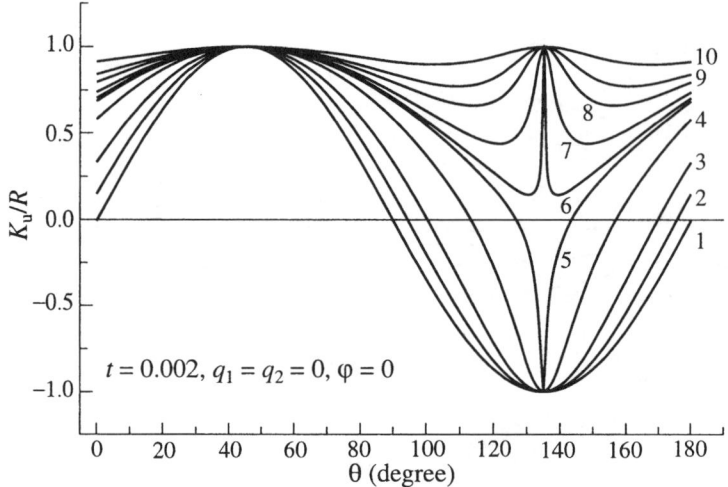

*Figure 2.* Angle dependence of $K_u/R$ ratio for $t = 0.02$, $q_1 = q_2 = 0$, $\varphi = 0$. Curve 1–10 corresponds to $h$ values 0.01, 0.08, 0.12, 0.17, 0.195, 0.20, 0.21, 0.23, 0.25, 0.3 respectively.

in Table 1 are

$$K_u = -(q_2 - 2t)\sin[2(\theta - \varphi)] + q_1 \cos[4\theta] - t^2 \cos[4(\theta - \varphi)] + h^2$$
$$K_d = (q_2 - 2t)\cos[2(\theta - \varphi)] + q_1 \sin[4\theta] - t^2 \sin[4(\theta - \varphi)]. \tag{8}$$

TABLE I.

| $ij$ | $\rho_{0ij}$ | $\rho_{45ij}$ |
|---|---|---|
| 11 | $-K_u/R - t(1+h/2)\sin[2(\theta-\varphi)] - h^2 + \ldots$ | $K_d/R - t(1+h/2)\cos[2(\theta-\varphi)] + \ldots$ |
| 12 | $K_u/R - t(1+h/2)\sin[2(\theta-\varphi)] - h^2 + \ldots$ | $-K_d/R - t(1+h/2)\cos[2(\theta-\varphi)] + \ldots$ |
| 21 | $K_u/R - t(1-h/2)\sin[2(\theta-\varphi)] - h^2 + \ldots$ | $-K_d/R - t(1-h/2)\cos[2(\theta-\varphi)] + \ldots$ |
| 22 | $-K_u/R - t(1-h/2)\sin[2(\theta-\varphi)] - h^2 + \ldots$ | $K_d/R - t(1-h/2)\cos[2(\theta-\varphi)] + \ldots$ |

Let we note that calculations presented at Table 1 has a physical meaning if each from transitions under consideration can be separately observed despite of possible approaches of transition energies $E_{e,i} - E_{hh,1}$ and $E_{e,i} - E_{hh,2}$. The main contributions in the polarizations $\rho_{0ij}$ and $\rho_{45ij}$ expect to be the $K_u/R$ or $K_d/R$. Next field-independence terms, $t\sin[2(\theta - \varphi)]$ and $t\cos[2(\theta - \varphi)]$, arise due to HH–LH mixing by $T$ type $C_{2v}$ invariant only. They allow finding the t parameter from angle dependence of $\rho_{0ij}$ and $\rho_{45ij}$ in zero magnetic field.

Authors of Ref. [8] introduce the $q_2$ invariant as an equivalent form of $T$ – invariant and Zeeman term interference in 2-th perturbation theory. Such equivalence is manifested itself in expressions for $R$, $K_u$, $K_d$, where $q_2$-parameter appears only as combination $q_2 - 2t$. We note here that $T$-invariant itself generates also other contributions (like field-independence terms and $4\theta$-dependent terms) to the $\rho_{0ij}$ and $\rho_{45ij}$, which cannot be reduced to $q_2$-parameter manifestation.

Note also some particularity of the $\rho_{0ij}$ and $\rho_{45ij}$ that appears near the conditions provided the HH sublevels crossing. These particularities are shown as angle dependencies of $K_u/R$ and $K_d/R$ terms in Figs. 2 and 3.

Let we enter into discussion of photo-luminescence (PL) PA. Now the intensities of PL transitions should be account with relative spin sublevel populations of electron $P_{ei}$ and HH $P_{hj}$:

$$P_{e1} = \frac{\exp(-\beta_e h_e/2)}{2\cosh(\beta_e h_e/2)};$$

$$P_{e2} = \frac{\exp(\beta_e h_e/2)}{2\cosh(\beta_e h_e/2)};$$

$$P_{h1} = \frac{\exp(-\beta_h G_{hh}/2\Delta)}{2\cosh(\beta_h G_{hh}/2\Delta)};$$

$$P_{h2} = \frac{\exp(\beta_h G_{hh}/2\Delta)}{2\cosh(\beta_h G_{hh}/2\Delta)}, \quad (9)$$

where $h_e = G_e/\Delta$, $\beta_e = \Delta/T_e$ and $\beta_h = \Delta/T_h$, $T_e$ and $T_h$ are spin temperature for electrons in 1e state and holes in 1hh state respectively.

The PL polarizations $\rho_0$ and $\rho_{45}$ include all four transition accounted with

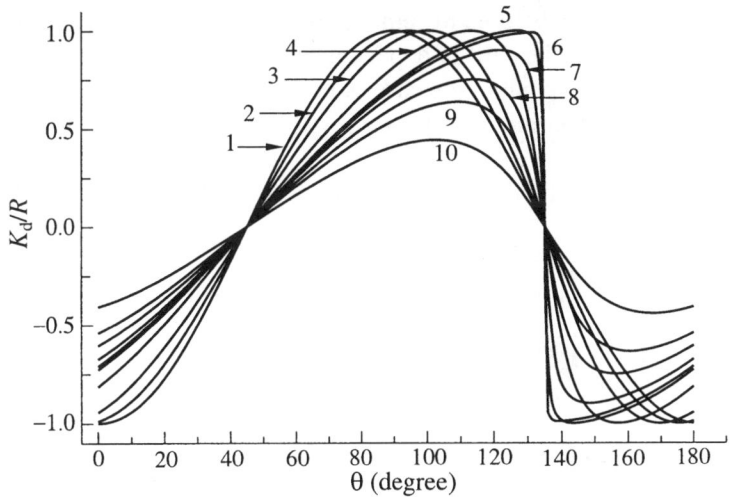

*Figure 3.* Angle dependence of $K_d/R$ ratio for $t = 0.02$, $q_1 = q_2 = 0$, $\varphi = 0$. Curve 1–10 corresponds to $h$ values 0.01, 0.08, 0.12, 0.17, 0.195, 0.20, 0.21, 0.23, 0.25, 0.3 respectively.

taking into account of population (9) and transition probabilities is:

$$\rho_0 = -t\sin[2(\theta-\varphi)] - h^2 - \frac{K_u}{R}\tanh\left[\frac{h_e\beta_e}{2}\right]\tanh\left[\frac{G_{hh}\beta_h}{2\Delta}\right]$$

$$+ h\left\{\frac{t}{2}\sin[2(\theta-\varphi)]\tanh\left[\frac{h_e\beta_e}{2}\right] - \frac{K_u t}{R}\text{sech}^2\left[\frac{h_e\beta_e}{2}\right]\right.$$

$$\left.\times\tanh\left[\frac{G_{hh}\beta_h}{2\Delta}\right]\right\} + O(h^2) \quad (10)$$

$$\rho_{45} = -t\cos[2(\theta-\varphi)] + \frac{K_d}{R}\tanh\left[\frac{h_e\beta_e}{2}\right]\tanh\left[\frac{G_{hh}\beta_h}{2\Delta}\right]$$

$$+ h\left\{\frac{t}{2}\cos[2(\theta-\varphi)]\tanh\left[\frac{h_e\beta_e}{2}\right] + \frac{K_d}{R}\text{sech}^2\left[\frac{h_e\beta_e}{2}\right]\right.$$

$$\left.\times\tanh\left[\frac{G_{hh}\beta_h}{2\Delta}\right]\right\} + O(h^2). \quad (11)$$

Now the ratio $h\beta_h/2$ (i.e. termalization of split components of HH doublet) controls the dependence of $\rho_0$ and $\rho_{45}$ on magnetic field direction. If the $T_h$ is low enough, $G_{hh}/2T_h > 1$, $\tanh[G_{hh}\beta_h/2\Delta] \to 1$, the PL polarizations reflect the PA of transitions to lowest energy level of HH, that look like the ratios $K_u/R$ and $K_d/R$. In opposite case, $\tanh[G_{hh}\beta_h/2\Delta] \approx G_{hh}/2T_h = (3/4)hR\beta_h$, the angle dependencies of the dominant terms in Eqs. (11), (12) transform to $\rho_0 \propto K_u$ and $\rho_{45} \propto K_d$ that are not sensitive to effect of HH sublevel crossing. For QWs with semimagnetic barriers based on $A^2B^6$ compounds, we expect the possibility to trace the transition from low field to high field regime of PA in the case of narrow QW (with strong magnetic field exchange enhancement) and at low enough temperature. More sound effect expects to be in semimagnetic QW provided rather strong HH spin splitting. Any way, the shift of dependence $\rho_0(\theta)$ being proportional to $h^2$ has to be detected. Note also that in the case of congruent HH splitting and linewidth, one can expect manifestation of $\rho_{0ij}(\theta)$, $\rho_{45ij}(\theta)$ peculiarities within of contour of luminescence line.

In the case of weak field, $G_{hh}/2T_h < 1$ and $G_e/2T_e < 1$, the Eqs. (10), (11) take the form:

$$\rho_0 \approx -t\sin[2(\theta-\varphi)] - h^2\left\{1 + \frac{3}{8}K_u\beta_h\left(2 + \frac{G_e}{G_h}\beta_h\right)\right.$$

$$\left.+ \frac{t}{4}\frac{G_e}{G_h}\beta_e\sin[2(\theta-\varphi)]\right\} + O(h^4)\ldots \quad (12)$$

$$\rho_{45} \approx -t\cos[2(\theta-\varphi)] + h^2\left\{\frac{3}{8}K_d\beta_h\left(2 + \frac{G_e}{G_h}\beta_e\right)\right.$$

$$\left.+ \frac{t}{4}\frac{G_e}{G_h}\beta_e\cos[2(\theta-\varphi)]\right\} + O(h^4)\ldots \quad (13)$$

The ratio $G_e/G_h$ in Eqs. (12), (13) does not practically depends on $h$. One would wait this ratio is in order of magnitude $\sim -(0.15-0.25)$ for structures on the base of $A^2B^6$ compound.

Our estimations show that the $\beta_e$ and $\beta_h$ can vary from 10 to 150. So the terms in (13), (14), which are proportional to product $\beta_e\beta_h$, can control the angle dependences of $\rho_0$ and $\rho_{45}$.

Eqs. (12), (13) are in accordance with the similar expressions (10), (11) from the paper [2, 3] in same characteristics. Meanwhile, there are important differences. Eqs. (12), (13) include additionally (i) the angle dependence of $\rho_0$ and $\rho_{45}$ in zero magnetic field, (ii) the $\pi/2$ periodical terms of $\rho_0(\theta)$ and $\rho_{45}(\theta)$ and (iii) spin-temperature dependences of PA.

Let us discuss the relation of the $\pi$- and $\pi/2$-periodical contributions in angle dependence of $\rho_0$ and $\rho_{45}$. The $\pi$- and $\pi/2$-periodical functions cannot satisfactorily describe $\theta$-dependences in the vicinity of HH sublevel crossing at low enough temperature (see Figs. 2, 3). If we consider the case $G_{hh}/2T_h < 1$, the $\pi/2$-contributions presented by the expressions $K_u$ and $K_d$ (8) are essentially less than $\pi$-contributions. Nevertheless, in real heterostructures, the strain induced $C_{2v}$ contributions may have the local nature while the averaging over structure plane results in relatively small their values. The interface roughness in nanostructures can generate the fluctuations of $t$ parameters over QW plane too. If the $t$-parameter is a random value in structure plane, the averaging over this fluctuations leads to change $tf_1(2\theta')$ by $\langle t \rangle f_1(2\theta')$ and $t^2 f_2(4\theta')$ by $\langle t^2 \rangle f_2(4\theta')$, where $f_1$ and $f_2$ are some trigonometric functions on some angle $\theta'$. In quantum structures that look as a whole like quite symmetrical, one can expect rather small value of $\langle t \rangle$ while dispersion of fluctuations can exceed this value: $\langle t \rangle < \langle t^2 \rangle$. Last inequality can lead to significant contribution of $f_2(4\theta')$-terms to the PA.

Note also, that $\pi/2$-periodical terms in expressions $K_u$, $K_d$ are proportional to $q_1$. We have not information about magnitude of the $q_1$ in $A^2B^6$ compounds. In accordance with [11], the $q_1$ parameter is small enough.

## 3. Experiment

The experiments were performed on the [001] oriented CT-866 structure grown in the Institute of Physics of Würzburg University. The structure includes few CdTe QWs with one barrier $Cd_{0.80}Mg_{0.20}Te$ and other one $Cd_{0.78}Mn_{0.22}Te$. In this paper we will discuss measurement of 1e–1hh exciton luminescence polarization $\rho_0$ and $\rho_{45}$ from QW of 24 Å width only. The measurements were carried out at temperature 2 K in the magnetic field up to 3.6 T directed in structure plane. Surprisingly, the structure had initially in-plane strain. After first series of measurements the structure had cracked on two part and the strain in each of them was reduced essentially. Then we perform second series of measurements on the same structure

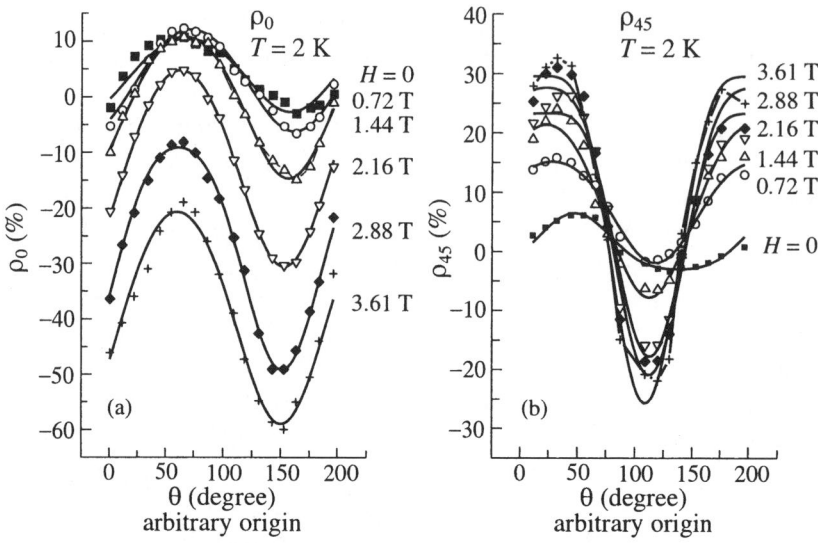

*Figure 4.* PA of exciton 1e–1hh luminescence in stressed 24 Å QW. (a) $\rho_0$; (b) $\rho_{45}$. The points — experiment for indicated magnetic field values; lines — fitting.

with reduced strain. The results of first and second series of measurements of $\rho_0$ and $\rho_{45}$ are shown in Fig. 4(a,b) and Fig. 5(a,b) correspondingly.

From the data presented at Fig. 4 we can estimate the t value in strained structure on angle dependence of $\rho_0$ and $\rho_{45}$ in zero magnetic field as 0.05±0.02. This value was decreased up to 0±0.02 after strain reducing (see Fig. 5). The data presented at Fig. 4 are very well fitted by the equations, which include constant part (for $\rho_0$) and $2\theta$ and $4\theta$ periodical terms. Contribution of the $4\theta$ terms is relatively larger for the case of reduced strain (Fig. 5). Figures 4 and 5 demonstrate additional peculiarity of "high field" regime of PA, the magnitude of angle dependence saturated in the big field. Unfortunately the obtained data do not give possibility to distinguish the interface and strain contributions. More detail analyses of the experimental results will be published separately.

## 4. Conclusion

The PA of 1e–1hh luminescence from [001] oriented QW with semimagnetic barriers in the in-plane magnetic field studied theoretically with taking into account two type of $C_{2v}$ invariants in the hole Hamiltonian and Zeeman terms. It is shown that $J^3B$ — invariant, without other terms, does not describe some important properties of PA. It is found the important role of $C_{2v}$ anisotropy and Zeeman term interplay in the PA. This effect must drastically change the angle dependence of PA at some range of in-plane magnetic field. The effect smooth out for PL line that is composed from unresolved transitions between spin components of 1e and 1hh

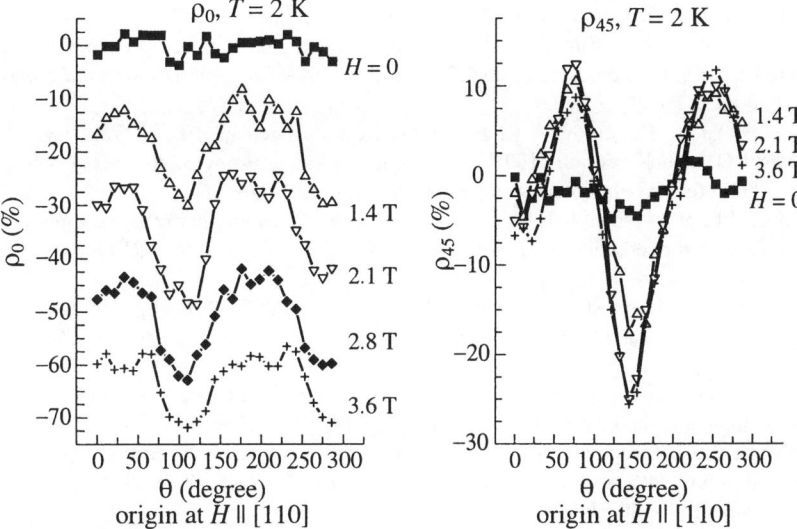

*Figure 5.* Angle dependence of $\rho_0$ (a) and $\rho_{45}$ (b) in the same structure after reducing of strain. The lines are drawn for a best view.

states. The transformation of this effect is considered with taking into account the spin temperatures of 1e and 1hh states. Relation of $\pi(2\theta)$- and $\pi/2(4\theta)$-periodical contributions to the angle dependence of PA is discussed and shown that $4\theta$-terms can be relatively enhanced due to averaging of the local strain effects and roughness in real QW.

The measurement of PA of 1e-1hh exciton luminescence carried out on the 24 Å width $Cd_{0.80}Mg_{0.20}Te/CdTe/Cd_{0.78}Mn_{0.22}Te$ QW with initial strain and after its reduction. It is shown that main peculiarity of obtained data agreed with qualitative predication of theoretical consideration.

## Acknowledgments

The authors are grateful to Dr D. R. Yakovlev for an allotment of the sample for measurement.

The work was partly supported by INTAS 99-015 grant and by grant of State Fundamental Research Foundation of Ukraine.

## References

1. Koudinov, A.V., Kusraev, Yu.G., Kavokin, K.V., Merkulov, I.A., and Zakharchenya, B.P. (1997) Magnetic anisotropy in $Cd_{1-x}Mn_xTe$ alloys revealed by polarized luminescence, *Inst. Phys. Conf.* **Ser. 155**: Chapter 10, 747–750 (*Proc. of 23-th Int. Symp. Compound Semicond.*, St. Petersburg, Russia, 23-27 September 1996, IOP Publishing Ltd. 1997);

2. Aksyakov, I.G., Koudinov, A.V., Kusraev, Yu.G., Zakhachenya, B.P., Woitowicz, T., Karczewski, G., and Kossut, J. (1999) Orthorhombic symmetry of valence-band states in CdTe/Cd$_{1-x}$Mn$_x$Te quantum wells, *Fiz. Tverd. Tela* **41**, 903–906 (*Phys. Solid State* (1999) **41**, 820-823).
3. Kusraev, Yu.G., Koudinov, A.V., Aksyanov, I.G., Zakhachenya, B.P., Woitowicz, T., Karczewski G., and Kossut, J. (1999) Extreme in-plane anisotropy of the heavy hole g-factor in (001) CdTe/MnTe quantum wells, *Phys. Rev. Lett.* **82**, 3176–3179.
4. Aleiner, I.L. and Ivchenko, E.L., (1992) The nature of anisotropic exchange splitting in type II GaAs/AlAs superlattices, *Pis'ma Zh. Eksp. Teor. Fiz.* **55**, p.662–664 (*JETP Lett.* (1992) **55**, 692-694);
   Ivchenko, E.L., Kaminskii, A.Yu., and Aleiner, I.L. (1993) The exchange splitting of exciton levels in the type I and II superlattices, *Zh. Eksp. Teor. Fiz.* **104**, 3401–3415 (*JETP* (1993) **77**, 609–623).
5. Krebs, O. and Voisin, P. (1996) Giant optical anisotropy of semiconductor heterostructures with no common atom and the quantum-confined Pockels effect, *Phys. Rev. Lett.* **77**, 1829–1832.
6. Ivchenko, E.L., Toropov, A.A., and Voisin, P. (1998) Interface induced optical anisotropy in heterostructure with non-common cations and anions, *Fiz. Tverd. Tela* **40**, 1925–1931 (*Phys. Solid State* (1998) **40**,1748–1754).
7. Gourdon, C. and Lavallard, P. (1992) Fine structure of heavy excitons in GaAs/AlAs superlattices, *Phys. Rev. B* **46**, 4644–4650.
8. Pikus, G.E. and Pikus, F.G. (1994) The mechanism of heavy and light hole mixing in GaAs/AlAs superlattices, *Solid State Commun.* **89**, 319–322.
9. Gaj, J.A., Grieshaber, W., Bodin-Deshayes, C., Cibert, J., Feuillet, G., Merle d'Aubigne, Y., and Wasiela, A. (1994) Magneto-optical study of interface mixing in the CdTe-(Cd,Mn)Te system, *Phys. Rev. B* **50**, 5512–5527.
10. Grieshaber, W., Haury, A., Cibert, J., Merle d'Aubigne, Y., Wasiela, A., and Gaj, J. (1996) Magneto-optical study of the interface in semimagnetic semiconductor heterostructures: Intrinsic effect and interface profile in CdTe-Cd$_{1-x}$Mn$_x$Te, *Phys. Rev. B* **53**, 4891–4904.
11. Bir, G.L. and Pikus, G.E. (1974) *Symmetry and Strain Induced Effects in Semiconductors*, Willey and Sons, NY.

# MODEL OF STRONGLY CORRELATED 2D FERMI LIQUIDS BASED ON FERMION-CONDENSATION QUANTUM PHASE TRANSITION

V.R. SHAGINYAN
*Petersburg Nuclear Physics Institute, Russian Academy of Sciences, Gatchina, 188300, Russia*

Abstract. A model of strongly correlated electron or hole liquids with the fermion condensate is presented and applied to the consideration of quasiparticle excitations in high temperature superconductors, in their superconducting and normal states. Within our model the appearance of the fermion condensate presents a quantum phase transition, that separates the regions of normal and strongly correlated electron liquids. Beyond the fermion condensation quantum phase transition point the quasiparticle system is divided into two subsystems, one containing normal quasiparticles and the other — fermion condensate localized at the Fermi surface. In the superconducting state the quasiparticle dispersion in systems with FC can be presented by two straight lines, characterized by effective masses $M^*_{FC}$ and $M^*_L$, respectively, and intersecting near the binding energy which is of the order of the superconducting gap. This same quasiparticle picture persists in the normal state, thus manifesting itself over a wide range of temperatures as new energy scales. Arguments are presented that fermion systems with FC have features of a "quantum protectorate". A theory of high temperature superconductivity based on the combination of the fermion-condensation quantum phase transition and the conventional theory of superconductivity is presented. This theory describes maximum values of the superconducting gap which can be as big as $\Delta_1 \sim 0.1\varepsilon_F$, with $\varepsilon_F$ being the Fermi level. We show that the critical temperature $2T_c \simeq \Delta_1$. If there exists the pseudogap above $T_c$ then $2T^* \simeq \Delta_1$, and $T^*$ is the temperature at which the pseudogap vanishes. A discontinuity in the specific heat at $T_c$ is calculated. The transition from conventional superconductors to high-$T_c$ ones as a function of the doping level is investigated. The single-particle excitations and their lineshape are also considered. Analyzing experimental data on the high temperature superconductivity in different materials induced by field-effect doping, we show that all these facts can be understood within a theory of the superconductivity based on the fermion condensation quantum phase transition, which can be conceived of as a universal cause of the superconductivity. The main features of a room-temperature superconductor are considered.

Key words: strongly correlated electrons, quantum phase transistor

## 1. Introduction

One of the most challenging problems of modern physics is the problem of systems with a big coupling constant. This problem is of crucial importance particularly in the quantum field theory, making even the quantum electrodynamics to be not a self-consistent theory [1]. It is well-known that a consideration of strongly correlated liquids is close to the problem of systems with the big coupling constant which persists in many-body physics. A solution of this problem has been offered in the context of the Landau theory of normal Fermi liquids by introducing into the theory a notion of the quasiparticles and parameters which characterize the effective interaction between the quasiparticles [2]. As a result, the Landau theory has removed high energy degrees of freedom and kept a sufficiently large number of relevant low energy degrees of freedom to treat liquid's low energy properties. Usually, it is assumed that the breakdown of the Landau theory is defined by the Pomeranchuk stability conditions and occurs when the Landau amplitudes being negative reach its critical value. Note that the new phase at which the stability conditions are restored can in principle be again described in the framework of the theory. To describe a strongly correlated electron liquid, taking place when the coupling constant becomes sufficiently big, a conventional way can be used, assuming that the correlated regime is connected to the noninteracting Fermi gas by adiabatic continuity in the same way as in the framework of the Landau normal Fermi liquid theory [2]. But a question exists whether this is possible at all. Most likely, the answer is negative. Therefore, we direct our attention to a model, in the frame of which a strongly correlated electron liquid is separated from conventional Fermi liquid by a phase transition related to the onset of the fermion condensate (FC) [3,4]. It was demonstrated rather recently [3] that the Pomeranchuk conditions are covering not all possible limitations: one is overlooked, being connected with the situation when, at temperature $T = 0$, the effective mass can become infinitely big. It has been demonstrated that such a situation, leading to profound consequences, can take place when the corresponding amplitude being positive reaches the critical value, producing a completely new class of strongly correlated Fermi liquids with FC [3,4] which is separated from a normal Fermi liquid by the fermion condensation quantum phase transition (FCQPT) [5]. In this case we are dealing with the strong coupling limit where an absolutely reliable answer cannot be given, based on pure theoretical first principle ground. Therefore, the only way to verify that FC occurs is to consider experimental facts which bear witness to the existence of such a state. We assume that these facts can be find in two-dimensional (2D) systems with interacting electrons or holes, which can be presented by modulation doped quantum wells, by high mobility metal-oxide-semiconductor field effect transistors, or by high-$T_c$ superconductors.

The aim of our report is to show that within the frameworks of our model of a strongly correlated electron (hole) liquid based on FCQPT the main properties of such liquids observed in the high-temperature superconductors can be understood. In Sec. 2, we review the general features of Fermi liquids with FC, showing that an electron liquid of low density in the high-$T_c$ materials inevitably undergoes FCQPT. In Sec. 3 we consider the high-temperature superconductivity, which takes place in the presence of FC. In Sec. 4 we describe the quasiparticle dispersion and lineshape.

Finally, in Sec. 5, we summarize our main results.

## 2. The General Features of Fermi Liquids with FC

Let us start by considering the key points of the FC theory. FC is a new solution of the Fermi liquid theory equations [2] for the quasiparticle occupation numbers $n(p, T)$

$$\frac{\delta(F - \mu N)}{\delta n(p, T)} = \varepsilon(p, T) - \mu(T) - T \ln \frac{1 - n(p, T)}{n(p, T)} = 0, \tag{1}$$

which depends on the momentum $p$ and temperature $T$. Here $F$ is the free energy, and $\mu$ is the chemical potential, while

$$\varepsilon(p, T) = \frac{\delta E[n(p)]}{\delta n(p, T)}, \tag{2}$$

is the quasiparticle energy. This energy is a functional of $n(p, T)$ just like the total energy $E[n(p)]$ and the other thermodynamic functions. Eq. (1) is usually presented as the Fermi–Dirac distribution,

$$n(p, T) = \left\{ 1 + \exp\left[\frac{(\varepsilon(p, T) - \mu)}{T}\right] \right\}^{-1}. \tag{3}$$

At $T \to 0$ one gets from Eqs. (1), (3) the standard solution $n_F(p, T \to 0) \to \theta(p_F - p)$, with $\varepsilon(p \simeq p_F) - \mu = p_F(p - p_F)/M_L^*$, where $p_F$ is the Fermi momentum, and $M_L^*$ is the Landau effective mass [2],

$$\frac{1}{M_L^*} = \frac{1}{p} \frac{d\varepsilon(p, T = 0)}{dp}\bigg|_{p=p_F}. \tag{4}$$

It is implied that $M_L^*$ is positive and finite at the Fermi momentum $p_F$. As a result, the $T$-dependent corrections to $M_L^*$, to the quasiparticle energy $\varepsilon(p)$, and other quantities, start with $T^2$-terms. But this solution is not the only one possible. There exist "anomalous" solutions of Eq. (1) associated with the so-called fermion condensation [3, 6]. Being continuous and satisfying the inequality $0 < n(p) < 1$ within some region in $p$, such solutions $n(p)$ admit a finite limit for the logarithm in Eq. (1) at $T \to 0$ yielding,

$$\varepsilon(p) = \frac{\delta E[n(p)]}{\delta n(p)} = \mu; \quad p_i \leq p \leq p_f. \tag{5}$$

At $T = 0$ Eq. (5) determines FCQPT, possessing solutions at some $r_s = r_{FC}$ as soon as the effective inter-electron interaction becomes sufficiently strong [7]. From here on we shall call a hole or electron system as electron one provided this will not lead to confusion. In a simple electron liquid, the effective inter-electron interaction is proportional to the dimensionless average interparticle distance $r_s = r_0/a_B$, with $r_0 = \sqrt{2}/p_F$ being the average distance, and $a_B$ is the Bohr radius. Equation (5) leads to the minimal value of $E$ as a functional of $n(p)$ when in system under consideration a strong rearrangement of the single particle spectra can take place. We see from Eq. (5) that the occupation numbers $n(p)$ become variational parameters: the solution $n(p)$ takes place if the energy $E$ can be lowered by alteration of the occupation numbers. Thus, within the region $p_i < p < p_f$, the solution $n(p) = n_F(p) + \delta n(p)$ deviates from the Fermi step function $n_F(p)$ in such a way that the energy $\varepsilon(p)$ stays constant while outside this region $n(p)$ coincides with $n_F(p)$. It is pertinent to note that the above general consideration was verified by inspecting simple models. As the result, it was shown that the onset of the FC does lead to lowering the free energy [6, 8]. It follows from the above consideration that the superconductivity order parameter $\kappa(\mathbf{p}) = \sqrt{n(\mathbf{p})(1 - n(\mathbf{p}))}$ has a nonzero value over the region occupied by FC. The superconducting gap $\Delta(\mathbf{p})$ being linear in the coupling constant of the particle-particle interaction $V_{pp}$ gives rise to the high value of $T_c$ because one has $2T_c \simeq \Delta$ [8] within the standard Bardeen–Cooper–Schrieffer (BCS) theory [9]. As it is shown in Sec. 3, if the superconducting gap $\Delta \neq 0$, the quasiparticle effective mass becomes finite. In consequence of these features the density of states at the Fermi level becomes finite and the involved quasiparticles are not localized. On the other hand, even at $T = 0$, $\Delta$ can vanish, provided $V_{pp}$ is repulsive or absent. Then, as it is seen from Eq. (5), the Landau quasiparticle system becomes separated into two subsystems. The first contains the Landau quasiparticles, while the second, related to FC, is localized at the Fermi surface and formed by dispersionless quasiparticles. As a result, the standard Kohn–Sham scheme for the single particle equations is no longer valid beyond the point of the FC phase transition [10]. Such a behavior of systems with FC is clearly different from what one expects from the well known local density calculations. Therefore these calculations are hardly applicable to describe systems with FC. It is also seen from Eq. (5) that a system with FC has a well-defined Fermi surface.

Let us assume that FC has just taken place, that is $p_i \to p_f \to p_F$, and the deviation $\delta n(p)$ is small. Expanding functional $E[n(p)]$ in Taylor's series with respect to $\delta n(p)$ and retaining the leading terms, one obtains from Eq. (5),

$$\mu = \varepsilon(\mathbf{p}) = \varepsilon_0(\mathbf{p}) + \int F_L(\mathbf{p}, \mathbf{p}_1) \delta n(\mathbf{p}_1) \frac{d\mathbf{p}_1}{(2\pi)^2}; \quad p_i \leq p \leq p_f, \qquad (6)$$

where $F_L(\mathbf{p}, \mathbf{p}_1) = \delta^2 E/\delta n(\mathbf{p})\delta n(\mathbf{p}_1)$ is the Landau interaction. Both the Landau interaction and the single-particle energy $\varepsilon_0(p)$ are calculated at $n(p) = n_F(p)$. It is seen from Eq. (6) that the FC quasiparticles forms a collective state, since their energies are defined by the macroscopical number of quasiparticles within the region $p_i - p_f$, and vice versa. The shape of the spectra is not effected by the Landau interaction, which, generally speaking, depends on the system's properties, including the collective states, impurities, etc. The only thing defined by the interaction is the width of the region $p_i - p_f$, provided the interaction is sufficiently strong to produce the FC phase transition at all. Thus, we can conclude that the spectra related to FC are of universal form, being dependent, as we will see below, mainly on temperature $T$, if $T > T_c$, or on the superconducting gap at $T < T_c$.

According to Eq. (1), the single-particle excitations within the interval $p_i - p_f$ have at $T_c \leq T \ll T_f$ the shape $\varepsilon(p, T)$ linear in T [8, 11], which can be simplified at the Fermi level,

$$\varepsilon(p, T) - \mu(T) = T \ln \frac{1 - n(p)}{n(p)} \simeq T \left. \frac{1 - 2n(p)}{n(p)} \right|_{p \simeq p_F}. \quad (7)$$

$T_f$ is the temperature, above which FC effects become insignificant [8],

$$\frac{T_f}{\varepsilon_F} \sim \frac{p_f^2 - p_i^2}{2M\varepsilon_F} \sim \frac{\Omega_{FC}}{\Omega_F}. \quad (8)$$

Here $\Omega_{FC}$ is the FC volume, $\varepsilon_F$ is the Fermi energy, and $\Omega_F$ is the volume of the Fermi sphere. We note that at $T_c \leq T \ll T_f$ the occupation numbers $n(p)$ are approximately independent of $T$, being given by Eq. (5). One can imagine that at these temperatures dispersionless plateau $\varepsilon(p) = \mu$ given by Eq. (5) is slightly turned counter-clockwise about $\mu$. As a result, the plateau is just a little tilted and rounded off at the end points. According to Eq. (7) the effective mass $M^*_{FC}$ related to FC is given by,

$$M^*_{FC} \simeq p_F \frac{p_f - p_i}{4T}. \quad (9)$$

To obtain Eq. (9) an approximation for the derivative $dn(p)/dp \simeq -1/(p_f - p_i)$ was used. Having in mind that $p_f - p_i \ll p_F$, and using (8) and (9) the following estimates for the effective mass $M^*_{FC}$ are obtained,

$$\frac{M^*_{FC}}{M_0} \sim \frac{N(0)}{N_0(0)} \sim \frac{T_f}{T}. \quad (10)$$

Eqs. (9) and (10) show the temperature dependence of $M^*_{FC}$. In (10) $M_0$ denotes the bare electron mass, $N_0(0)$ is the density of states of noninteracting electron gas, and $N(0)$ is the density of states at the Fermi level. Multiplying both sides of Eq. (9) by $p_f - p_i$ we obtain the energy scale $E_0$ separating the slow dispersing low energy part, related to the effective mass $M^*_{FC}$, from the faster dispersing relatively high energy part, defined by the effective mass $M^*_L$ [5, 12],

$$E_0 \simeq 4T. \quad (11)$$

It is seen from Eq. (11) that the scale $E_0$ does not depend on the condensate volume. The single particle excitations are defined according to Eqs. (7) and (9) by the temperature and by $n(p)$, given by Eq. (5). Thus, we are led to the conclusion that the one-electron spectrum is negligible disturbed by thermal excitations, impurities, etc, so that one observes the features of the quantum protectorate [13, 14].

It is seen from Eq. (5) that at the point of FC phase transition $p_f \to p_i \to p_F$, $M_{FC}^*$ and the density of states, as it follows from Eqs. (5), (10), tend to infinity. One can conclude that at $T = 0$ and as soon as $r_s \to r_{FC}$, FCQPT takes place being connected to the absolute growth of $M_L^*$. It is essential to have in mind, that the onset of the charge density wave instability in a many-electron system, such as electron liquid, which takes place as soon as the effective inter electron constant reaches its critical value $r_s = r_{cdw}$, is preceded by the unlimited growth of the effective mass. Therefore, the FC occurs before the onset of the charge density wave. Hence, at $T = 0$, when $r_s$ reaches its critical value $r_{FC} < r_{cdw}$, the FCQPT inevitably takes place [7]. It is pertinent to note that this growth of the effective mass with decreasing electron density was observed experimentally in a metallic 2D electron system in silicon at $r_s \simeq 7.5$ [15]. Therefore we can take $r_{FC} \sim 7.5$. On the other hand, there exist charge density waves or strong fluctuations of charge ordering in underdoped high-$T_c$ superconductors [16]. Thus, the formation of FC in high-$T_c$ compounds can be thought as a general property of an electron liquid of low density which is embedded in these solids, rather then an uncommon and anomalous solution of Eq. (1) [7]. Beyond the point of FCQPT the condensate volume is proportional to $(r_s - r_{FC})$ as well as $T_f/\varepsilon_F \sim (r_s - r_{FC})/r_{FC}$ at least when $(r_s - r_{FC})/r_{FC} \ll 1$. Note, that such a behavior is in accordance with the general properties of second order phase transitions. Therefore, we can accept a model relating systems with FC to high-$T_c$ compounds, assuming that the effective coupling constant $r_s$ increases with decreasing doping, exceeding its critical value $r_{FC}$ at the levels corresponding to optimal doped samples. We remark, that this critical value $r_{FC}$ corresponds to the $r_s$ values of highly overdoped samples [7]. As the result, our quite natural model suggests that both quantities, $T_f$ and condensate volume $\Omega_{FC}$, increase with decrease of doping. Thus, these values are higher in underdoped samples as compared to overdoped ones provided $r_s$ meets the mentioned above conditions. While, in the highly overdoped regime only slight deviations from the normal Fermi liquid are observed [17]. All these peculiar properties are naturally explained within a model proposed in [5, 12, 18, 19] and allow to relate the doping level $x$ regarded as the density of mobile charge carriers (holes or electrons) per unit area to the density of Fermi liquid with FC. We assume that $x_{FC}$ corresponds to the highly overdoped regime at which FCQPT takes place, and introduce the effective coupling constant $g_{eff} \sim (x - x_{FC})/x_{FC}$. In our model, the doping level $x$ at $x \leq x_{FC}$ in metals is related to $(p_f - p_i)$ in the following way:

$$g_{eff} \sim \frac{(x_{FC} - x)}{x_{FC}} \sim \frac{(p_f - p_i)(p_f + p_i)}{p_F^2} \sim \frac{p_f - p_i}{p_F}. \quad (12)$$

According to experimental facts the large density of states at the Fermi level reaches its maximum in the vicinity of the Hove singularities, that is around the point $(\pi, 0)$ of the Brillouin zone, or $\bar{M}$, in high-$T_c$ compounds. The density of states reaches its minimum value at the intersection of the so called nodal direction of the Brillouin zone with the Fermi surface (see, e.g., [20]). The FC sets in around the van Hove singularities [11], causing, according to Eqs. (9) and (10), large density of states and large value of the difference $(p_f - p_i)$ at the point $\bar{M}$. Then, the volume $\Omega_{FC}$ and difference $(p_f - p_i)$ start to depend on the point of the Fermi surface, say, on the angle $\phi$ along the Fermi surface, which we count from the point $\bar{M}$ to the nodal direction. Nonetheless, as it is seen from Eq. (11), $E_0$ remains constant, being independent of the angle. It is not the case for the effective mass $M_{FC}^*$, that can strongly depend upon the angle via the difference $(p_f(\phi) - p_i(\phi))$ increasing from the nodal direction towards $\bar{M}$, as it follows from Eq. (9). It is pertinent to note that outside the FC region the single particle spectrum is negligible affected by the temperature, being defined by $M_L^*$. Thus, we come to the conclusion that a system with FC is characterized by two effective masses: $M_{FC}^*$ that is related to the single particle spectrum at lower energy scale, and $M_L^*$ describing the spectrum at higher energy scale. These two effective masses manifest itself as a break in the quasiparticle dispersion, which can be approximated by two straight lines intersecting at the energy $E_0$. This break takes place at temperatures $T_c \leq T \ll T_f$ in accordance with the experimental findings [21], and, as we will see, at $T \leq T_c$ corresponding to the experimental facts [21, 22]. As to the quasiparticle formalism, it is applicable to this problem since the width $\gamma$ of single particle excitations is not large compared to their energy being proportional $\gamma \sim T$ at $T > T_c$ [8]. The lineshape can be approximated by a simple Lorentzian [12], being in accordance with experimental data obtained from scans at a constant binding energy [23], see Sec. 4. Then, FC serves as a stimulating source of new phase transitions which lift the degeneration of the spectrum. For example, FC can generate the spin density wave, or antiferromagnetic phase transition, thus leading to a whole variety of the system's properties. Then, the onset of the charge density wave is preceded by FCQPT, and both of these phases can coexist at the sufficiently low density when $r_s \geq r_{cdw}$. The simple consideration presented above explains extremely large variety of properties of high-$T_c$ superconductors. We have seen above that the superconductivity is strongly aided by FC, because both of the phases are characterized by the same order parameter. As a result, the superconductivity, removing the spectrum degeneration, "wins the competition" with the other phase transitions up to the critical temperature $T_c$. We turn now to a consideration of both the superconducting state and quasiparticle dispersions at $T \leq T_c$.

## 3. The Superconducting State

The explanation of the large values of the critical temperature $T_c$, of the maximum value of the superconducting gap $\Delta_1$, of the relation between $\Delta_1$ and the temperature $T^*$ at which the pseudogap vanishes are, as years before, among the main problems in the physics of high-temperature superconductivity. To solve them, one needs to know the single-particle spectra of corresponding metals. Recent studies of photoemission spectra in copper oxide based compounds discovered an energy scale in the spectrum of low-energy electrons in copper oxides, which manifests itself as a kink in the single-particle spectra [21–24]. As a result, the spectra in the energy range (-200—0) meV can be described by two straight lines intersecting at the binding energy $E_0 \sim (50-70)$ meV [21, 22]. The existence of the energy scale $E_0$ could be attributed to the interaction between electrons and the collective excitations, for instance, phonons [24]. On the other hand, the analysis of the experimental data on the single-particle electron spectra demonstrates that the perturbation of the spectra by phonons or other collective states is in fact very small, therefore, the corresponding state of electrons has to be described as a strongly collectivized quantum state and was named "quantum protectorate" [13, 14]. Thus, the interpretation of the above mentioned kink as a consequence of electron-phonon interaction can very likely be in contradiction with the quantum protectorate concept. To describe the large values of $T_c$, the single-particle spectra and the kink, the assumption can be used that the electron system of high-$T_c$ superconductor has undergone FCQPT.

The compounds are extremely complex materials having a great number of competing degrees of freedom which produce a great variety of physical properties. In turn, these properties can compete and coexist with the superconductivity hindering the understanding of the universal cause of the superconductivity. As a result, it was suggested that the unique superconducting properties in these compounds are defined by the presence of the Cu-O planes, by the $d$-wave pairing symmetry, and by the existence of the pseudogap phenomena in optimally doped and underdoped cuprates, see e.g. [25–27]. However, recent studies of quasiparticle tunneling spectra of cuprates have revealed that the pairing symmetry may change from the $d$ to $s$-wave symmetry, depending on the hole, or electron, doping level [28–30]. Then, the high temperature $s$-wave superconductivity has been observed in electron doped infinite layer cuprates [31] with a sharp transition at $T = 43$ K and the absence of pseudogap [32]. Therefore, we can conclude that the $d$-wave symmetry and the pseudogap phenomena are not integral parts of . After all, recent studies of high-$T_c$ superconductivity in $C_{60}$ crystals [33] with the use of field-induced doping, when an increase in the superconducting transition temperature $T_c$ to 52 K was achieved in hole doped samples, have shown that the presence of the Cu-O planes is not the necessary condition to observe high-$T_c$ superconductivity. Then, in lattice expanded $C_{60}$ by intercalating $CHCl_3$ and $CHBr_3$ into the lattice, the higher $T_c$ of 80 K in hole doped $C_{60}$/$CHCl_3$ and of 117 K in $C_{60}$/$CHBr_3$ were observed [34]. In the electron-doped case, $T_c = 11$ K is reached for $C_{60}$ crystals, $T_c = 18$ K and $T_c = 26$ K were observed in samples intercalated with $CHCl_3$ and $CHBr_3$ respectively [34, 35]. The described above technique, when the corresponding dopant densities $x$ of electrons or holes are induced by gate doping in a field-effect transistor geometry, permits constructing the variation in $T_c^{\alpha\gamma}(x)$ as a function of $x$ in a wide region of the doping variation [33–35]. Here $\alpha$ denotes the material, say $C_{60}$ or intercalated $C_{60}$, etc., and $\gamma$ denotes hole or electron doping. One important point to remember is that in case of electron doped metals we have to treat $x$ as the density of the mobile charge carriers, which corresponds to a narrow variation region of $x$ around half-filling of the conduction band. While the limitation of the hole doping variation is defined by the electric breakdown strength of the gate oxide taking place at a sufficiently high level of the doping [34]. This technique allows the study of properties of metals in question as a function of the doping level $x$ without inducing disorder or possible defects which could have a strong impact on the superconductivity. It is very essential to note that the shape of the functions $T_c^{\alpha\gamma}(x)$ is similar in samples with or without intercalation, that is the shape does not depend on both $\alpha$ and $\gamma$ [34]. Moreover, this observation is also valid in the case of the field induced superconductivity in both a spin-ladder cuprate [$CaCu_2O_3$]$_4$ [36] and $CaCuO_2$ [37]. Thus, we can use a simple approximation

$$T_c^{\alpha\gamma}(x) = T_1^{\alpha} T_2^{\gamma}(x_1 - x)x, \tag{13}$$

where the coefficients $T_1^{\alpha}$ and $T_2^{\gamma}$ define the transition temperature $T_c$ for a given hole (or electron) metal, and $x$ is the density of the mobile charge carriers with $x$ is obviously tending continuously to zero at the insulator-metal transition. It is directly follows from Eq. (13) that the transition temperature reaches its maximum value $T_c^M$ at the optimal doping level $x_{opt}$

$$T_c^M = T_c^{\alpha\gamma}(x_{opt}) = T_1^{\alpha} T_2^{\gamma}\left(\frac{x_1}{2}\right)^2. \tag{14}$$

Now we can calculate the value of $x_{opt}$ for the different hole and electron metals studied in [33–37] in terms of the dimensionless parameter $r_s$, $\pi r_s^2 a_B^2 = 1/x$. We have that $r_s^{opt} \sim 10$ and, thus, $r_s^{opt}$ is approximately independent of the metals. As a result we can recognize that these striking experimental facts, the general shape of the function $T_c(x)$ and the constant value of $r_s^{opt}$, point to a fact, that the generic properties of high-$T_c$ superconductivity are defined by the 2D charge (electron or hole) strongly correlated liquid rather then by solids which hold this liquid. While the solids arrange the presence of the pseudogap phenomena, the $s$ or $d$-wave pairing symmetry, the electron-phonon coupling constant defining $T_c$, the variation region of $x$, and so on.

At $T = 0$, the ground state energy $E_{gs}[\kappa(\mathbf{p}), n(\mathbf{p})]$ of 2D electron liquid is a functional of the order parameter of the superconducting state $\kappa(\mathbf{p})$ and of the quasiparticle occupation numbers $n(\mathbf{p})$ and is determined by the known equation of the weak-coupling theory of superconductivity, see e.g. [38]

$$E_{gs} = E[n(\mathbf{p})] + \int \lambda_0 V(\mathbf{p}_1, \mathbf{p}_2) \kappa(\mathbf{p}_1) \kappa^*(\mathbf{p}_2) \frac{d\mathbf{p}_1 d\mathbf{p}_2}{(2\pi)^4}. \quad (15)$$

Here $E[n(\mathbf{p})]$ is the ground-state energy of normal Fermi liquid, $n(\mathbf{p}) = v^2(\mathbf{p})$ and $\kappa(\mathbf{p}) = v(\mathbf{p})\sqrt{1 - v^2(\mathbf{p})}$. It is assumed that the pairing interaction $\lambda_0 V(\mathbf{p}_1, \mathbf{p}_2)$ is weak. Minimizing $E_{gs}$ with respect to $\kappa(\mathbf{p})$ we obtain the equation connecting the single-particle energy $\varepsilon(\mathbf{p})$ to $\Delta(\mathbf{p})$,

$$\varepsilon(\mathbf{p}) - \mu = \Delta(\mathbf{p}) \frac{1 - 2v^2(\mathbf{p})}{2\kappa(\mathbf{p})}, \quad (16)$$

here the single-particle energy $\varepsilon(\mathbf{p})$ is determined by the Landau equation (2). The equation for the superconducting gap $\Delta(\mathbf{p})$ takes form

$$\Delta(\mathbf{p}) = -\int \lambda_0 V(\mathbf{p}, \mathbf{p}_1) \kappa(\mathbf{p}_1) \frac{d\mathbf{p}_1}{4\pi^2}$$
$$= -\frac{1}{2} \int \lambda_0 V(\mathbf{p}, \mathbf{p}_1) \frac{\Delta(\mathbf{p}_1)}{\sqrt{(\varepsilon(\mathbf{p}_1) - \mu)^2 + \Delta^2(\mathbf{p}_1)}} \frac{d\mathbf{p}_1}{4\pi^2}. \quad (17)$$

If $\lambda_0 \to 0$, then, the maximum value $\Delta_1 \to 0$, and Eq. (16) reduces to Eq. (5) [3]

$$\varepsilon(\mathbf{p}) - \mu = 0, \text{ if } 0 < n(\mathbf{p}) < 1; \quad p_i \leq p \leq p_f. \quad (18)$$

Now we can study relationships between the state defined by Eq. (18) and the superconductivity. At $T = 0$, Eq. (18) defines a particular state of Fermi liquid with FC for which the modulus of the order parameter $|\kappa(\mathbf{p})|$ has finite values in the $L_{FC}$ range of momenta $p_i \leq p \leq p_f$, and $\Delta_1 \to 0$ in the $L_{FC}$. Such a state can be considered as superconducting, with infinitely small value of $\Delta_1$ so that the entropy of this state is equal to zero. It is obvious, that this state, being driven by the quantum phase transition, disappears at $T > 0$ [5]. When $p_i \to p_F \to p_f$, Eq. (18) determines the point $r_{FC}$ at which the FCQPT takes place. It follows from Eq. (18) that the system brakes into two quasiparticle subsystems: the first subsystem in the $L_{FC}$ range is occupied by the quasiparticles with the effective mass $M_{FC}^* \propto 1/\Delta_1$, while the second one is occupied by quasiparticles with finite mass $M_L^*$ and momenta $p < p_i$. If $\lambda_0 \neq 0$, $\Delta_1$ becomes finite, leading to a finite value of the effective mass $M_{FC}^*$ in $L_{FC}$, which can be obtained from Eq. (16) [5, 12]

$$M_{FC}^* \simeq p_F \frac{p_f - p_i}{2\Delta_1}. \quad (19)$$

As to the energy scale, it is determined by the parameter $E_0$:

$$E_0 = \varepsilon(\mathbf{p}_f) - \varepsilon(\mathbf{p}_i) \simeq 2\frac{(p_f - p_F)p_F}{M_{FC}^*} \simeq 2\Delta_1. \quad (20)$$

It is reasonably safe to suggest that we have come back to the Landau theory by integrating out high energy degrees of freedom and introducing the quasiparticles. The sole difference between the Landau Fermi liquid and Fermi liquid undergone FCQPT is that we have to expand the number of relevant low energy degrees of freedom by adding both a new type of quasiparticles with the effective mass $M_{FC}^*$, given by Eq. (19), and the energy scale $E_0$ given by Eq. (20). We have also to bear in mind that the properties of these new quasiparticles of a Fermi liquid with FC cannot be separated from the properties of the superconducting state, as it follows from Eqs. (16), (19) and (20). We may say that the quasiparticle system in the range $L_{FC}$ becomes very "soft" and is to be considered as a strongly correlated liquid. On the other hand, the system's properties and dynamics are dominated by a strong collective effect having its origin in FCQPT and determined by the macroscopic number of quasiparticles in the range $L_{FC}$. Such a system cannot be disturbed by the scattering of individual quasiparticles and has features of a quantum protectorate [5, 13, 14].

We assume that the range $L_{FC}$ is small, $(p_f - p_F)/p_F \ll 1$, and $2\Delta_1 \ll T_f$ so that the order parameter $\kappa(\mathbf{p})$ is governed mainly by the FC [5, 18]. To solve Eq. (17) analytically, we take the Bardeen–Cooper–Schrieffer (BCS) approximation for the interaction [9]: $\lambda_0 V(\mathbf{p}, \mathbf{p}_1) = -\lambda_0$ if $|\varepsilon(\mathbf{p}) - \mu| \le \omega_D$, the interaction is zero outside this region, with $\omega_D$ being the characteristic phonon energy. As a result, the gap becomes dependent only on the temperature, $\Delta(\mathbf{p}) = \Delta_1(T)$, being independent of the momentum, and Eq. (17) takes the form

$$1 = N_{FC}\lambda_0 \int_0^{E_0/2} \frac{d\xi}{\sqrt{\xi^2 + \Delta_1^2(0)}} + N_L\lambda_0 \int_{E_0/2}^{\omega_D} \frac{d\xi}{\sqrt{\xi^2 + \Delta_1^2(0)}}. \quad (21)$$

Here we set $\xi = \varepsilon(\mathbf{p}) - \mu$ and introduce the density of states $N_{FC}$ in the $L_{FC}$, or $E_0$, range. As it follows from Eq. (19), $N_{FC} = (p_f - p_F)p_F/2\pi\Delta_1(0)$. The density of states $N_L$ in the range $(\omega_D - E_0/2)$ has the standard form $N_L = M_L^*/2\pi$. If the energy scale $E_0 \to 0$, Eq. (21) reduces to the BCS equation. On the other hand, assuming that $E_0 \le 2\omega_D$ and omitting the second integral in the right hand side of Eq. (21), we obtain

$$\Delta_1(0) = \frac{\lambda_0 p_F(p_f - p_F)}{2\pi} \ln\left(1 + \sqrt{2}\right) = 2\beta\varepsilon_F \frac{p_f - p_F}{p_F} \ln\left(1 + \sqrt{2}\right), \quad (22)$$

where the Fermi energy $\varepsilon_F = p_F^2/2M_L^*$, and dimensionless coupling constant $\beta = \lambda_0 M_L^*/2\pi$. Taking the usual values of the dimensionless coupling constant $\beta \simeq 0.3$, and $(p_f - p_F)/p_F \simeq 0.2$, we get from Eq. (22) the large value of $\Delta_1(0) \sim 0.1\varepsilon_F$, while for normal metals one has $\Delta_1(0) \sim 10^{-3}\varepsilon_F$. Taking into account the omitted integral, we obtain

$$\Delta_1(0) \simeq 2\beta\varepsilon_F \frac{p_f - p_F}{p_F} \ln\left(1 + \sqrt{2}\right)\left(1 + \beta \ln \frac{2\omega_D}{E_0}\right). \quad (23)$$

It is seen from Eq. (23) that the correction due to the second integral is small, provided $E_0 \simeq 2\omega_D$. Below we show that $2T_c \simeq \Delta_1(0)$, which leads to the conclusion that there is no isotope effect since $\Delta_1$ is independent of $\omega_D$. But this effect is restored as $E_0 \to 0$. Assuming $E_0 \sim \omega_D$ and $E_0 > \omega_D$, we see that Eq. (21) has no standard solutions $\Delta(p) = \Delta_1(T = 0)$ because $\omega_D < \varepsilon(p \simeq p_f) - \mu$ and the interaction vanishes at these momenta. The only way to obtain solutions is to restore the condition $E_0 < \omega_D$. For instance, we can define the momentum $p_D < p_f$ such that

$$\Delta_1(0) = 2\beta\varepsilon_F \frac{p_D - p_F}{p_F} \ln\left(1 + \sqrt{2}\right) = \omega_D, \qquad (24)$$

while the other part in the $L_{\text{FC}}$ range can be occupied by a gap $\Delta_2$ of the different sign, $\Delta_1(0)/\Delta_2 < 0$. It follows from Eq. (24) that the isotope effect is presented, while the both gaps can have $s$-wave symmetry. A more detailed analysis will be published elsewhere.

At $T \simeq T_c$, Eqs. (19) and (20) are replaced by the equation, which is valid also at $T_c \leq T \ll T_f$ in accord with Eq. (9) [5]

$$M^*_{\text{FC}} \simeq p_F \frac{p_f - p_i}{4T_c}, \quad E_0 \simeq 4T_c; \quad \text{if } T_c \leq T: \quad M^*_{\text{FC}} \simeq p_F \frac{p_f - p_i}{4T}, \quad E_0 \simeq 4T. \qquad (25)$$

Equation (21) is replaced by its conventional finite temperature generalization

$$1 = N_{\text{FC}}\lambda_0 \int_0^{E_0/2} \frac{d\xi}{\sqrt{\xi^2 + \Delta_1^2(T)}} \tanh\frac{\sqrt{\xi^2 + \Delta_1^2(T)}}{2T}$$

$$+ N_L\lambda_0 \int_{E_0/2}^{\omega_D} \frac{d\xi}{\sqrt{\xi^2 + \Delta_1^2(T)}} \tanh\frac{\sqrt{\xi^2 + \Delta_1^2(T)}}{2T}. \qquad (26)$$

Putting $\Delta_1(T \to T_c) \to 0$, we obtain from Eq. (26)

$$2T_c \simeq \Delta_1(0), \qquad (27)$$

with $\Delta_1(T = 0)$ being given by Eq. (21). By comparing Eqs. (19), (25) and (27), we see that $M^*_{\text{FC}}$ and $E_0$ are almost temperature independent at $T \leq T_c$. In the same way, as it was done in Sec. 2, we can conclude that $E_0$ does not change along the Fermi surface, while $M^*_{\text{FC}}$ increases when moving from the nodal direction to the point $\bar{M}$. Now a few remarks are in order. One can define $T_c$ as the temperature when $\Delta_1(T_c) \equiv 0$. At $T \geq T_c$, Eq. (26) has only the trivial solution $\Delta_1 \equiv 0$. On the other hand, $T_c$ can be defined as a temperature at which the superconductivity vanishes. Thus, we deal with two different definitions, which can lead to two different temperatures $T_c$ and $T^*$ in case of the $d$-wave symmetry of the gap. It was shown [12, 39] that in the case of the d-wave superconductivity, taking place in the presence of the FC, there exist a nontrivial solutions of Eq. (26) at $T_c \leq T \leq T^*$ corresponding to the pseudogap state. It happens when the gap occupies only such a part of the Fermi surface, which shrinks as the temperature increases. Here $T^*$ defines the temperature at which $\Delta_1(T^*) \equiv 0$ and the pseudogap state vanishes. The superconductivity is destroyed at $T_c$, and the ratio $2\Delta_1/T_c$ can vary in a wide range and strongly depends upon the material's properties, as it follows from consideration given in [12, 19, 39]. Therefore, provided there exists the pseudogap above $T_c$, then $T_c$ is to be replaced by $T^*$, and Eq. (27) takes the form

$$2T^* \simeq \Delta_1(0). \tag{28}$$

The ratio $2\Delta_1/T_c$ can reach very high values. For instance, in the case of $Bi_2Sr_2CaCu_2O_{6+\delta}$, where the superconductivity and the pseudogap are considered to be of the common origin, $2\Delta_1/T_c$ is about 28, while the ratio $2\Delta_1/T^* \simeq 4$, which is also valid for various cuprates [40]. Thus, Eq. (28) gives good description of the experimental data. We remark that Eq. (21) gives also good description of the maximum gap $\Delta_1$ in the case of the d-wave superconductivity, because the different regions with the maximum absolute value of $\Delta_1$ and the maximal density of states can be considered as disconnected [41]. Therefore, the gap in this region is formed by attractive phonon interaction which is approximately independent of the momenta.

Consider now two possible types of the superconducting gap $\Delta(\mathbf{p})$ given by Eq. (17) and defined by interaction $\lambda_0 V(\mathbf{p}, \mathbf{p}_1)$. If this interaction is dominated by a phonon-mediated attraction, the even solution of Eq. (17) with the $s$-wave, or the $s + d$ mixed waves, will have the lowest energy. Provided the pairing interaction $\lambda_0 V(\mathbf{p}_1, \mathbf{p}_2)$ is the combination of both the attractive interaction and sufficiently strong repulsive interaction, the $d$-wave odd superconductivity can take place, see e.g. [41]. But both the $s$-wave even symmetry and $d$-wave odd one lead to the approximately same value of the gap $\Delta_1$ in Eq. (21) [18]. Therefore, the non-universal pairing symmetries in high-$T_c$ superconductivity is likely the result of the pairing interaction, and the $d$-wave pairing symmetry cannot be considered as essential to high-$T_c$ in keeping with experimental findings [28–32]. In case, if there were only the $d$-wave pairing, the crossover from superconducting gap to pseudogap can take place, so that the superconductivity is destroyed at the temperature $T_c$, with the superconducting gap being smoothly transformed into the pseudogap which closes at some temperature $T^* > T_c$ [19, 39]. In the case of the $s$-wave pairing we can expect the absence of the pseudogap phenomena in accordance with the experimental observation, see [32] and references therein.

We turn now to a consideration of the maximum value of the superconducting gap $\Delta_1$ as a function of the density $x$ of mobile charge carriers. Being rewritten in terms of $x$ and $x_{FC}$ related to the variables $p_i$ and $p_f$ by Eq. (12), Eq. (22) takes

the form
$$\Delta_1 \propto \beta(x_{FC} - x)x. \tag{29}$$
Here we take into account that the Fermi level $\varepsilon_F \propto p_F^2$, the density $x \propto p_F^2$, and thus, $\varepsilon_F \propto x$. Considering the field induced superconductivity, we can safely assume that $T_c \propto \Delta_1$ because this technique allows the study of properties of metal in question as a function of $x$ without inducing additional defects or disorder which can have a dramatic impact on the transition temperature. Then, instead of Eq. (29) we have
$$T_c^{\alpha\gamma} \propto \beta^\alpha \beta^\gamma (x_{FC} - x)x. \tag{30}$$
In Eq. (30), we made the natural change $\beta = \beta^\alpha \beta^\gamma$ since the coupling constant $\beta$ is fixed by the properties of metal in question. Following reference [42], we take that hole doped metals differ from electron doped ones only in the magnitude of the coupling constant $\beta^\gamma$ which is smaller in case of electron doped metals. Now it is seen that Eq. (30) coincides with Eq. (13) producing the universal optimal doping level $x_{opt} = x_{FC}/2 = x_1/2$ in line with the experimental facts. In our model, we have $x_{opt}/x_{FC} = (r_s^{opt}/r_{FC})^2 = 2$, taking the value $r_s^{opt} \simeq 10$, we obtain $r_{FC} \simeq 7.0$. This result is in a reasonable agreement with the experimental value $r_{FC} \sim 7.5$ corresponding to sharp increase of the effective mass [15]. In line with facts [34], it follows from Eq. (30) that among the hole doped fullerides, the $T_c$ ratios for $C_{60}$/CHBr$_3$-$C_{60}$/CHCl$_3$-$C_{60}$ have to be the same as in the case of the respective electron doped fullerides because the factor $\beta^\gamma$ drops out of the ratios.

As an example of the implementation of the previous analysis let us consider the main features of a room-temperature superconductor [43]. The superconductor has to be a quasi two-dimensional structure, presented by infinite-layer compounds or by field-induced superconductivity in gated structures. As it follows from Eq. (22), $\Delta_1 \sim \beta\varepsilon_F \propto \beta/r_s^2$. Noting that FCQPT takes place in 3D systems at $r_s \sim 20$ and in 2D systems at $r_s \sim 8$ [7], we can expect that $\Delta_1$ of 3D system comprises 10% of the corresponding maximum value of 2D superconducting gap, reaching values as high as 60 meV for underdoped crystals with $T_c = 70$ [44]. On the other hand, it is seen form Eq. (22), that $\Delta_1$ can be even large, $\Delta_1 \sim 75$ meV, and one can expect $T_c \sim 300$ K in the case of the $s$ wave pairing as it follows from the simple relation $2T_c \simeq \Delta_1$. In fact, we can safely take $\varepsilon_F \sim 300$ meV, $\beta \sim 0.5$ and $(p_f - p_i)/p_F \sim 0.5$. Thus, we can conclude that a possible room-temperature superconductor has to be the $s$-wave superconductor in order to get rid of the pseudogap phenomena, which tremendously reduces the transition temperature. The density $x$ of the mobile charge carriers must satisfy the condition $x \leq x_{FC}$ and be flexible to reach the optimal doping level. It is worth noting that the coupling constant $\beta$ has to be sufficiently big because FC giving rise to the order parameter $\kappa(\mathbf{p})$ does not produce the gap $\Delta$ by itself. For instance, the coupling constant can be enhanced by an intercalation as it is done for fullerides [34, 42].

Now we turn to the calculations of the gap and the specific heat at the temperatures $T \to T_c$. It is worth noting that this consideration is valid provided $T^* = T_c$, otherwise the considered below discontinuity is smoothed out over the temperature range $T^* \div T_c$. For the sake of simplicity, we calculate the main contribution to the gap and the specific heat coming from the FC. The function $\Delta_1(T \to T_c)$ is found from Eq. (26) upon expanding the right hand side of the first integral in powers of $\Delta_1$ and omitting the contribution from the second integral on the right hand side of Eq. (26). This procedure leads to the following equation [18]

$$\Delta_1(T) \simeq 3.4 T_c \sqrt{1 - \frac{T}{T_c}}. \tag{31}$$

Thus, the gap in the spectrum of the single-particle excitations has quite usual behavior. To calculate the specific heat, the conventional expression for the entropy $S$ [9] can be used

$$S = 2 \int \left[ f(\mathbf{p}) \ln f(\mathbf{p}) + (1 - f(\mathbf{p})) \ln(1 - f(\mathbf{p})) \right] \frac{d\mathbf{p}}{(2\pi)^2}, \tag{32}$$

where

$$f(\mathbf{p}) = \frac{1}{1 + \exp[E(\mathbf{p})/T]}; \quad E(\mathbf{p}) = \sqrt{(\varepsilon(\mathbf{p}) - \mu)^2 + \Delta_1^2(T)}. \tag{33}$$

The specific heat $C$ is determined by

$$C = T\frac{dS}{dT} \simeq 4\frac{N_{FC}}{T^2} \int_0^{E_0} f(E)[1 - f(E)] \left[ E^2 + T\Delta_1(T)\frac{d\Delta_1(T)}{dT} \right] d\xi$$
$$+ 4\frac{N_L}{T^2} \int_{E_0}^{\omega_D} f(E)[1 - f(E)] \left[ E^2 + T\Delta_1(T)\frac{d\Delta_1(T)}{dT} \right] d\xi. \tag{34}$$

When deriving Eq. (34) we again use the variable $\xi$ and the densities of states $N_{FC}$, $N_L$, just as before in connection to Eq. (21), and use the notation $E = \sqrt{\xi^2 + \Delta_1^2(T)}$. Equation (34) predicts the conventional discontinuity $\delta C$ in the specific heat $C$ at $T_c$ because of the last term in the square brackets of Eq. (34). Upon using Eq. (31) to calculate this term and omitting the second integral on the right hand side of Eq. (34), we obtain

$$\delta C \simeq \frac{3}{2\pi}(p_f - p_i)p_F. \tag{35}$$

In contrast to the conventional result when the discontinuity is a linear function of $T_c$, $\delta C$ is independent of the critical temperature $T_c$ because the density of state varies inversely with $T_c$ as it follows from Eq. (25). Note, that deriving Eq. (35) we take into account the main contribution coming from the FC. This contribution vanishes as soon as $E_0 \to 0$ and the second integral of Eq. (34) gives the conventional result.

## 4. The Lineshape of the Single-Particle Spectra

Consider the lineshape $L(q, \omega)$ of the single-particle spectrum which is a function of two variables. Measurements carried out at a fixed binding energy $\omega = \omega_0$, where $\omega_0$ is the energy of a single-particle excitation, determine the lineshape $L(q, \omega = \omega_0)$ as a function of the momentum $q$. We have shown above that $M_{FC}^*$ is finite and constant at $T \leq T_c$. Therefore, at excitation energies $\omega \leq E_0$ the system behaves like an ordinary superconducting Fermi liquid with the effective mass given by Eq. (19) [5, 12]. At $T_c \leq T$ the low energy effective mass $M_{FC}^*$ is finite and is given by Eq. (9). Once again, at the energies $\omega < E_0$, the system behaves as a Fermi liquid, the single-particle spectrum is well defined, while the width of single-particle excitations is of the order of $T$ [5, 8]. This behavior was observed in experiments on measuring the lineshape at a fixed energy [23]. It is pertinent to note that recent measurements of the lineshape suggest that quasiparticle excitation even in the $(\pi, 0)$ region of the Brillouin zone of $Bi_2Sr_2CaCu_2Q_{8+\delta}$ (Bi2212) are much better defined then previously believed from earlier Bi2212 data [45]. We remark that our model is in accordance with these measurements suggesting that well-defined quasiparticles exist at the Fermi level.

The lineshape can also be determined as a function $L(q = q_0, \omega)$ at a fixed $q = q_0$. At small $\omega$, the lineshape resembles the one considered above, and $L(q = q_0, \omega)$ has a characteristic maximum and width. At energies $\omega \geq E_0$, quasiparticles with the mass $M_L^*$ come into play, leading to a growth of the function $L(q = q_0, \omega)$. As a result, the function $L(q = q_0, \omega)$ possesses the known peak-dip-hump structure [46] directly defined by the existence of the two effective masses $M_{FC}^*$ and $M_L^*$ [5, 12]. To have more quantitative and analytical insight into the problem we use the Kramers–Krönig transformation to construct the imaginary part $\text{Im}\Sigma(\mathbf{p}, \varepsilon)$ of the self-energy $\Sigma(\mathbf{p}, \varepsilon)$ starting with the real one $\text{Re}\Sigma(\mathbf{p}, \varepsilon)$ which defines the effective mass [47]

$$\frac{1}{M^*} = \left(\frac{1}{M} + \frac{1}{p_F}\frac{\partial \text{Re}\Sigma}{\partial p}\right) \bigg/ \left(1 - \frac{\partial \text{Re}\Sigma}{\partial \varepsilon}\right). \tag{36}$$

Here $M$ is the bare mass, while the relevant momenta $p$ and energies $\varepsilon$ are subjected to the conditions: $|p - p_F|/p_F \ll 1$, and $\varepsilon/\varepsilon_F \ll 1$. We take $\text{Re}\Sigma(\mathbf{p}, \varepsilon)$ in the simplest form which accounts for the change of the effective mass at the energy scale $E_0$:

$$\text{Re}\Sigma(\mathbf{p}, \varepsilon) = -\varepsilon\frac{M_{FC}^*}{M} + \left(\varepsilon - \frac{E_0}{2}\right)\frac{M_{FC}^* - M_L^*}{M}\left[\theta(\varepsilon - E_0/2) + \theta(-\varepsilon - E_0/2)\right]. \tag{37}$$

Here $\theta(\varepsilon)$ is the step function. Note that in order to ensure a smooth transition from the single-particle spectrum characterized by $M_{FC}^*$ to the spectrum defined by $M_L^*$ the step function is to be substituted by some smooth function. Upon inserting Eq. (37) into Eq. (36) we can check that inside the interval $(-E_0/2, E_0/2)$ the effective mass $M^* \simeq M^*_{FC}$, and outside the interval $M^* \simeq M^*_L$. By applying the Kramers–Krönig transformation to $\mathrm{Re}\,\Sigma(\mathbf{p},\varepsilon)$, we obtain the imaginary part of the self-energy [18]

$$\mathrm{Im}\,\Sigma(\mathbf{p},\varepsilon) \sim \varepsilon^2 \frac{M^*_{FC}}{\varepsilon_F M} + \frac{M^*_{FC} - M^*_L}{M}\left(\varepsilon \ln\left|\frac{\varepsilon + E_0/2}{\varepsilon - E_0/2}\right| + \frac{E_0}{2}\ln\left|\frac{\varepsilon^2 - E_0^2/4}{E_0^2/4}\right|\right). \tag{38}$$

We can see from Eq. (38) that at $\varepsilon/E_0 \ll 1$ the imaginary part is proportional to $\varepsilon^2$; at $2\varepsilon/E_0 \simeq 1$ $\mathrm{Im}\,\Sigma \sim \varepsilon$; at $E_0/\varepsilon \ll 1$ the main contribution to the imaginary part is approximately constant. This is the behavior that gives rise to the known peak-dip-hump structure. Then, it is seen from Eq. (38) that when $E_0 \to 0$ the second term on the right hand side tends to zero, the single-particle excitations become better defined resembling that of a normal Fermi liquid, and the peak-dip-hump structure eventually vanishes. On the other hand, the quasiparticle amplitude $a(\mathbf{p})$ is given by [47]

$$\frac{1}{a(\mathbf{p})} = 1 - \frac{\partial \mathrm{Re}\,\Sigma(\mathbf{p},\varepsilon)}{\partial \varepsilon}. \tag{39}$$

It follows from Eq. (36) that the quasiparticle amplitude $a(\mathbf{p})$ rises as the effective mass $M^*_{FC}$ decreases. Since, as it follows from Eq. (12), $M^*_{FC} \sim (p_f - p_i)/p_F \sim (x_{FC} - x)/x_{FC}$, we are led to a conclusion that the amplitude $a(\mathbf{p})$ rises as the doping level rises, and the single-particle excitations become better defined in highly overdoped samples. It is worth noting that such a behavior was observed experimentally in so highly overdoped Bi2212 that the gap size is about 10 meV [17]. Such a small size of the gap testifies that the region occupied by the FC is small since $E_0/2 \simeq \Delta_1$. Quasiparticles located at the intersection of the nodal direction of the Brillouin zone with the Fermi level should also be well-defined comparatively to quasiparticles located at the point $\bar{M}$ because of the decrease of the effective mass $M^*_{FC}$ when moving from the point $\bar{M}$ to the nodal direction, as it was discussed in Sections 2 and 3. This is especially true in regard to strongly underdoped samples.

## 5. Summary

In conclusion, we have shown that the theory of high temperature superconductivity based on the fermion-condensation quantum phase transition and on the conventional theory of superconductivity permits to describe high values of $T_c$, $T^*$ and of the maximum value of the gap $\Delta_1$, which may be as big as $\Delta_1 \sim 0.1\varepsilon_F$ or even larger. We have also traced the transition from conventional superconductors to high-$T_c$ and demonstrated that in the highly overdoped cuprates the single-particle excitations become much better defined, resembling that of a normal Fermi liquid.

We have also shown by a simple, self-consistent analysis, that the general features of the shape of the critical temperature $T_c$ as a function of the density $x$ of the mobile carriers in the metals and the value of the optimal doping $x_{opt}$ can be understood within the framework of the theory of the high-$T_c$ superconductivity based on FCQPT. We have demonstrated that neither the $d$-wave pairing symmetry, nor the pseudogap phenomenon, nor the presence of the Cu-O planes are of importance for the existence of the high-temperature superconductivity. As a result, we can conclude that the generic properties of high-temperature superconductors are defined by the 2D charge (electron or hole) strongly correlated liquid rather then by solids which hold this liquid. While the solids arrange the presence of the pseudogap phenomena, the $s$-wave pairing symmetry or $d$-wave one, the electron-phonon coupling constant defining $T_c$, the variation region of $x$, and so on. The main features of a room-temperature superconductor have also been outlined.

## Acknowledgements

This work was supported by the Russian Foundation for Basic Research, project 01-02-17189.

## References

1. Feynman, R.P. (1985) *QED The Strange Theory of Light and Matter*, Princeton, New Jersey, Princeton University Press.
2. Landau, L.D. (1956) *Zh. Eksp. Teor. Fiz.* **30**, 1058. [*Sov. Phys. JETP* **3**, 920 (1956)].
3. Khodel, V.A. and Shaginyan, V.R. (1990) *Pis'ma Zh. Eksp. Teor. Fiz.* **51**, 488 [*JETP Lett.* **53**, 51, 553 (1990)].
4. Volovik, G.E. (1991) *Pis'ma Zh. Eksp. Teor. Fiz.* **53**, 208 [*JETP Lett.* **53**, 222 (1991)].
5. Amusia, M.Ya. and Shaginyan, V.R. (2001) *Pis'ma Zh. Eksp. Teor. Fiz.* **73**, 268 [*JETP Lett.* **73**, 232 (2001)];
Amusia, M.Ya. and Shaginyan, V.R. (2001) *Phys. Rev. B* **63**, 224507.
6. Khodel, V.A., Shaginyan, V.R. and Khodel, V.V. (1994) *Phys. Rep.* **249**, 1.
7. Khodel, V.A., Shaginyan, V.R. and Zverev, M.V. (1997) *Pis'ma Zh. Eksp. Teor. Fiz.* **65**, 242 [*JETP Lett.* **65**, 253 (1997)].
8. Dukelsky, J., Khodel, V.A., Schuck, P. and Shaginyan, V.R. (1997) *Z. Phys.* **102**, 245;
Khodel, V.A. and Shaginyan, V.R. (1997) *Condensed Matter Theories*, **12**, 222.
9. Bardeen, J., Cooper, L.N. and Schrieffer, J.R. (1957) *Phys. Rev.* **108**, 1175.
10. Shaginyan, V.R. (1998) *Phys. Lett. A* **249**, 237.
11. Khodel, V.A., Clark, J.W. and Shaginyan, V.R. (1995) *Solid Stat. Comm.* **96**, 353.
12. Artamonov, S.A. and Shaginyan, V.R. (2001) *Zh. Eksp. Teor. Fiz.* **119**, 331 [*JETP* **92**, 287 (2001)].
13. Laughlin, R.B. and Pines, D. (2000) *Proc. Natl. Acad. Sci. USA* **97**, 28.
14. Anderson, P.W. cond-mat/0007185; cond-mat/0007287.
15. Shashkin, A.A., Kravchenko, S.V., Dolgopolov V.T. and Klapwijk, T.M. *Phys. Rev. B* in press; cond-mat/0111478.
16. Grüner, G. (1994) *Density Waves in Solids* Addison-Wesley, Reading, MA.
17. Yusof, Z. et al. (2002) *Phys. Rev. Lett.* **88**, 167006;
Yusof, Z. et al. cond-mat/01044367.
18. Amusia, M.Ya., Artamonov, S.A. and Shaginyan, V.R. (1998) *Pis'ma Zh. Eksp. Teor. Fiz.* **74**, 396 [*JETP Lett.* **74**, 435 (2001)].

19. Amusia, M.Ya. and Shaginyan, V.R. (2002) *Phys. Lett. A* **298**, 193.
20. Shen, Z.-X. and Dessau, D. (1995) *Phys. Rep.* **253**, 1.
21. Bogdanov, P.V. *et al.* (2000) *Phys. Rev. Lett.* **85**, 2581.
22. Kaminski, A. *et al.* (2001) *Phys. Rev. Lett.* **86**, 1070.
23. Valla, T. *et al.* (1999) *Science* **285**, 2110;
    Valla, T. *et al.* (2000) *Phys. Rev. Lett.* **85**, 828.
24. Lanzara, A. *et al.* (2001) *Nature* **412**, 510.
25. Timusk, T. and Statt, B. (1999) *Rep. Prog. Phys.* **62**, 61.
26. Tsuei, C.C. and Kirtley, J.R. (2000) *Rev. Mod. Phys.* **72**, 969.
27. Varma, C.M., Nussinov, Z. and Wim van Saarloos (2002) *Phys. Rep.* **361**, 267.
28. Yeh, N.-C. *et al.* (2001) *Phys. Rev. Lett.* **87**, 087003.
29. Biswas, A. *et al.* (2002) *Phys. Rev. Lett.* **88**, 207004.
30. Skinta, J.A. *et al.* (2002) *Phys. Rev. Lett.* **88**, 207005.
31. Skinta, J.A. *et al.* (2002) *Phys. Rev. Lett.* **88**, 207003.
32. Chen, C.-T. *et al.* (2002) *Phys. Rev. Lett.* **88**, 227002.
33. Schön, J.H., Kloc, Ch. and Batlogg, B. (2000) *Nature* **408**, 549.
34. Schön, J.H., Kloc, Ch. and Batlogg, B.(2001) *Science* **293**, 2432.
35. Schön, J.H., Kloc, Ch., Haddon, R.C. and Batlogg, B. (2000) *Science* **288**, 656.
36. Schön, J.H. *et al.* (2001) *Science* **293**, 2430.
37. Schön, J.H. *et al.* (2001) *Nature* **414**, 434.
38. Tilley, D.R. and Tilley, J. (1975) *Superfluidity and Superconductivity*, Bristol, Hilger.
39. Shaginyan, V.R. (1998) *Pis'ma Zh. Eksp. Teor. Fiz.* **68**, 491. [JETP Lett. **68**, 527 (1998)].
40. Kugler, M. *et al.* (2001) *Phys. Rev. Lett.* **86**, 4911.
41. Abrikosov, A.A. (1995) *Phys. Rev. B* **52**, R15738; Abrikosov, A.A. cond-mat/9912394.
42. Bill, B. and Kresin, V.Z. (2002) *Eur. Phys. J. B* **26**, 3.
43. Shaginyan, V.R. cond-mat/0206575.
44. Miyakawa, N. *et al.* (1999) *Phys. Rev. Lett.* **83**, 1018.
45. Feng, D.L. *et al.* cond-mat/0107073.
46. Dessau, D.S. *et al.* (1991) *Phys. Rev. Lett.* **66**, 2160.
47. Migdal, A.B. *Theory of Finite Fermi Systems and Applications to Atomic Nuclei* (Benjamin, Reading, MA, 1977).

# THE HEAVY-HOLE $X^+$ TRION IN DOUBLE QUANTUM WELLS

R.A. SERGEEV AND R.A. SURIS
*Ioffe Physico-Technical Institute RAS,*
*26 Polytekhnicheskaya, St Petersburg 194021, Russia*

Abstract. The system of two heavy holes in 2D Quantum Well bound by an electron in the adjacent 2D Quantum Well is considered. Simple variational wave function with a few variational parameters is suggested to calculate the ground state energy of such $X^+$ trion in the whole range of the distances between the wells. The trion appears to be bound up to unexpectedly large distance values. The critical value of the latter is found to be more than 34 (in atomic units) which is 68 times of the distance between the holes in 2D $X^+$ trion. The resonant state of the trion at the distances between QWs greater than the critical is also considered. A few simple estimations of the trion binding energy for GaAs and ZnSe double quantum well structures are proposed.

Key words: trions, charged excitons, double quantum wells

## 1. Introduction

Three particle electron-hole complexes (trions) in bulk semiconductors were predicted by Lampert in 1958 [1]. However, the similar system – $H^-$ ion was firstly considered by Bethe as early as 1929 [2]. Nevertheless, the experimental investigation of trions in bulk semiconductors was difficult by their small binding energy.

Recently the interest in experiment and theory of trions has risen due to considerable progress in the semiconductor heterostructure fabrication. The theoretical calculations performed in the 1980s [3–6] predicted a considerable (up to tenfold) increase of the trion binding energy in quantum well heterostructures as against bulk trions. This made possible experimental studies, which were firstly performed by K. Kheng *et al.* [7].

In recent years, the trion energies versus the effective mass ratio were intensively studied in a two-dimensional case or in a single quantum well [8–15]. It is of interest to consider the double quantum well heterostructures with the spatial separation of the carriers [16–21]. Variation of the distance between QWs gives us an additional possibility to control the trion parameters. In the present paper the simplest model of spatially indirect heavy-hole $X^+$ trion in such a heterostructure is examined.

In Section 2 the choice of the indirect exciton model is discussed. In Section 3 the model of the trion is proposed and the ground state energy dependence versus the value of the spatial separation is variationally calculated. The geometrical structure of the trion is discussed in Section 4. A few simple estimations of the trion binding energy for GaAs and ZnSe double quantum well structures are proposed in Section 5. The last section contains the conclusions.

## 2. The Indirect Exciton Variational Calculations

We consider the heterostructure with spatially separated electron and holes using the oversimplified model of two 2D quantum wells: one – for the holes and the other – for the electrons. In order to obtain the binding energy of the trion we have to find the full trion energy and subtract it from the exciton energy. Therefore, if we intend to calculate the trion binding energy with reasonable accuracy, we should calculate the exciton energy in the same way as the trion energy. Consequently, the trion function should be based on the exciton function, transforming to the latter if the distance between the holes tends to the infinity. Therefore, at first, we have to find the simplest trial exciton wave function with the minimum number of the variational parameters which gives plausible results for the exciton binding energy and for the shape of the wave function in the whole range of the distances between the quantum wells.

The Schrödinger Equation for the in-plane motion of the exciton on the $s$-state in neighboring 2D quantum wells is:

$$-\frac{1}{\rho}\frac{\partial}{\partial \rho}\rho\frac{\partial}{\partial \rho}\Psi_{ex}(\rho) - \frac{2}{\sqrt{\rho^2+d^2}}\Psi_{ex}(\rho) = -E_{ex}\Psi_{ex}(\rho). \tag{1}$$

Here $\rho$ is the 2D projection of the vector from the hole to the electron, $d$ is the distance between the wells, $E_{ex}$ is the exciton binding energy. Here and after the atomic units are used: the energy scale is bulk exciton Bohr energy, $\mu e^4/2\varepsilon^2\hbar^2$, and the length scale is the Bohr radius, $\hbar^2\varepsilon/\mu e^2$, where $\mu$ is the reduced mass of the electron and the hole, $\varepsilon$ is the dielectric constant. It is easily to see, that at small $d \to 0$ Eq. (1) transforms into the 2D exciton Schrödinger equation, and therefore the wave function $\Psi_{ex}(\rho)$ should have the exponential form:

$$\Psi_{ex}(\rho) = \exp(-2\rho). \tag{2}$$

The energy of 2D exciton in chosen units is: $E_{ex} = 4$. In the opposite limit, when $d$ becomes much larger than 2D Bohr radius, the potential in (1) can be approximately written as:

$$V(\rho) = -\frac{2}{\sqrt{\rho^2+d^2}} \approx -\frac{2}{d} + \frac{\rho^2}{d^3}. \tag{3}$$

Therefore, the hole potential transforms into the oscillator-like potential and the wave function of the exciton tends to take the gaussian form:

$$\Psi_{ex}(\rho) = \exp\left(-\frac{\rho^2}{2d^{3/2}}\right). \tag{4}$$

The exciton binding energy in this approximation is:

$$E_{ex} = \frac{2}{d} - \frac{2}{d^{3/2}}. \tag{5}$$

It makes reasonable the choice of the simple one-parameter trial wave function for exciton energy calculations in the following way:

$$\Psi_{ex}(\rho) = \exp(-\beta\sqrt{\rho^2 + d^2}). \tag{6}$$

Here $\beta$ is the only variational parameter, nontrivially corresponding to the length scale of the exciton wavefunction.

It is easily to see that function (6) produces the plausible results for the exciton binding energy and the shape of the wave function for all values of $d$. Indeed, at $d \to 0$ it takes the form of the function (2) (and the parameter $\beta$ gets the value 2). In the other limit, $d \to \infty$, function (6) transforms into gaussian (4) at $\rho \ll d$; the $\beta$ parameter in such case is asymptotically equal to:

$$\beta = \frac{1}{\sqrt{d}}. \tag{7}$$

However, it should be noted, at $\rho \sim d$ the approximation (3) becomes incorrect and the real behavior of the exciton function at $\rho \to \infty$ is exponential. The chosen function (6) satisfies this condition, but with the wrong exponential factor.

The binding energy and the parameter $\beta$, calculated with the function (6) are shown in the Fig. 1. For comparison, the approximations (5) and (7) are also plotted.

## 3. The Indirect Trion Variational Calculations

The X$^+$ trion in coupled 2D quantum wells can be represented as two heavy holes in the same quantum well bound together by an electron in the adjacent quantum well. The holes are considered to be infinitely heavy (compared to the electron) so the adiabatic approximation is used and the electron and the holes wave functions are separated. The electron in-plane wave function is proposed to be the symmetrical sum of two exciton-like wave functions (6) matched to each of the holes and shifted toward each other due to the polarization effect:

$$\Psi_{tr}(\rho) = \exp\left[-\beta\sqrt{\left(\rho + \frac{1-c}{2}R\right)^2 + d^2}\right]$$

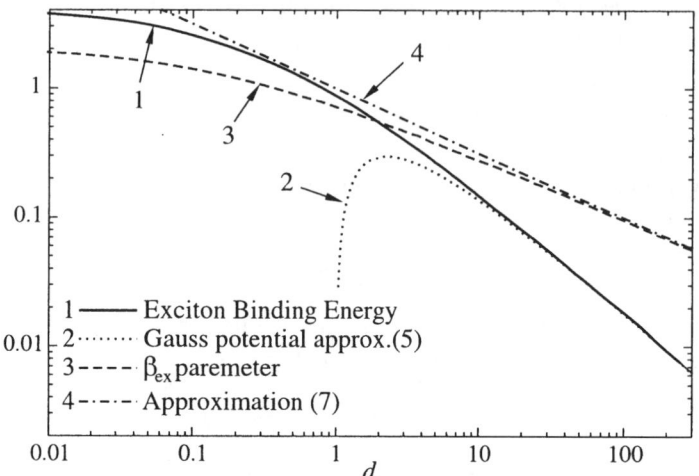

*Figure 1.* The bold curves are: the binding energy $E_{ex}$ (1, solid) and the scaling variational parameter $\beta_{ex}$ (3, dash) of indirect exciton versus the distance between the quantum wells $d$. The curves 2 and 4 are the approximations (5) and (7) respectively.

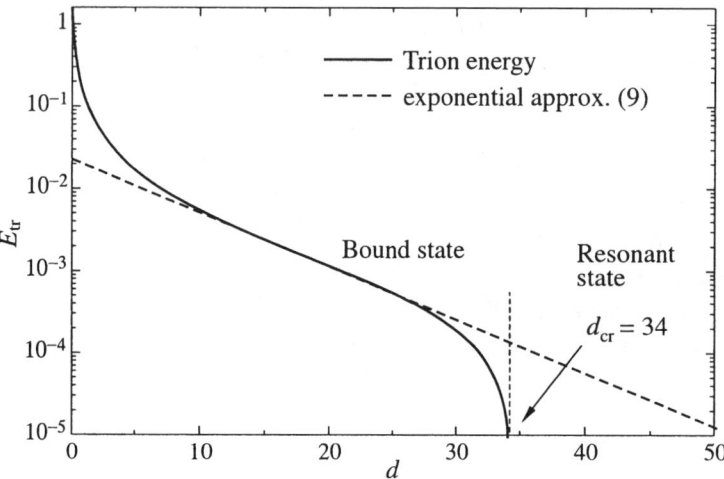

*Figure 2.* The solid curve is the indirect trion binding energy $E_{tr}$ in the units of 3D exciton binding energy versus the distance between the quantum wells $d$. The dashed one is the approximaion (9).

$$+ \exp\left[-\beta\sqrt{\left(\rho - \frac{1-c}{2}R\right)^2 + d^2}\right]. \qquad (8)$$

Here $\rho$ is the 2D projection of the vector from the center of mass of the holes to the electron, $R$ is the 2D vector from one hole to the other, $d$ is the distance between the quantum wells; $\beta$ is the same variational parameter as in the exciton function

(6), and $c$ is a parameter corresponding to polarization. The equality $c = 0$ means that both exciton parts of function (8) are strongly centered on the corresponding holes, $c = 1$ refers to the configuration of the electron wave function concentrated opposite the center of distance between the holes. $R$, $\beta$ and $c$ we consider as the fitting parameters.

The spatially-indirect trion binding energy $E_{tr}$ (in 3D exciton energy units) calculated with the function (8) versus the distance between the quantum wells is plotted in Fig. 2. It can be seen, that trion binding energy quickly decreases at small $d$. As it is shown in the next section, the configuration of the electron function qualitatively changes at $d \ll 1$.

In the region of $d \in 10-25$, the energy decreases strictly exponentially. It is because the binding energy in this interval is formed mainly by the term caused by overlapping of the exciton functions in (8), which decrease exponentially with $d$. The approximation for the energy in the foregoing range is:

$$E_{appr}(d) \approx 2.28 \cdot 10^{-2} \exp(-0.15d), \qquad (9)$$

and it is also plotted in Fig. 2.

At $d \approx 34$ the trion binding energy becomes negative, as it is easily to see in Fig. 3, where the spatially-indirect trion and exciton binding energy ratio versus the distance between the quantum wells is plotted. It means, that at $d > 34$ the trion transforms into the exciton and the free hole. However the effective potential between the holes has the minimum on relatively small $R$, and the energy corresponding to the minimum exceeds the exciton binding energy. The existence of the minimum signifies that there is a metastable state separated from the ground state of the system ($R \to \infty$) by a maximum. With the holes of finite masses, when we have to apply quantum mechanics to the holes, it means that we deal with a

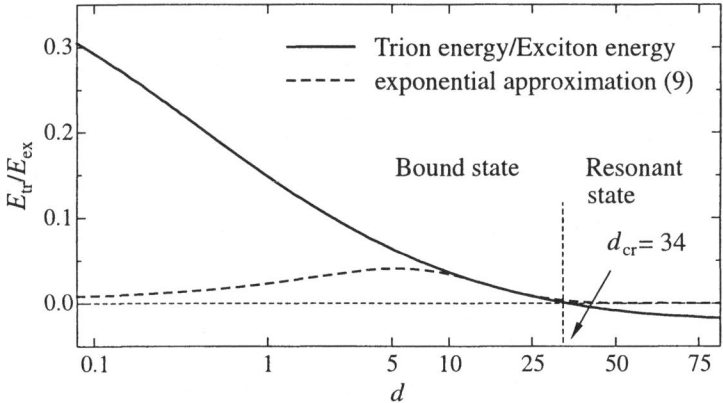

*Figure 3.* The solid curve is the indirect trion and exciton binding energy ratio $E_{tr}/E_{ex}$ versus the distance between the quantum wells $d$. For comparison, the approximaion (9) normalized to $E_{ex}$ is also plotted (the dashed one).

resonant state. The minimum of the effective potential of the hole-hole interaction is found in a significant range of $d$ values up to $d \sim 80$.

Finally, it is necessary to note, that the function (8) produces the considerable mistake in the trion binding energy value due to the oversimplified form of the exciton function (6) at small values of the distance between the quantum wells ($d \ll 1$). However, at $d = 0$ the function (8) produces the plausible result $E_{tr}/E_{ex} = 0.40$, whereas the exact result [10] is $E_{tr}/E_{ex} = 0.41$.

## 4. The Structure of the Indirect Trion

In Fig. 4 the trion and the exciton scaling parameter ratio is shown. One can see that the ratio becomes less than unity at the distance between the wells $d \approx 26$ where the trion is still bound. This fact implicitly indicates that the effective potential binding the holes together has the negative minimum at small $R$, due to the overlap of the exponential parts in the electron function (8) and the polarization, but it is positive at $R \to \infty$ due to the Coulomb interaction.

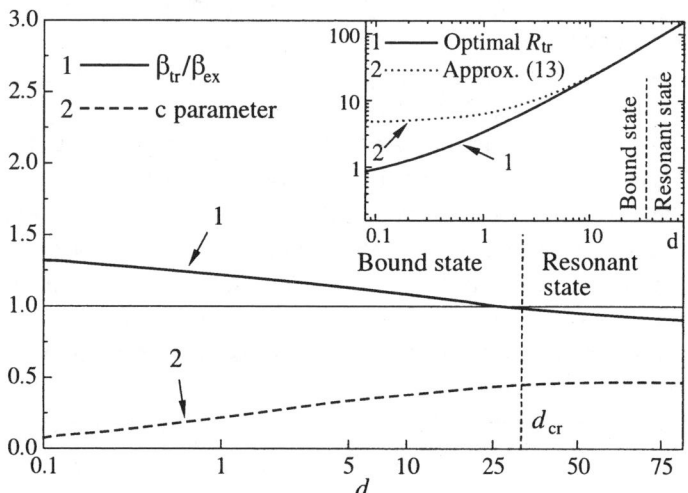

*Figure 4.* The solid curve is the indirect trion and exciton scaling parameter ratio versus the distance between the quantum wells $d$. The dashed one is the polarization parameter $c$ dependence. In the inset the optimal distance between the holes $R$ and the approximation (13) versus $d$ are plotted.

Indeed, the Coulomb potential between the indirect exciton and the adjacent hole, is:

$$V(R) \approx \frac{1}{R} - \int \frac{\Psi_{ex}(\rho)^2}{\sqrt{(R-\rho)^2 + d^2}} d\rho. \quad (10)$$

At large distances between the holes: $R \gg d, \rho$, the integration element can be expanded in series of $1/R$ and for the main term we have:

$$V(R) \approx -\frac{\langle \rho^2 \rangle - 2d^2}{4R^3}, \quad (11)$$

where

$$\langle \rho^2 \rangle = \int \rho^2 \Psi_{ex}(\rho)^2 d\rho. \tag{12}$$

In (11) the exciton function is taken to be axially symmetric. It is easily to derive that the anisotropy arising from the polarization leads only to the forth power term and is negligible. Having $\langle \rho^2 \rangle$ being estimated with the function (6), it can be fount that the potential between the exciton and the hole is repulsive on $R \to \infty$ at $d > 1.87$. It leads to existence of the resonant state of the trion at $d > 34$.

The polarization coefficient $c$ versus $d$ dependence is also shown in Fig. 3. The polarization quickly increases at small $d$ and nearly saturates at $c \approx 4.5$. It means, that at large $d$ the trion has the stable configurations with the distance between the holes nearly two times larger than the distance between the maxima of the electron wave function. In contrast to, for example, the biexciton $X_2$, the trion $X^+$ is charged system. Therefore, if we mentally divide the trion on the exciton and the adjacent hole we find out that the exciton should be strongly polarized by the external strong coulomb field, especially if the optimal distance between the holes is comparable to the average distance between the hole and the electron in the indirect exciton. As it is shown in the inset of Fig. 4, the latter statement holds at all distances between the quantum wells (up to $d \leq 80$). At large values of $d$, the electron-hole potential takes the quadratic form (3), and the addition of the external field of the adjacent hole (which we can consider as nearly uniform one, because the lateral size of the indirect exciton increases slower than $d$) leads to the ordinary shift of the potential minimum without any modification of the shape. This reasoning is not correct at small values of $d$, ($d \ll 1$), when the strong coulomb potential of the holes prevents the shift of the electron wave function as a single whole and reduce the effect of the polarization. Therefore the polarization coefficient $c$ is negligible at $d \to 0$. In this limit, the polarization is usually considered as a perturbation of the function shape, but at considerable $d$ it is small compared to the shift polarization and is beyond the scope of the paper.

Finally, in the inset of Fig. 3 the optimal distance between the holes $R$ versus the $d$ is shown. It can be seen that at $d > 5$ (and up to $d \sim 80!$) the curve nearly merges with the linear approximation:

$$R_{appr}(d) \approx 4.61 + 1.856d, \tag{13}$$

which is also plotted. One of the explanations of such plain dependence may be as follows: At large $d$ the exciton function transforms into the gaussian function (4). The scale of the latter is $d^{3/4}$ which is smaller than $d$. Therefore, at large values of $d$ only one scale may be left. More accurate explanation of this fact is to be done.

To illustrate the evolution of the trion structure with $d$, in Fig. 5 the contour diagrams of the probability density of the electron in-plane function (8) for different values of the distance between the quantum wells ($d = 0$, 1 and 30) are shown. For

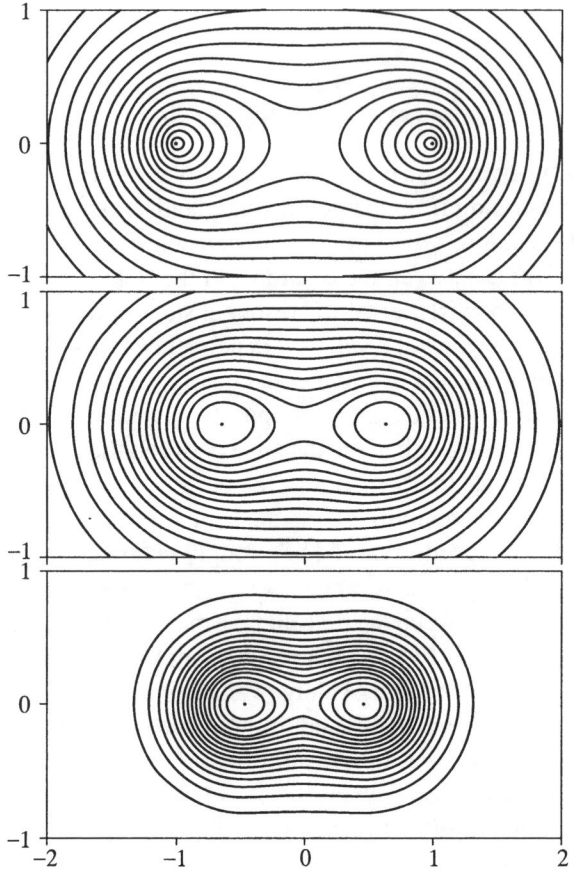

*Figure 5.* The contour diagrams of the probability density of the electron in-plane function (8) for different values of the distance between the quantum wells (from top to bottom $d = 0$, 1 and 30 respectively).

comparison, the diagrams are plotted in corresponding normalized coordinates. Thus, the coordinates of the infinitely heavy holes are (1,0) and (-1,0) for all plots. It is clearly seen, that increasing of $d$ leads to the shift of the maxima of the electron density toward each other corresponding to increasing of the polarization parameter $c$. Increasing of $d$ also leads to the more compact form of the density function due to decreasing the exciton in-plane scale ($d^{3/4}$) compared to the distance between the wells ($d$).

## 5. The Trion Binding Energy Estimations

One of the possible ways to create the spatially-indirect trion is to build the double quantum well heterostructure with the external electric field applied parallel to the growth direction. The applied bias introduces an asymmetry between the wells and each type of the carriers is collected preferably in corresponding quantum well. If the electric field is strong enough to neglect the overlapping of the electron and the holes wavefunctions and the localization length of the carriers in the growth direction is smaller than the effective distance between electron and holes layers, the above-mentioned approximation of two 2D quantum wells with different types of carriers can be used. In such model the trion binding energy does not depend on the bias and localization length of the carriers. Nevertheless, the former implicitly changes the trion energy varying the effective distance between the layers of holes and electrons and the localization length of the carriers. The latter, in its turn, leads to the comparable small correction and can be simply taken into account by means of the perturbation theory.

Thus, in order to estimate the binding energy of the trion, we have to know the in-plane Bohr radius and 3D Bohr energy for the material that the quantum wells are build from. For example, the typical values for GaAs quantum wells are:

$$E_B = \frac{m_e m_h}{m_e + m_h} \frac{e^4}{2\varepsilon^2 \hbar^2} \approx 4.5 \text{ meV}, \quad a_B = \frac{m_e + m_h}{m_e m_h} \frac{\hbar^2 \varepsilon}{e^2} \approx 120 \text{ Å}. \quad (14)$$

By using the dependences shown in Figs. 1 and 2 it is easy to obtain the spatially-indirect exciton ($E_{ex}$) and trion ($E_{tr}$) binding energies for any value of the effective distance between the layers of the electrons and holes ($d_{eff}$, it should be normalized to $a_B$). However, all calculations in the present paper are performed in the infinite hole mass approximation, but in GaAs the hole is only 6-8 times heavier than the electron. Therefore, for the plausible estimation of the trion energy the energy of zero oscillations of the holes should be taken into account. The simple estimation of the latter can be obtained in the same way as it was performed in [15]:

$$\Delta E_{tr} = \frac{\hbar \omega}{2} \approx \sqrt{\frac{m_e}{m_e + m_h} \frac{E_B E_{tr}}{R^2}}, \quad (15)$$

where the optimal distance between the holes $R$ in units of $a_B$ can be taken from the inset of Fig. 3.

The final results for the exciton and the trion binding energies for some values of $d_{eff}$ are shown in Table 1. The corresponding results for ZnSe QWs ($E_B \approx 20$ meV, $a_B \approx 40$ Å) are also presented. The exciton in-plane radius ($R_{ex} = \sqrt{\langle \rho^2 \rangle}$) calculated with the formulae (6) and (12) and the optimal distance between the holes in the trion ($R_{tr}$) are also given in the table.

## 6. Conclusions

The energy and the geometrical structure dependence of indirect trion versus the distance between the coupled 2D Quantum Wells are calculated variationally. The trion is found to be bound at the well distances $d < 34$. At the distance values

TABLE I. Examples of Calculated Parameters of Spatially-Indirect Exciton and $X^+$ Trion for GaAs and ZnSe QWs

|  | $d_{\text{eff}}$, Å | $E_{\text{ex}}$, meV | $E_{\text{tr}}$, meV | $R_{\text{ex}}$, Å | $R_{\text{tr}}$, Å |
|---|---|---|---|---|---|
| GaAs | 50 | 6.5 | 0.9 | 162 | 227 |
|  | 100 | 4.4 | 0.5 | 214 | 358 |
| ZnSe | 50 | 15.0 | 1.3 | 86 | 160 |
|  | 100 | 9.2 | 0.6 | 123 | 274 |

$d > 34$ the system of the exciton and the hole seems to have a resonant state. The parameters of the resonant state have been traced up to $d \sim 80$.

## Acknowledgements

This work was supported by Russian Foundation for Basic Research, grants 02-02-17610, 02-02-06243 and by Russian Federal Program on Support of Leading Scientific Schools, grant 00-15-96812.

## References

1. Lampert, M.A. (1958) *Phys. Rev. Lett.* **1**, 450.
2. Bethe, H.A. (1929) *Z. Phys.* **57**, 815.
3. Stebe, B. and Comte, C. (1977) *Phys. Rev. B* **15**, 3967.
4. Schilling, R and Mattis, D.C. (1982) *Phys. Rev. Lett.* **49**, 808.
5. Stebe, B. and Ainane, A. (1989) *Superlatt. Microstruct.* **5**, 545.
6. Chen, Z. and Lin, C.D. (1990) *Phys. Rev. A* **42**, 18.
7. Kheng, K. *et al*, (1993) *Phys. Rev. Lett.* **71**, 1752.
8. Thilagam, A. (1996) *Phys. Rev. B* **55**, 7804.
9. Stebe, B., Munschy, G., Stauffer, L., Dujardin, F., and Murat, J. (1997) *Phys. Rev. B* **56**, 12454.
10. Usukura, J., Suzuki, Y., and Varga, K. (1999) *Phys. Rev. B* **59**, 5652.
11. Ruan, W.Y., Chan, K.S., Ho, H.P., Zhang, R.Q. and Pun, E.Y.B. (1999) *Phys. Rev. B* **60**, 5714.
12. Stebe, B., Moradi, A., and Dujardin, F. (2000) *Phys. Rev. B* **61**, 7231.
13. Esser, A., Runge, E., and Zimmermann, R. (2000) *Phys. Rev. B* **62**, 8232.
14. Riva, C., Peeters, F.M., and Varga, K. (2000) *Phys. Status Solidi A* **178**, 513.
15. Sergeev, R.A. and Suris, R.A. (2001) *Phys. Solid State* **43**, 746.
16. Dignam, M.M. and Sipe, J.E. (1991) *Phys. Rev. B* **43**, 4084.
17. Peeters, F.M. and Golub, J.E. (1991) *Phys. Rev. B* **43**, 5159.
18. Binder, E., Kuhn, T., and Mahler, G. (1994) *Phys. Rev. B* **50**, 18319.
19. Bigenwald, P. and Gil, B. (1995) *Phys. Rev. B* **51**, 9780.
20. Shields, A.J., *et al.* (1997) *Phys. Rev. B* **55**, 1318
21. Timofeev, V.B., *et al.* (1999) *Phys. Rev. B* **60**, 8897.

# Contributors

**Ashkinadze** Boris
Physics Department,
Technion-Israel Institute of Technology
Haifa 32000
borisa@tx.technion.ac.il
ISRAEL

**Bronold** Franz
Institut für Theoretische Physik,
Otto-von-Güricke-Universität
D-39016 Magdeburg
Franz.Bronold@physik.uni-magdeburg.de
GERMANY

**Combescot** Monique
Universite Paris 6
2 Place Jussieu, 75005 Paris
combescot@gps.jussieu.fr
FRANCE

**Cox** Ronald
CEA-Grenoble
SP2M, CEA-Grenoble,
17 rue des Martyrs,
F-38054 Grenoble-Cedex 9
rcox@cea.fr
FRANCE

**Deveaud** Benoit
l'Institut de Photonique et Electronique
Quantiques, Ecole Polytechnique
Federal
EPFL CE Ecublens, CH-1015 Lausanne
deveaud@dpmail.epfl.ch
SWITZERLAND

**Esser** Axel
AG Halbleitertheorie,
Humboldt-Universität zu Berlin
Hausvogteiplatz 5-7, Berlin D-10117
axel.esser@physik.hu-berlin.de
GERMANY

**Hawrylak** Pawel
Institute for Microstructural Sciences,
National Research Council Canada
Bldg. M-50, Rm. 107, Ottawa K1A OR6
pawel.hawrylak@nrc.ca
CANADA

**Ivchenko** Eougueniious
Ioffe Physico-Technical Institute, RAS
194021 St Petersburg,
Polytechnicheskaya 26
ivchenko@coherent.ioffe.rssi.ru
RUSSIA

**Keller** Dirk
Physikalisches Institut EP III,
Universität Würzburg
Am Hubland, 97074, Würzburg
dkeller@physik.uni-wuerzburg.de
GERMANY

**Kochereshko** Vladimir
Ioffe Physico-Technical Institute, RAS
194021 St Petersburg,
Polytechnicheskaya 26
vladimir.kochereshko@pop.ioffe.rssi.ru
RUSSIA

**Korenev** Vladimir
Ioffe Physico-Technical Institute, RAS
194021 St Petersburg,
Polytechnicheskaya 26
korenev@orient.ioffe.rssi.ru
RUSSIA

**Kukushkin** Igor
Institute for Solid State Physics, RAS
142432 Chernogolovka,
Moscow Region
kukush@issp.ac.ru
RUSSIA

**Lavallard** Philippe
Groupe de Physique des Solides,
Universite Paris 7
Tour 23, 2 Place Jussieu,
75251 Paris Cedex 05
lavallar@gps.jussieu.fr
FRANCE

**Marie** Xavier
Laboratoire de Physique de la Matiere
Condensee, INSA-CNRS
135 avenue de Rangueil, 31077 Toulouse
marie@insa-tlse.fr
FRANCE

**McCombe** Bruce
The State University of New York,
University at Buffalo
4260 Gunville Road,
Clarence NY 14031
mccombe@mccombe.physics.buffalo.edu
USA

**Ossau** Wolfgang
Physikalisches Institut EP III,
Universität Würzburg
Am Hubland, 97074, Würzburg
ossau@physik.uni-wuerzburg.de
GERMANY

**Ryabchenko** Sergiy
Institute of Physics, NAS of Ukraine
Saperne pole Str. 26-20, 01042 Kiev
ryabch@iop.kiev.ua
UKRAINE

**Shaginyan** Vasily
Petersburg Nuclear Physics Institute,
RAS
Gatchina
vasily@vs3381.spb.edu
RUSSIA

**Sergeev** Rinat
Ioffe Physico-Technical Institute, RAS
194021 St Petersburg,
Polytechnicheskaya 26
rinat@theory.ioffe.rssi.ru
RUSSIA

**Suris** Robert
Ioffe Physico-Technical Institute, RAS
194021 St Petersburg,
Polytechnicheskaya 26
suris@theory.ioffe.rssi.ru
RUSSIA

**Timofeev** Vladislav
Institute for Solid State Physics, RAS
142432 Chernogolovka,
Moscow Region
timofeev@issp.ac.ru
RUSSIA

**Viña** Luis
Departamento Fisica de La Materiales,
Universidad Autonoma, Cantoblanco
Madrid E-28049
luis.vina@uam.es
SPAIN

**Yakovlev** Dmitry
Experementelle Physik II,
Universit"at Dortmund
Otto-Hahn-Str. 4, D-44227 Dortmund
Dmitry.Yakovlev@physik.uni-dortmund.de
GERMANY

# Index

2D electron gas, 112
2DEG, 125, 126, 128–131, 135, 136, 193, 194

Absorption spectrum, 101
Anderson-like equation, 107
Asymmetric trion lineshape, 100

Band-gap renormalization, 151
Binding energy of trions, 54
Boson, 89
Bound-to-bound transition, 26
Bound-to-continuum transition, 27

Carrier localization, 151
CdTe, 42, 159–162, 164–166, 168
CdTe modulation-doped quantum well, 205
CdTe/CdMgTe, 207
Charged exciton, 41, 151, 157
Charged excitons, 193
Charged magnetoexciton, 25
Circular polarized light, 103
Cluster meanfield approximation, 169
Coherent property, 24
Combined exciton cyclotron resonance, 126
Combined exciton cyclotron resonance line, 129
Combined optical process, 125
Combined trion cyclotron resonance line, 129
Composite fermion, 4
Correlation length, 107
Coulomb form factor, 101
Cyclotron resonance, 4, 9

D'yakonov–Perel' spin relaxation, 181, 183
Dark, 31, 152
Dark state, 104
Density-matrix approach, 100
Diffusion, 207
Diluted magnetic semiconductor, 217
Double quantum well, 13, 279, 280, 286
Dynamical response, 169
Dynamical screening, 169

Electron cyclotron resonance, 6
Electron–electron collision, 181
Electron–electron interaction, 25

Electron–electron scattering, 189
Electrostatic disorder, 99, 106
Exchange interaction, 81, 86, 159, 163, 168, 239–245
Exciton, 89, 112, 125, 126, 132, 135, 136, 151, 160, 170, 205, 217, 233
Excitonic correction, 163, 164, 166

Family of states, 28, 29
Far infrared, 32
Fermi edge singularity, 170
Fermi energy, 112
Fermi-edge singularity, 26
Filling factor, 151
Four-wave mixing, 205, 207
Fractional quantum Hall effect, 5
Free exciton, 152
Free trion, 206

GaAs, 42, 206
GaAs/AlAs, 233
GaAs/AlGaAs, 32
Generalized truncation scheme, 101
Green function, 122

Hartree–Fock approximation, 101
Hidden symmetry, 29
High density regime, 170
Hole, 111, 112

Inhomogeneous electron gas, 151
Integer quantum Hall effect, 6
Interacting many-body systems, 5
Interacting two-dimensional electron gas, 151
Interband optical properties, 25
Interband polarization, 100
Internal exciton transitions, 26
Internal transition, 25, 27
Interwell exciton, 13, 14

Kohn's theorem, 6

Ladder diagram, 103
Landau level, 26, 130, 152
Linear polarized light, 103

Localization, 99
Localization length, 100, 107
Low densities, 169
Low density limit, 90
Low symmetry effect, 248
Low-energy tail, 108
Luminescence polarization, 248, 255

Magnetic field, 247, 248, 250–252, 254–256
Magnetic translational invariance, 29
Magneto-exciton, 44
Magneto-optics, 159
Magnetophotoluminescence, 193
Magnetoplasma-cyclotron resonance, 8
Magnetoplasmon, 25
Many-body, 90
Many-body system, 26
Many-electron effect, 25
Microcavities, 64
Microcavities, 74
Microwave-modulated photoluminescence, 193
Motional narrowing, 108

Nanocrystal, 239, 242
Negatively charged exciton, 42
Non-linear effect, 206

Optical spectrum, 112
Optical susceptibility, 103, 104, 170
Optical trion density, 108
Optically detected resonance, 25, 32
Oscillator strength transfer, 100

Pauli blocking, 169
Pauli exclusion, 89
Photoluminescence, 46, 151
Photon, 112
Polariton, 65, 70
Polarization anisotropy, 248
Polarized photoluminescence, 218
Positively charged exciton, 42

Quantum dot, 79
Quantum phase transistor, 259
Quantum well, 111, 112, 159, 162, 168, 217, 247
Quantum wire, 233
Quasiparticle, 152

Radiative decay, 207
Recoil energy, 102

Recoil process, 100
Reflectivity, 46, 218
Resonance excitation, 15
Rough interface, 99

Scattering amplitude, 114
Scattering state, 169
Semiconductor, 239, 241, 245, 248
Semimagnetic, 247, 248, 254, 256
Singlet, 26, 27, 49, 152
Singularity, 115, 118
Spin dynamics, 65, 73
Spin polarized 2DEG, 217
Spin property, 79
Spin splitting of electron subbands, 182
Spin-flip Raman scattering, 233, 234
Stimulated emission, 69, 74
Strongly correlated electron, 259, 260

Three-particle bound, 100
Time-resolved, 65
Time-resolved emission, 66
Trion, 41, 111, 112, 169, 170, 205, 206, 217, 279–287
Trion amplitude, 101
Trion continuum, 106
Trion Green's function, 102
Triplet, 26, 27, 49, 152
Triplet state, 31
Two-dimensional electron gas, 151
Two-dimensional electrons, 193

Wave function, 119
Weak disorder, 100

ZnSe, 41, 42